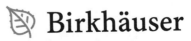

Frontiers in Mathematics

This series is designed to be a repository for up-to-date research results which have been prepared for a wider audience. Graduates and postgraduates as well as scientists will benefit from the latest developments at the research frontiers in mathematics and at the "frontiers" between mathematics and other fields like computer science, physics, biology, economics, finance, etc. All volumes are online available at SpringerLink.

More information about this series at http://www.springer.com/series/5388

Bruno Bianchini • Luciano Mari • Patrizia Pucci •
Marco Rigoli

Geometric Analysis of Quasilinear Inequalities on Complete Manifolds

Maximum and Compact Support Principles and Detours on Manifolds

 Birkhäuser

Bruno Bianchini
Dipartimento di Matematica Pura e Applicata
Università degli Studi di Padova
Padova, Italy

Luciano Mari
Dipartimento di Matematica
Università degli Studi di Torino
Torino, Italy

Patrizia Pucci
Dipartimento di Matematica e Informatica
Università degli Studi di Perugia
Perugia, Italy

Marco Rigoli
Dipartimento di Matematica
Università degli Studi di Milano
Milano, Italy

ISSN 1660-8046   ISSN 1660-8054   (electronic)
Frontiers in Mathematics
ISBN 978-3-030-62703-4   ISBN 978-3-030-62704-1   (eBook)
https://doi.org/10.1007/978-3-030-62704-1

Mathematics Subject Classification: Primary 58J05, 53C42, 53C21; Secondary 35B08, 35B50, 35B53, 58J32, 58J65

This book is published under the imprint Birkhäuser, www.birkhauser-science.com, by the registered company Springer Nature Switzerland AG.

The registered company address is: Gewerbestrasse 11, 6330 Cham, Switzerland

Preface

The use of typical tools from Analysis to shed light on geometric properties is at the core of modern Global Differential Geometry. The aim of the book is to discuss the influence of the structure of a manifold M on the qualitative behaviour of solutions to certain classes of elliptic PDEs arising in a Riemannian context. The differential inequalities that we shall consider are elliptic, possibly degenerate, with the left-hand side described by a quasilinear operator Δ_φ, shortly called the φ-Laplacian. Depending on the choice of φ, the class encompasses well-known examples, including, for instance, the p-Laplacian and the mean curvature operator. To illustrate our results, in the first two chapters we have chosen to discuss some geometric and physical problems where such inequalities naturally appear, and in what follows we shall often pay special attention to the mean curvature operator, which arises when considering the graph associated to a smooth function u on the manifold. This prototype case leads to investigate a large class of differential inequalities whose right-hand side depends not only on $x \in M$ and on the solution u but also on the gradient of u. Having motivated our study, a key part of the book deals with various generalized forms of the maximum principle for Δ_φ, which serve as a bridge to relate the Geometry of M to the analytical properties of the PDE under consideration. Our technical achievements will provide a bulk of flexible methods to describe qualitative properties of solutions to a variety of problems. In the last chapters, Liouville type theorems and compact support principles are investigated in detail, with special emphasis on the role played by the integrability requirements that, in the literature, are known as the Keller–Osserman conditions. Geometric applications, among others, touch upon Bernstein type theorems for graphs with prescribed mean curvature (notably, minimal or solitons for the mean curvature flow), Yamabe and capillarity equations. We refer to the "Contents" for a more exhaustive list.

The book presents the most recent points of view and trends in the field and is therefore to be considered at an advanced level suitable to researchers and senior post-graduate students with a solid knowledge in Differential Geometry and elliptic PDEs. Besides the presentation of new theorems, we collect and organize sharp refinements of some research results obtained by the authors and their collaborators in a period of about twenty years. In this respect, we mention the monographs of P. Pucci and J. Serrin [194], of S. Pigola, M. Rigoli and A.G. Setti [182], and of L.J. Alías, P. Mastrolia and M. Rigoli [6], which

v

could serve as a starting introduction for the present work. In fact, we will consistently refer to them as well as to the references therein for the previous literature. Non-expert readers who are interested in the topic and have an eye on geometric applications can profit from the introductory chapters in [6]. The monograph aims to fulfil two purposes. First, to provide an exposition of some fundamental tools in Geometric Analysis and elliptic PDEs on complete manifolds, a subject that has become of general interest in the last decades. Second, to enable researchers to get familiar with the necessary results and concepts to proceed towards a further study of the specialized literature on the subject. We included an extensive and commented bibliography in the attempt to make it a reference book for future research.

We end by expressing our deep gratitude to the colleagues and students who made this book possible. Special thanks goes also to Sabrina Hoecklin from Springer Nature Switzerland AG, Birkhäuser, for her useful help at all stages of the editing process, and to Daniel Jagadisan, Sarah Annette Goob and Mahalakshmi Balamurugan for taking care of the production process.

Padova, Italy

Bruno Bianchini

Torino, Italy

Luciano Mari

Perugia, Italy

Patrizia Pucci

Milano, Italy

Marco Rigoli

August 2020

Contents

List of Symbols

The following symbols are frequently used along the manuscript:

$\mathbb{R}_0^+$	$\mathbb{R}_0^+ = [0, \infty)$;
$\mathbb{R}^+$	$\mathbb{R}^+ = (0, \infty)$;
$\mathbb{R}^m$	Flat m-dimensional Euclidean space;
$\mathbb{S}^m$	Unit sphere in $\mathbb{R}^{m+1}$ endowed by the induced Euclidean metric;
$\mathbb{H}^m$	Hyperbolic space of sectional curvature -1;
$d\theta^2$	Round metric of curvature 1 on the unit sphere $\mathbb{S}^{m-1}$;
ω_{m-1}	Volume of the unit sphere $(\mathbb{S}^{m-1}, d\theta^2)$;
$\mathbb{R} \times_h M$	Warped product of $\mathbb{R}$ and $(M, \langle, \rangle)$ with metric $ds^2 + h(s)^2\langle, \rangle$;
$\text{Lip}_{\text{loc}}(\Omega)$	Locally Lipschitz functions on Ω;
Δ_p	p-Laplacian;
$\Delta_{p,q}$	(p, q)-Laplacian;
Δ_φ	φ-Laplacian;
$\mathcal{O}$	Origin of M;
$D_{\mathcal{O}}$	Maximal domain of normal coordinates centred at $\mathcal{O}$;
$\text{cut}(\mathcal{O})$	Cut-locus of $\mathcal{O}$;
Ric	Ricci curvature;
Sec	Sectional curvature;
Sec_{rad}	Radial sectional curvature;
M_g	Radially symmetric model manifold with warping function g;
$\mathcal{G}$	Green kernel of Δ_p;
$\mathcal{G}^g(r)$	Green kernel of Δ_p on a model M_g;
$\text{cap}_p(K)$	p-capacity of a set K;
ϱ	Fake distance from a given origin;
(FMP)	Finite maximum principle;
(WMP$_\infty$)	Weak maximum principle at infinity;
(SMP$_\infty$)	Strong maximum principle at infinity;
(OWMP$_\infty$)	Open form of the weak maximum principle at infinity;
(L)	Liouville property;
(SL)	Strong Liouville property;

(CSP)	Compact support principle;
(FE)	Feller property;
(KO_∞)	Keller–Osserman condition at infinity;
(KO_0)	Keller–Osserman condition at zero;
$(\neg KO_\infty)$	Failure of the Keller–Osserman condition at infinity;
$(\neg KO_0)$	Failure of the Keller–Osserman condition at zero;
$(\mathcal{WS})$	Weak Sard property.

Some Geometric Motivations

Let $(M, \langle\,,\,\rangle)$ be a complete Riemannian manifold of dimension $m \geq 2$, and let Δ_φ be a quasilinear operator depending on a real function φ, to be detailed in the next chapter. For instance, the family Δ_φ includes the Laplace–Beltrami operator Δ, defined with the sign agreement that

$$\Delta u = \mathrm{div}(\nabla u) = \sum_{j=1}^{m} \frac{\partial^2 u}{\partial x_j^2} \qquad \text{on } \mathbb{R}^m.$$

Since the birth of Geometric Analysis, several geometric applications lead to the study of differential inequalities of the type

$$\Delta_\varphi u \geq b(x) f(u) \qquad \text{on } M, \tag{1.1}$$

for given $b : M \to \mathbb{R}$ and $f : \mathbb{R} \to \mathbb{R}$. Here, u is a function encoding some property of the problem under consideration, and the determination of the qualitative properties of u can therefore be read in terms of constraints for the original geometric problem. Hence, it is fundamental to understand how the geometry of M influences the behaviour of u. By now, (1.1) is fairly well understood for most of the operators Δ_φ. A thorough investigation, with applications ranging from Lorentzian Geometry to the theory of Ricci and mean curvature flow solitons, to Kähler manifolds, Yang-Mills fields and minimal submanifolds, among many others, can be found in the monograph [182] and in the book [6]. However, some relevant problems are also characterized by the appearing of gradient terms in the right-hand side of (1.1), notably those related to the mean curvature operator

$$\mathrm{div}\left(\frac{\nabla u}{\sqrt{1 + |\nabla u|^2}}\right), \tag{1.2}$$

B. Bianchini et al., *Geometric Analysis of Quasilinear Inequalities on Complete Manifolds*, Frontiers in Mathematics, https://doi.org/10.1007/978-3-030-62704-1_1

and consequently to graphs with prescribed mean curvature. Despite the many works devoted to the study of (1.2) (some of them will be described in the next section), the existing literature is not sufficiently complete from the point of view that we are going to consider in this book. In particular, the results in the above mentioned references are not sufficient to provide a full picture of how geometry influences the behaviour of u. Although the mean curvature operator should be regarded as the main focus of the present work from the geometric point of view, the techniques that we present apply to a wide class of quasilinear elliptic inequalities of the type

$$\Delta_\varphi u \geq b(x) f(u) l(|\nabla u|) \qquad \text{on } M \tag{1.3}$$

for rather general b, f, l. This book can be considered as a natural continuation of [6, 182] as well as of [194], that are an invaluable source of material and whose results will be repeatedly commented herein. However, both our focus and our results significantly depart from those in [6, 182, 194], and in various instances, they give new insight even for equations of the type (1.1). It will be later underlined how the study of equations with a non-constant gradient term l will be also crucial to achieve sharp Liouville theorems for (1.4), where "apparently" no gradient term appears. This is remarkably the case of the capillarity equation, as we shall detail in Chap. 2. The purpose of this chapter is to motivate, via some natural examples, the study of (1.3) with a nontrivial gradient term.

1.1   Prescribing the Mean Curvature of a Graph

The mean curvature operator arises when considering the graph immersion

$$\varphi : M \to \mathbb{R} \times M, \qquad \varphi(x) = \big(u(x), x\big)$$

associated to a smooth function $u : M \to \mathbb{R}$. Precisely, we endow $\mathbb{R} \times M$ with the product metric $ds^2 + \langle \, , \, \rangle$, and M with the induced graph metric

$$g = \varphi^*\big(ds^2 + \langle \, , \, \rangle\big),$$

so that the image $\varphi(M)$ is isometric to (M, g). Then, the unnormalized mean curvature H of $\varphi(M)$ in the upward-pointing normal direction satisfies the differential equation

$$\text{div}\left(\frac{\nabla u}{\sqrt{1 + |\nabla u|^2}}\right) = H \qquad \text{on } M,$$

where the gradient ∇ and the divergence operator are those of the metric $\langle\,,\,\rangle$. For instance, if $\varphi(M)$ is minimally immersed, that is, $H = 0$, the equation becomes

$$\mathrm{div}\left(\frac{\nabla u}{\sqrt{1+|\nabla u|^2}}\right) = 0 \qquad\qquad \text{(MSE)}$$

and, already in the case M is the Euclidean plane $\mathbb{R}^2$, its study has brought to deep connections between Geometry and Analysis. We refer the reader to the old, but still actual, monumental work of J.C.C. Nitsche [172], to the book of E. Giusti [102] and to the informative survey of L. Simon [223], to appreciate the breadth of viewpoints and the depth of the techniques necessary to understand equation (MSE).

Hereafter, a solution of a differential equation or inequality is said to be *entire* if it is defined on the whole of M. In particular, differently from part of the existing literature, the term "entire" does not indicate a specific behaviour of u at infinity.

A basic question related to (MSE) is whether there exist entire solutions possibly with some restrictions such as, for instance, a controlled growth at infinity or the boundedness on one side. Classically, the problem has first been considered in Euclidean space $\mathbb{R}^m$, and the efforts to solve it favoured the flourishing of Geometric Analysis. In 1915, S. Bernstein [18] proved that the only entire solutions of (MSE) over $\mathbb{R}^2$ are affine functions (a topological gap in his highly nontrivial argument has been pointed out and corrected in [125, 159]). Various other proofs later appeared in the literature, [76, 97, 117, 160, 171, 188, 223], each one inferring the validity of Bernstein's theorem from some peculiarities of $\mathbb{R}^2$ and thus not extendible to higher dimensions. We suggest to consult Farina's survey [83] for more details. The first proof to allow for a higher-dimensional generalization has been given by W.H. Fleming [99], and since then, in few years, fundamental contributions of E. De Giorgi ([68, 69], $m = 3$), F. Almgren ([8], $m = 4$) and J. Simons ([224], $m \le 7$) extended the validity of Bernstein theorem up to dimension $m = 7$. In 1969, E. Bombieri, De Giorgi and E. Giusti [30] gave the first example of entire minimal graph over $\mathbb{R}^m$ for $m \ge 8$ (further examples were later given by L. Simon [222]), thus leading to the following remarkable result:

$$\text{each minimal graph over } \mathbb{R}^m \text{ is affine} \qquad \Longleftrightarrow \qquad m \le 7. \qquad (\mathscr{B}1)$$

If the graph is subjected to some a priori bound, then further tools are available and more rigidity is expected. We mention that

$(\mathscr{B}2)$ by work of Bombieri, de Giorgi and M. Miranda [29], positive solutions of (MSE) over $\mathbb{R}^m$ are constant for every $m \ge 2$;

$(\mathscr{B}3)$ coupling [29] with works of De Giorgi [68, 69] and J. Moser [163], every entire solution of (MSE) over $\mathbb{R}^m$ with at most linear growth is affine.

We also mention that an entire solution of (MSE) having $m - 7$ derivatives bounded on one side is necessarily affine. This result was shown by A. Farina [85], improving on [28, 84].

If $M \neq \mathbb{R}^m$, the set of solutions of (MSE) may drastically change. For instance, if M is the hyperbolic space $\mathbb{H}^m$ of curvature -1, then (MSE) has plenty of bounded solutions: for every ϕ continuous on the boundary at infinity $\partial_\infty \mathbb{H}^m \approx \mathbb{S}^{m-1}$, there exists u solving (MSE) on $\mathbb{H}^m$ and approaching ϕ at infinity. A proof of this result is contained in [169] for $m = 2$, and in [77] for larger m. Up to defining $\partial_\infty M$ appropriately, the same Plateau's problem at infinity can be considered on any Cartan–Hadamard manifold, that is, on any complete, simply connected manifold whose sectional curvature satisfies

$$\mathrm{Sec} \leq 0 \qquad \text{on } M.$$

Since the problem is solvable on $\mathbb{H}^m$ but not on $\mathbb{R}^m$ because of $(\mathscr{B}2)$, it is natural to wonder about the sharp thresholds on Sec, the sectional curvature, that guarantee its solvability. The question has recently been addressed in a series of works [39–41, 44, 123, 210]. The picture is indeed subtle, and in general a negative upper bound on Sec does not suffice to guarantee that the Plateau's problem has a solution. A thorough discussion of the above results would lead us a bit too far from the main focus of this introductory chapter, so we refer the interested reader to the above mentioned works, as well as to the surveys [21, 114]. Similarly, it is natural to consider the following

Question Which geometric conditions on M can guarantee the validity of results analogous to those in $(\mathscr{B}1)$, $(\mathscr{B}2)$ and $(\mathscr{B}3)$?

To present, a thorough answer to the above question is still far from reaching, and for instance we are aware of no results addressing $(\mathscr{B}1)$ on manifolds different from $\mathbb{R}^m$. In view of the structure theory initiated by J. Cheeger and T. Colding in [46–49], it is reasonable to hope that results similar to those holding in $\mathbb{R}^m$ could be obtainable on manifolds with non-negative Ricci curvature:

$$\mathrm{Ric} \geq 0.$$

Nevertheless, a complete solution of the question would be relevant even on manifolds with $\mathrm{Sec} \geq 0$. The guess might also be motivated by the fact that (MSE) rewrites as

$$\Delta_g u = 0 \qquad \text{on } M,$$

with Δ_g the Laplace–Beltrami operator of the graph metric g, and by the theory of harmonic functions on manifolds with non-negative Ricci or sectional curvature. Nevertheless, by Gauss equation it should be stressed that no control on any of these curvatures on the graph (M, g) would easily follow from a control on the corresponding one on the base manifold $(M, \langle \, , \, \rangle)$, so the analogy does not allow to directly apply techniques from the theory of harmonic functions to the realm of minimal graphs. To the best of our knowledge, some interesting results regarding $(\mathscr{B}2)$ and $(\mathscr{B}3)$ are known under both Ricci and sectional curvature requirements, see [44, 73, 212]. Very recently in [59, 72], the authors independently obtained the following extension of ($\mathscr{B}2$) without the need of sectional curvature bounds:

Theorem 1.1 ([59, 72]) *If M is a complete manifold with* Ric ≥ 0, *then every positive solution of* (MSE) *is constant.*

A complete solution to the above question is likely to significantly improve the current understanding of the geometry of the mean curvature operator on Riemannian manifolds.

Together with the minimal case, it is of interest to study graphs whose mean curvature H is a prescribed function of the height, possibly via a non-homogeneous weight, namely we investigate solutions to

$$\operatorname{div}\left(\frac{\nabla u}{\sqrt{1+|\nabla u|^2}}\right) = b(x) f(u) \tag{1.4}$$

on M, for some $f \in C(\mathbb{R})$ and $0 < b \in C(M)$. Equation (1.4) models hypersurfaces that, on every $\Omega \Subset M$, are stationary for the functional

$$u \mapsto \int_\Omega \left[\sqrt{1+|\nabla u|^2} + b(x) F(u)\right] dx,$$

where

$$F(t) = \int_0^t f(s)\,ds, \tag{1.5}$$

with respect to variations compactly supported inside Ω. A notable example is that of the *capillarity equation*, for which $b \equiv 1$ and $f(t) = \kappa t$ for some constant $\kappa > 0$. In this case, the graph of u describes the surface of a fluid, and the equation arises from balancing forces of tension (proportional to the mean curvature of the capillarity surface) with the weight of the fluid supported. The physical constant κ is positive or negative depending on whether the surface is an upper or lower boundary of the fluid. Various aspects of capillarity surfaces were studied in detail in R. Finn's book [96].

In Euclidean space and if f is non-decreasing, the behaviour of solutions of (1.4) is well understood thanks to the following beautiful Liouville theorem, due to V. Tkachev [225] for $b \equiv 1$ and later improved by Y. Naito and H. Usami [168] and J. Serrin [219], see also [83, Thm. 10.4].

Theorem 1.2 ([168, 219, 225]) *Assume that*

$$b(x) \geq C\left(1+|x|\right)^{-\mu} \qquad on\ \mathbb{R}^m,$$

for some constants $C > 0$ and $\mu < 1$. If $u \in C^2(\mathbb{R}^m)$ solves (1.4) with f non-decreasing and $f \not\equiv 0$, then u is constant.

In particular, the result guarantees that the only solution of the capillarity equation on the entire $\mathbb{R}^m$ is $u \equiv 0$, and guarantees that entire graphs with constant mean curvature H must be minimal. This last property of $\mathbb{R}^m$ was first proved by E. Heinz [118] if $m = 2$, and by S.S. Chern [54] and H. Flanders [98] for general m, see also improvements by I. Salavessa [213]. Condition $f \not\equiv 0$ is necessary, as the example of affine functions shows. As we shall detail later, the restriction $\mu < 1$ is sharp.

It is instructive to observe that the proof of Theorem 1.2 splits into two parts:

(a) first, the author shows the identity $f(u) \equiv 0$ on $\mathbb{R}^m$, thus u solves (MSE);
(b) then, since $f \not\equiv 0$ and is non-decreasing, u must be bounded from one side, so the constancy of u follows from ($\mathscr{B}2$) above.

Let us comment on the above two steps with a simple, heuristic argument: hereafter, we define

$$u^* \doteq \sup_M u, \qquad u_* \doteq \inf_M u.$$

Suppose that M is compact, and that u solves (1.4). Evaluating the equation at a point x_0 where u achieves its maximum u^*, and recalling that the mean curvature operator is elliptic, we deduce $f(u^*) \leq 0$. Similarly, evaluating at a minimum point gives $f(u_*) \geq 0$. Since f is non-decreasing, we therefore deduce $f(u) \equiv 0$ on M. Step (a) can therefore be obtained as a consequence of a maximum principle for u. Assume for convenience that $b \equiv 1$. If M is non-compact and we assume that u is bounded, even though u may fail to attain its maximum or minimum, the same conclusion $f(u) \equiv 0$, with the same argument, can be achieved provided that there exist sequences $\{x_k\}, \{y_k\}$ of points of M such that

$$u(x_k) \to u^*, \qquad \mathrm{div}\left(\frac{\nabla u}{\sqrt{1 + |\nabla u|^2}}\right)(x_k) < \tfrac{1}{k},$$

$$u(y_k) \to u_*, \qquad \mathrm{div}\left(\frac{\nabla u}{\sqrt{1 + |\nabla u|^2}}\right)(y_k) > -\tfrac{1}{k}.$$

The existence of $\{x_k, y_k\}$ is not guaranteed on each manifold, but depends on the geometry of M at infinity: if such sequences can be found for each bounded $u \in C^2(M)$, then we say that the mean curvature operator satisfies the (weak) maximum principle at infinity. Maximum principles at infinity have first been introduced in the celebrated papers by H. Omori [174] and S.T. Yau ([239] and [50], the second with S.Y. Cheng), for the Laplace–Beltrami operator, and proved to be an essential tool in Geometric Analysis.

Their use, in suitable forms, is ubiquitous in the present monograph, and identifying the optimal thresholds for their validity is one of the central problems to reach sharp existence-nonexistence results for equations like (1.4) and many more. In the setting of Theorem 1.2, however, the function u is not known to be bounded a priori, so one needs further arguments.

The Seeming Lack of a Keller–Osserman Condition

Even though the actual proof of Step (a), for Theorem 1.2, is rather special, it is convenient for the moment to think that the strategy to show $f(u) \equiv 0$ proceeds via the following substeps:

– first, prove that a solution of (1.4) is bounded;
– then, show that a maximum principle at infinity holds for the mean curvature operator on the underlying manifold.

In fact, this will be a general strategy applied to classes of PDEs that we are going to study, that includes (1.4). It is illustrative to compare Theorem 1.2, say in the case $b \equiv 1$, to the situation occurring for solutions of

$$\Delta u = f(u) \qquad \text{on } \mathbb{R}^m, \tag{1.6}$$

with $f \in C(\mathbb{R})$. It is known that positive solutions of $\Delta u = 0$ are still constant, that is, the corresponding of Step (b) holds; indeed, by Yau's theorem [239], positive solutions of $\Delta u = 0$ are constant on every complete manifold M with non-negative Ricci curvature. However, it is false that every solution of (1.6) enjoys $f(u) \equiv 0$ on $\mathbb{R}^m$: for instance, by a result of D. Fisher-Colbrie and R. Schoen [97], and W.F. Moss and J. Piepenbrink [165], there are plenty of positive solutions of

$$\Delta u = \kappa u \qquad \text{on } \mathbb{R}^m$$

for constants $\kappa > 0$. As we show in a moment, the choice $f(t) = \kappa t$ is sharp for the existence of nontrivial solutions. In two seminal works, J.B. Keller and R. Osserman [135, 175] independently studied conditions on f to guarantee that solutions of (1.6) satisfy L^∞ bounds which, in many instances, make them trivial. Their investigation arose in connection to the type problem for Riemann surfaces [175] and to uniqueness problems for charged fluids in a container [134]. If $f(t) > 0$ for $t \gg 1$, both of the authors identified the integrability requirement

$$\frac{1}{\sqrt{F}} \in L^1(\infty), \tag{1.7}$$

where F is defined in (1.5), that hereafter will be called a Keller–Osserman condition. Note that f satisfies (1.7) if, loosely speaking, it grows faster than linearly at infinity; for instance, $f(t) = t^\sigma$ satisfies (1.7) if and only if $\sigma > 1$. Under the validity of (1.7) and assuming that

$$f \text{ is non-decreasing on } \mathbb{R},$$

the authors prove that a solution of the inequality

$$\Delta u \geq f(u) \qquad \text{on } \mathbb{R}^m$$

must satisfy

$$u^* < \infty, \qquad f(u^*) \leq 0. \tag{1.8}$$

Furthermore, if (1.7) fails (the non-decreasing assumption on f is dropped here), the authors constructed a radial, positive unbounded solution of (1.6) on $\mathbb{R}^m$. It should be stressed that claim (1.8) is a reformulation of the results in [135, 175] in a form more suited to our purposes. For convex f, a characterization of solutions to (1.6) depending on the validity of (1.7) was previously obtained by E.K. Haviland [110]. Claim (1.8) easily implies the following Liouville theorem via a "reflection" argument that we shall repeatedly use hereafter. Assume that

$$tf(t) > 0 \qquad \text{on } \mathbb{R}\backslash\{0\},$$

and that Keller–Osserman conditions hold both at ∞ and $-\infty$, that is,

$$\frac{1}{\sqrt{F}} \in L^1(\infty) \cap L^1(-\infty), \tag{1.9}$$

(note that, in our assumptions, F is positive on $\mathbb{R}\backslash\{0\}$). Then, the only solution u of (1.6) is $u \equiv 0$. Indeed, Claim (1.8) implies that $f(u^*) \leq 0$, so $u^* \leq 0$. On the other hand, $\bar{u} \doteq -u$ solves

$$\Delta \bar{u} = \bar{f}(\bar{u}) \qquad \text{on } \mathbb{R}^m, \text{ where } \bar{f}(t) \doteq -f(-t),$$

and the integrability at $-\infty$ in (1.9) guarantees the validity of (1.7) with $\bar{f}$ replacing f. Thus, $\bar{f}(\bar{u}^*) \leq 0$, equivalently $f(u_*) \geq 0$. Concluding, $u^* \leq 0 \leq u_*$, that is, $u \equiv 0$.

Turning to the mean curvature operator, in Theorem 1.2 no growth condition of Keller–Osserman type is required on f, a fact that shows a further, striking difference with the Laplacian. In the next section and in Chap. 10, we aim to clarify the reasons for their absence, and to investigate the behaviour of solutions of (1.4) on more general manifolds.

As we shall see, the picture will be quite interesting and Keller–Osserman conditions will appear again in a somehow hidden way for Eq. (1.4), as soon as the volume of geodesic balls centred at a fixed origin of M grows faster than polynomially. The case of the capillarity equation reveals quite instructive.

Geodesic Graphs in $\mathbb{R} \times_h M$

Although sufficiently general to be analytically challenging, ambient manifolds of the type $\mathbb{R} \times M$ leave aside various cases of interest, notably that of the hyperbolic space $\mathbb{H}^{m+1}$. For this reason, it is useful to consider ambient manifolds that can be written as the warped product

$$\bar{M}^{m+1} = \mathbb{R} \times_h M^m,$$

that is topologically $\mathbb{R} \times M$ endowed with the Riemannian metric

$$(\,,\,) = ds^2 + h(s)^2 \langle\,,\,\rangle,$$

for some positive $h \in C^\infty(\mathbb{R})$. Recall that the hyperbolic space $\mathbb{H}^{m+1}$ of constant sectional curvature -1 admits the following three different representations as a warped product of the above type:

– if we remove a fixed origin o, and we endow $\mathbb{H}^{m+1} \backslash \{o\}$ with polar coordinates $(r, \theta) \in \mathbb{R}^+ \times \mathbb{S}^m$, we can write

$$\mathbb{H}^{m+1} = \mathbb{R}^+ \times_{\sinh s} \mathbb{S}^m,$$

where $\mathbb{S}^m$ is the round sphere of curvature 1;
– consider the upper half-space model

$$\mathbb{H}^{m+1} = \left\{ (x_0, x) \in \mathbb{R} \times \mathbb{R}^m \ : \ x_0 > 0 \right\}$$

with metric

$$(\,,\,) = \frac{1}{x_0^2} \left(dx_0^2 + \langle\,,\,\rangle_{\mathbb{R}^m} \right).$$

With the change of variables $s = -\log x_0$ we express $(\,,\,)$ as the warped product

$$\mathbb{H}^{m+1} = \mathbb{R} \times_{e^s} \mathbb{R}^m, \qquad (\,,\,) = ds^2 + e^{2s} \langle\,,\,\rangle_{\mathbb{R}^m},$$

whose slices $\{s = \text{const}\}$ are horospheres;

– similarly, we can view $\mathbb{H}^{m+1}$ as a warped product

$$\mathbb{H}^{m+1} = \mathbb{R} \times_{\cosh s} \mathbb{H}^m, \qquad (,) = ds^2 + \cosh^2 s \langle , \rangle_{\mathbb{H}^m}$$

along totally umbilical hyperspheres, that in the upper half-space model correspond to Euclidean spheres having the same $(m-1)$-dimensional sphere as a trace on the boundary at infinity $\{x^0 = 0\}$.

Given a function $v : M \to \mathbb{R}$, one can consider the graph

$$\Sigma^m = \Big\{ (s, x) \in \mathbb{R} \times M, \ \ s = v(x) \Big\}.$$

Note that $X = h(s)\partial_s$ is a conformal field with geodesic flow lines, as a direct computation gives

$$\bar{\nabla}_v X = h'(s)v \qquad \forall v \in T\bar{M},$$

with $\bar{\nabla}$ the connection on $\bar{M}$. For this reason, we call Σ a *geodesic graph*. We let Φ_t be the flow of X, and note that the flow parameter satisfies

$$t = \int_0^s \frac{d\sigma}{h(\sigma)}, \qquad t : \mathbb{R} \to t(\mathbb{R}) = I. \tag{1.10}$$

We let $s(t)$ be its inverse. For convenience, we express the prescribed mean curvature equation in terms of the function $u(x) = t(v(x))$. Let ∇ be the connection on $(M, \langle , \rangle)$. If H is the normalized mean curvature of Σ with respect to the upward pointing unit normal

$$\nu = \frac{1}{\lambda(u)\sqrt{1 + |\nabla u|^2}} \Big(\partial_t - (\Phi_u)_* \nabla u \Big), \tag{1.11}$$

then a computation in [62] shows that $u : M \to I$ solves

$$\mathrm{div}\left(\frac{\nabla u}{\sqrt{1 + |\nabla u|^2}} \right) = m\lambda(u)H + m\frac{\lambda_t(u)}{\lambda(u)}\frac{1}{\sqrt{1 + |\nabla u|^2}}$$

$$= mh(v)H + \frac{mh'(v)}{\sqrt{1 + |\nabla u|^2}} \qquad \text{on } M, \tag{1.12}$$

where λ_t is the derivative of λ with respect to t, and h' is the derivative of h with respect to s. In particular, entire minimal graphs satisfy

$$\mathrm{div}\left(\frac{\nabla u}{\sqrt{1 + |\nabla u|^2}} \right) = \frac{mh'(v)}{\sqrt{1 + |\nabla u|^2}} \qquad \text{on } M, \tag{1.13}$$

that belongs to the family described in (1.3). The behaviour of solutions to (1.12), of course, depends on the relation between the geometry of M and the growth of H, h'. In the hyperbolic space $\mathbb{H}^{m+1}$, the existence of graphs with constant mean curvature (CMC) along horospheres or hyperspheres has been considered in a paper by M.P. Do Carmo and H.B. Lawson [75], who proved the following remarkable result:

Theorem 1.3 *Let* $\Sigma^m \to \mathbb{H}^{m+1}$ *be a graph over a horosphere or a hypersphere* M. *Suppose that* Σ *has constant mean curvature* $H \in [-1, 1]$. *Then, the following occurs:*

(i) *if* M *is a horosphere, then* $H = \pm 1$ *and* Σ *is a horosphere;*
(ii) *if* M *is a hypersphere, then* $H \in (-1, 1)$ *and* Σ *is a hypersphere.*

In other words, in the above setting the graph function is constant. Note that, differently from the Euclidean case, no restriction appears on the dimension m. The result applies in fact to the much more general setting of properly embedded hypersurfaces, but its proof, relying on the moving plane method, is not suited to ambient manifolds with few isometries or to graphs with non-constant mean curvature, and calls for new ideas and techniques. It is natural to ask for which classes of M and h a result like Theorem 1.3 holds for CMC graphs in $\mathbb{R} \times_h M$.

Mean Curvature Flow Solitons

A further example leading to an equation of the type (1.3) is that of solitons for the mean curvature flow. We recall that a smooth map

$$\varphi : [0, T) \times \Sigma^m \to \bar{M}^{m+1}$$

with $\varphi_t = \varphi(t, \cdot)$ an immersion for each t, is said to be a mean curvature flow (shortly, MCF) if

$$\frac{\partial \varphi}{\partial t} = \vec{H}_t,$$

where $\vec{H}_t$ is the unnormalized mean curvature vector of φ_t. The MCF φ is called a soliton if there exists a conformal vector field Y on $\bar{M}$ such that

$$\varphi_t(M) = \Psi_{\tau(t)}(\varphi_0(M)) \qquad \text{for every } t \in [0, T),$$

where Ψ_τ is the flow of Y and $\tau(t)$ is a time reparametrization. In other words, we require

$$\varphi(t, x) = \Psi\big(\tau(t), \eta(t, x)\big) \qquad \forall (x, t) \in \Sigma \times [0, T),$$

with η the flow of some time-dependent tangential vector field on M. Differentiating at $t = 0$, a soliton satisfies the identity $\vec{H}_0 = \tau'(0)Y^\perp$, where $\perp$ is the orthogonal projection on the normal bundle. Up to rescaling Y, we can assume that $\tau'(0) = 1$, obtaining

$$\vec{H} = Y^\perp. \tag{1.14}$$

Solitons in $\mathbb{R}^{m+1}$ with respect to the homothetically shrinking and to the translating vector fields Y give rise, respectively, to classical self-shrinkers and self-translators, that model the singularities developed under the MCF (cf. [149]). Bernstein type theorems for shrinkers that are graphs over $\mathbb{R}^m$ have been proved in [80, 234], and for translators in [15]. Although solitons in more general ambient spaces can no longer describe the blow-up picture near a singularity, nevertheless they are still relevant since they act as barriers for the MCF evolution. Suppose that Σ is a geodesic graph, with graph function $u : M \to I$ along the lines of the conformal field $X = \partial_t$, and note that $\vec{H} = mHv$. If the soliton field Y coincides with $\pm X$, Eq. (1.12) specifies to

$$\text{div}\left(\frac{\nabla u}{\sqrt{1 + |\nabla u|^2}}\right) = \left[\frac{m\lambda_t(u) \pm \lambda^3(u)}{\lambda(u)}\right] \frac{1}{\sqrt{1 + |\nabla u|^2}}$$
$$= \frac{mh'(v) \pm h^2(v)}{\sqrt{1 + |\nabla u|^2}}. \tag{1.15}$$

In particular, the equation for a self-translator in $\mathbb{R}^{m+1}$ that is a translating graph in the vertical direction is

$$\text{div}\left(\frac{\nabla u}{\sqrt{1 + |\nabla u|^2}}\right) = \frac{\pm 1}{\sqrt{1 + |\nabla u|^2}}.$$

Equidistant Graphs in $M \times_h \mathbb{R}$

Together with geodesic graphs in $\mathbb{R} \times_h M$, it is natural to study graphs in products $M \times_h \mathbb{R}$ endowed with the metric

$$(\,,\,) = \langle\,,\,\rangle + h(x)^2 ds^2,$$

for some $0 < h \in C^\infty(M)$. In this case, $X = \partial_s$ is Killing but not parallel, unless h is constant. Curves $\{x = \bar{x}\}$ for $\bar{x}$ constant have the property that the intersection of any two of them with a slice $\{s = \text{const}\}$ is a pair of points whose distance does not depend on the slice, and for this reason graphs of $u : M \to \mathbb{R}$ will be named *equidistant graphs*. For instance, $\mathbb{H}^{m+1}$ admits two such warped product decompositions, according to whether X has one or two fixed points at infinity: the first can be obtained by isolating the coordinate

$s = x_m$ in the upper half-space model, leading to

$$\mathbb{H}^{m+1} = \mathbb{H}^m \times_h \mathbb{R} \qquad \text{with } h(x_0, \dots, x_{m-1}) = \frac{1}{x_0}; \qquad (1.16)$$

the second can be written as

$$\mathbb{H}^{m+1} = \mathbb{H}^m \times_{\cosh r} \mathbb{R}, \qquad (\,,\,) = \langle\,,\,\rangle_{\mathbb{H}^m} + \left(\cosh^2 r(x)\right)ds^2, \qquad (1.17)$$

with $r : \mathbb{H}^m \to \mathbb{R}$ the distance from a fixed origin in $\mathbb{H}^m$, and corresponds, in the upper half-space model, to the fibration of $\mathbb{H}^{m+1}$ via Euclidean lines orthogonal to the totally geodesic hypersphere $\{x_0^2 + |x|^2 = 1\}$. Having defined the normal direction

$$\nu = \frac{1}{h\sqrt{1 + h^2|\nabla u|^2}}\left(\partial_t - h^2(\Phi_u)_*\nabla u\right),$$

a computation in [61] shows that u solves

$$\operatorname{div}\left(\frac{h\nabla u}{\sqrt{1 + h^2|\nabla u|^2}}\right) = mH - \left\langle\frac{h\nabla u}{\sqrt{1 + h^2|\nabla u|^2}}, \frac{\nabla h}{h}\right\rangle \qquad \text{on } M. \qquad (1.18)$$

If we consider the conformal deformation

$$\widehat{(\,,\,)} = h^{-2}\langle\,,\,\rangle,$$

and we denote with $\|\cdot\|$, D and $\widehat{\operatorname{div}}$, respectively, the norm, connection and divergence in the metric $\widehat{(\,,\,)}$, then (1.18) is equivalent to

$$\widehat{\operatorname{div}}_h\left(\frac{Du}{\sqrt{1 + \|Du\|^2}}\right) = mHh^2 \qquad \text{on } \left(M, \widehat{(\,,\,)}\right), \qquad (1.19)$$

where $\widehat{\operatorname{div}}_h$ is the following weighted divergence:

$$\widehat{\operatorname{div}}_h Y = h^{m-1}\widehat{\operatorname{div}}\left(h^{1-m}Y\right).$$

In particular, if φ is a soliton for $Y = \pm X$, then (1.19) becomes

$$\widehat{\operatorname{div}}_h\left(\frac{Du}{\sqrt{1 + \|Du\|^2}}\right) = \frac{\pm h(x)^3}{\sqrt{1 + \|Du\|^2}} \qquad \text{on } \left(M, \widehat{(\,,\,)}\right). \qquad (1.20)$$

Up to replacing the Riemannian divergence with a weighted one, (1.19) and (1.20) belong to the family of equations that are considered in the present book. Indeed, the proof of some of our major results proceed verbatim in the case of a weighted divergence in the left-hand side of (1.3). However, to apply our main nonexistence results we shall need that the metric $\widehat{\langle\,,\,\rangle}$ be complete, which impose restrictions on h and excludes graphs in $\mathbb{H}^{m+1}$ along the two decompositions mentioned above. In fact, Rigidity does not hold for such graphs in $\mathbb{H}^{m+1}$, at east in the CMC case: the Plateau's problem at infinity for graphs with constant $H \in (0, 1)$ is always solvable, by work of P. Guan and J. Spruck [108] for (1.17), and J. Ripoll and M. Telichevesky [211] for (1.16). Nevertheless, our theorems apply when $h \in L^\infty(M)$, notably for h vanishing at infinity, that is analytically the most subtle case of (1.20). Existence for the prescribed mean curvature equation on more general products $M \times_h \mathbb{R}$ is studied in [42].

1.2   A Problem from the Theory of Stochastic Control

Equation (1.12) for geodesic graphs with prescribed mean curvature suggests to consider more general problems of the type

$$\Delta_\varphi u \geq b(x)f(u) + \bar{b}(x)\bar{f}(u)\bar{l}(|\nabla u|), \tag{1.21}$$

for suitable $b, \bar{b}$ positive on M, $f, \bar{f} \in C(\mathbb{R})$ and $\bar{l} \in C(\mathbb{R}_0^+)$, where we set $\mathbb{R}_0^+ \doteq [0, \infty)$. Here, Δ_φ denotes a general quasilinear operator that depends on an increasing function φ on $\mathbb{R}_0^+$, whose properties will be listed in the next chapter. For suitable φ, the family includes the mean curvature operator a well as the Laplace–Beltrami and the p-Laplacian

$$\Delta_p u \doteq \mathrm{div}\big(|\nabla u|^{p-2}\nabla u\big), \qquad p > 1.$$

As a matter of fact, a prototype case appears in Stochastic Control Theory, and a detailed description of the model can be found in J.-M. Lasry and P.-L. Lions [140]. It can be roughly summarized as follows: given an open subset $\Omega \subset \mathbb{R}^m$, and given a continuous function $a : \Omega \to \mathbb{R}^m$, the state of a controlled system is assumed to be a diffusion process X_t valued in $\overline{\Omega}$ that satisfies the stochastic PDE

$$\mathrm{d}X_t = a(X_t)\mathrm{d}t + \mathrm{d}\mathscr{B}_t,$$

where $\mathscr{B}_t$ is a Brownian motion and $a(X_t)$ models a feedback control. The function a is assumed to lie a suitable class $\mathscr{A} \subset C(\Omega, \mathbb{R}^m)$ to guarantee that X_t is valued in Ω almost surely, so typically $|a(x)| \to \infty$ on $\partial\Omega$. In this case, no boundary conditions have to be imposed on $\partial\Omega$. For $b \in C(\overline{\Omega})\cap L^\infty(\Omega)$, one can consider a cost function $J : \Omega \times \mathscr{A} \to \mathbb{R}$ given by

$$J(x, a) = \mathbb{E}\left[\int_0^\infty \left(\frac{1}{q}|a(X_t)|^q - b(X_t)\right)e^{-\lambda t}\mathrm{d}t\right]$$

Here, $\lambda > 0$ represents a discount factor, and $q > 1$. In view of Bellman's dynamic programming principle, the Bellman function

$$u(x) = \inf_{a \in \mathscr{A}} J(x, a) \qquad \forall x \in \Omega$$

solves

$$\Delta u = \frac{1}{p}|\nabla u|^p + \lambda u + b(x) \qquad \text{on } \Omega, \text{ where } p = \frac{q}{q-1}. \qquad (1.22)$$

(at least heuristically: as pointed out in [140], the restrictions on a, X_t require nontrivial arguments to justify (1.22)). The case $q \geq 2$, that is, a running cost blowing up fast at infinity, is particularly interesting, and corresponds to $1 < p \leq 2$. It is also meaningful to consider the problem set in the entire space $\Omega = \mathbb{R}^m$, and in this case to consider *large* solutions, i.e. solutions satisfying $u(x) \to \infty$ as $|x| \to \infty$. Needless to say, the case of manifold-valued processes X is also natural, so is the study of (1.22) on Riemannian manifolds. Note that (1.22) implies an inequality of the type (1.21) under various reasonable assumptions on b, λ. In particular, the ergodic limit $\lambda \to 0$ leads to the study of

$$\Delta u = b(x) + \frac{1}{p}|\nabla u|^p \qquad \text{on } M.$$

As another example, assume $b \geq 0$ on $\mathbb{R}^m$ and $\lambda \geq \|b\|_\infty$, in which case

$$\Delta u \geq \frac{1}{p}|\nabla u|^p + b(x)(u+1)$$

is matched on the set $\{u > 0\}$ (open, if we assume $u \in C(\mathbb{R}^m)$). In particular, choosing $\varepsilon > 0$ and $f \in C(\mathbb{R})$ with $f(t) = t+1$ for $t > \varepsilon$, $f \equiv 0$ on $(-\infty, 0)$ and $0 \leq f(t) \leq t+1$ on $\mathbb{R}$, $v \doteq u_+$ solves

$$\Delta v \geq b(x)f(v) + \frac{1}{p}|\nabla v|^p \qquad \text{on } M.$$

The existence and nonexistence problem on the entire $\mathbb{R}^m$ for solutions to

$$\Delta u = b(x)f(u) + c|\nabla u|^p, \qquad \text{with } c > 0, \ p > 0,$$

where $0 < b \in C(\mathbb{R}^m)$, $f \in C(\mathbb{R})$, was studied by A.V. Lair and A.W. Wood [139], M. Ghergu and V. Radulescu [103], F. Toumi [227] and R. Filippucci, P. Pucci and M. Rigoli [94]. On complete manifolds, it was addressed in [157]. Although the main results of the present book are restricted to inequalities of the type

$$\Delta_\varphi u \geq b(x) f(u) l(|\nabla u|),$$

we mention that most of them can be applied in the more general setting of (1.21). However, a direct study of (1.21) would have lead to much more involved results, making hardly readable our attempt to describe in detail the influence of geometry on the behaviour of solutions.

An Overview of Our Results

2

The study of differential inequalities of the type

$$\operatorname{div} \mathcal{A}(x, u, \nabla u) \geq \mathcal{B}(x, u, \nabla u) \tag{2.1}$$

on Euclidean space $\mathbb{R}^m$ is a classical subject, and a great deal of work has been devoted to the analysis of the qualitative properties of solutions. The literature is vast, and, as said in Chap. 1, we restrict ourselves to the special case

$$\mathcal{B}(x, u, \nabla u) = b(x) f(u) l(|\nabla u|), \tag{2.2}$$

for continuous b, f, l. With no claim of completeness, we quote [14, 66, 86, 87, 92, 95, 105, 144, 162, 195], and for similar inequalities in the sub-Riemannian setting of Carnot groups, [5, 33–35, 63, 147]. The results in the references above will be related to those in our work in a more precise way later on.

To the best of our knowledge, only a few authors have analysed the influence of geometry on the behaviour of solutions of (2.1), (2.2) in a general setting, for instance see [6, 147, 157, 198], leaving however the picture still fragmentary, especially in case where l in (2.2) is a non-constant function. As one of the main purposes of the present work, we aim to give a detailed account of how geometry comes into play at the global level. Nevertheless, some interesting questions and problems remain open, and will be specified in due course in the book.

© The Author(s), under exclusive license to Springer Nature Switzerland AG 2021
B. Bianchini et al., *Geometric Analysis of Quasilinear Inequalities on Complete Manifolds*, Frontiers in Mathematics, https://doi.org/10.1007/978-3-030-62704-1_2

2.1   Setting and Main Properties Under Investigation

From now on, we let $(M, \langle \, , \, \rangle)$ be a Riemannian manifold of dimension $m \geq 2$. We shall assume throughout this book that M is non-compact. To avoid excessive technicalities, while still keeping a good amount of generality, we study the following subclass of (2.1): we consider a quasilinear operator Δ_φ, called the φ-Laplacian, weakly defined by

$$\Delta_\varphi u = \operatorname{div}\left(\frac{\varphi(|\nabla u|)}{|\nabla u|} \nabla u\right),$$

where we assume

$$\varphi \in C(\mathbb{R}_0^+), \quad \varphi(0) = 0, \quad \varphi(t) > 0 \text{ on } \mathbb{R}^+; \tag{2.3}$$

hereafter, $\mathbb{R}_0^+ = [0, \infty)$ and $\mathbb{R}^+ = (0, \infty)$. Different choices of φ give rise to well-known operators including, for instance, those presented in the examples below.

Example 2.1 The p-Laplace operator Δ_p for $p > 1$, where $\varphi(t) = t^{p-1}$:

$$\Delta_p u = \operatorname{div}\left(|\nabla u|^{p-2} \nabla u\right).$$

Besides its importance in Physics and Image Processing, the p-Laplacian is also interesting since it allows to efficiently bridge between two relevant operators obtained formally as (suitably normalized) limits as $p \to 1$ and $p \to \infty$:

$$\Delta_1 u \doteq \operatorname{div}\left(\frac{\nabla u}{|\nabla u|}\right), \qquad \Delta_\infty u \doteq \nabla^2 u\left(\frac{\nabla u}{|\nabla u|}, \frac{\nabla u}{|\nabla u|}\right).$$

The 1-Laplacian $\Delta_1 u$ describes the mean curvature of the level set $\{u = \text{const}\}$ in the normal direction $-\nabla u/|\nabla u|$, and is therefore important in the study of level set methods for the weak definition of geometric flows driven by the mean curvature. In fact, approximation with the p-Laplacian works well, for instance, to construct solutions of the inverse mean curvature flow, see [137, 156, 164]. On the other hand, the (normalized) ∞-Laplacian $\Delta_\infty u$ is tightly related to the metric geometry of the underlying manifold, and indeed the geodesic completeness of a Riemannian metric can be detected in terms of a potential theory for the ∞-Laplacian (see [153, 154]).

Example 2.2 The mean curvature operator, for which $\varphi(t) = t(1 + t^2)^{-1/2}$.

Example 2.3 The operator of exponentially harmonic functions, where $\varphi(t) = t \exp\left(t^2\right)$, was introduced in [79, 81]. The operator has connections with the nonlinear Hodge–De Rham theory developed in [221], and has interesting applications in gas dynamics: indeed, following [221], if u represents the velocity potential of a compressible fluid, and if the density of the fluid is assumed to be a positive function $\varphi(|\nabla u|)/|\nabla u|$ of the speed $|\nabla u|$, in view of the motion and continuity equations, and because of Bernoulli's law, u satisfies

$$\Delta_\varphi u = 0.$$

The ellipticity condition $\varphi' > 0$ characterizes the flow to be subsonic. A prototype case is that of the polytropic flow, corresponding to the choice

$$\varphi(t) = t\left(1 - \frac{\gamma - 1}{2}t^2\right)^{\frac{1}{\gamma-1}}$$

for a given adiabatic constant $\gamma > 1$, that is subsonic whenever

$$t < \sqrt{\frac{2}{\gamma + 1}}.$$

In this case, our techniques can be applied provided that we know, a priori, that $|\nabla u|^2$ is less than the sonic value $2/(\gamma + 1)$. More details on the physical interpretation can be found in [19].

Example 2.4 The (p, q)-Laplacian

$$\Delta_{p,q} u = \Delta_p u + \Delta_q u,$$

associated to $\varphi(t) = t^{p-1} + t^{q-1}$ with $1 < p < q$. This operator was independently introduced by V.V. Zhikov [240] and P. Marcellini [152], and can be regarded as a model example related to functionals with non-standard growth conditions. In particular, Zhikov introduced the operator in his study of anisotropic materials, homogenization and elasticity, cf. also [241]. The (p, q)-Laplacian also appears in models for quantum and plasma physics, and in chemical reaction design: for instance, in [71], G.H. Derrick proposed to use operators like $\Delta_{p,q}$ in an attempt to overcome a problem arising in the description of elementary particles by means of static solutions with finite energy to the generalized Klein–Gordon equation. The problem has been tackled in [16], see also [53].

We focus our attention on the differential inequalities

$$(P_\geq) \quad \Delta_\varphi u \geq b(x) f(u) l(|\nabla u|)$$
$$(P_\leq) \quad \Delta_\varphi u \leq b(x) f(u) l(|\nabla u|) \qquad\qquad (2.4)$$
$$(P_=) \quad \Delta_\varphi u = b(x) f(u) l(|\nabla u|)$$

in a *connected* open set, that is, a domain $\Omega \subset M$. Typically, we do not require Ω to be relatively compact, indeed Ω might coincide with M or with an end of M, that is, with a non-relatively compact connected component of $M \backslash K$, for some compact set K.

We fix the next basic assumptions on b, f, l:

$$
\begin{aligned}
b &\in C(M), \qquad b > 0 \text{ on } M, \\
f &\in C(\mathbb{R}), \\
l &\in C(\mathbb{R}_0^+), \qquad l > 0 \text{ on } \mathbb{R}^+.
\end{aligned}
\tag{2.5}
$$

Because of the positivity of b and l, if $f \geq 0$ the problems in (2.4) are called completely coercive in the recent literature (see [64, 86, 87]). Obviously, solutions of (2.4) are considered in the weak sense and, in view of geometric applications, we confine ourselves to locally Lipschitz or C^1 solutions. It should be stressed that relaxing their regularity class is by no means a trivial or just a technical issue. For instance, under our requirements on φ, b, l, we are not aware of the validity of weak Harnack inequalities for (2.4), and solutions may not even be locally bounded.

Definition 2.5 A function $u : \Omega \to \mathbb{R}$ is a C^1 *solution* (respectively, $\mathrm{Lip}_{\mathrm{loc}}$ *solution*) of $(P_\geq)$ in (2.4) if $u \in C^1(\Omega)$ (resp., $u \in \mathrm{Lip}_{\mathrm{loc}}(\Omega)$) and satisfies $(P_\geq)$ in the weak sense, that is,

$$
-\int_\Omega \frac{\varphi(|\nabla u|)}{|\nabla u|} \langle \nabla u, \nabla \psi \rangle \geq \int_\Omega b(x) f(u) l(|\nabla u|) \psi \qquad \text{for each } \psi \in C_c^\infty(\Omega), \ \psi \geq 0,
$$

where integration is performed with respect to the Riemannian measure. The analogous statement, with the reverse inequality, defines C^1 and $\mathrm{Lip}_{\mathrm{loc}}$ solutions of $(P_\leq)$.

We make a preliminary observation. Suppose that f has at least a zero on $\mathbb{R}$: then, by the translation invariance of (2.4) with respect to u, without loss of generality we can assume that $f(0) = 0$. The function $u \equiv 0$ is then a solution of $(P_=)$ on M, and the reduction principle in [64] (see Lemma 6.6 below, [141] and the appendix of [5]) guarantees that

$$
u_+ = \max\{u, 0\} \quad \text{solves } (P_\geq) \text{ weakly on } \Omega.
$$

Therefore, when f has a zero, without loss of generality we can restrict ourselves to investigate $(P_\geq)$ under the further assumption $f(0) = 0$ and, if $u > 0$ somewhere, we can also suppose $u \geq 0$ on Ω.

Definition 2.6 We say that:

- the *compact support principle* (shortly, (CSP)) holds for $(P_{\geq})$ if each non-negative C^1 solution of $(P_{\geq})$ on an end Ω of M, satisfying $u(x) \to 0$ as $x \in \Omega$, $\text{dist}(x, \partial\Omega) \to \infty$, has compact support, in $\overline{\Omega}$, that is, $u \equiv 0$ outside some compact set of $\overline{\Omega}$.
- the *finite maximum principle* (shortly, (FMP)) holds for $(P_{\leq})$ on the domain $\Omega \subset M$ if any non-negative C^1 solution of $(P_{\leq})$ for which $u(x_0) = 0$ at some $x_0 \in \Omega$, satisfies $u \equiv 0$ on Ω;
- the *strong Liouville property* (shortly, (SL)) holds for $(P_{\geq})$ if there exist no non-negative, non-constant C^1 solutions of $(P_{\geq})$ on all of M.
- the *Liouville property* (shortly, (L)) holds for $(P_{\geq})$ if there exist no non-negative, non-constant, *bounded* C^1 solutions of $(P_{\geq})$ on all of M.

Remark 2.7 We emphasize that the only difference between properties (L) and (SL) is that, in (L), we require that the solution of $(P_{\geq})$ be a priori bounded.

Remark 2.8 In some but not all of our results, we will indeed prove (SL), (L) when u is just assumed to be locally Lipschitz (actually, even an appropriate Sobolev regularity would suffice). If this were the case, we accordingly say that (SL), (L) hold for Lip_{loc} solutions.

Remark 2.9 (Constant Solutions) It is clear, by the properties of b, f, l in (2.5), that a constant $u = u^*$ solves $(P_{\geq})$ if and only if

$$l(0) = 0, \quad \text{independently of } f, \quad \text{or}$$

$$l(0) > 0, \quad f(u^*) \leq 0.$$

Therefore, in what follows we will always concentrate on *non-constant* solutions.

Note that (FMP) is of a local nature, and thus its validity should not depend on the considered manifold. On the other hand, (CSP), (L) and (SL) are *global* properties, and for this reason they are expected to depend on the geometry at infinity of M and not only on the structure of the operator related to φ, b, f, l. More precisely, the next scheme summarizes what occurs in general:

$$\begin{Bmatrix} \text{geometric conditions} \\ \text{related both to } b \\ \text{and to } \varphi, f, l \end{Bmatrix} + \begin{Bmatrix} \text{condition} \\ \text{on } \varphi, f, l \end{Bmatrix} \implies \begin{Bmatrix} \text{either (SL),} \\ \text{or (L),} \\ \text{or (CSP)} \end{Bmatrix}.$$

2.2   Keller–Osserman Conditions

We now describe the requirements on φ, f, l needed in order to possibly obtain (SL) or (FMP), (CSP), and next we will consider the role of geometry and of b. Our conditions will measure the combined growth/decay of f, l with respect to φ, respectively in a neighbourhood of zero (for (FMP) and (CSP)) and in a neighbourhood of infinity (for (SL)), and will extend in a nontrivial way the classical Keller–Osserman integrability assumption

$$\frac{1}{\sqrt{F}} \in L^1(\infty) \tag{2.6}$$

which has been recalled in Chap. 1. We assume

$$\begin{cases} \varphi \in C^1(\mathbb{R}^+), & \varphi' > 0 \text{ on } \mathbb{R}^+, \\ \dfrac{t\varphi'(t)}{l(t)} \in L^1(0^+). \end{cases}$$

Then, the function

$$K(t) = \int_0^t \frac{s\varphi'(s)}{l(s)} \, \mathrm{d}s \tag{2.7}$$

realizes a homeomorphism of $\mathbb{R}_0^+$ onto its image $[0, K_\infty)$, with inverse

$$K^{-1} : [0, K_\infty) \to \mathbb{R}_0^+.$$

Unless otherwise specified, we set

$$F(t) = \int_0^t f(s) \mathrm{d}s. \tag{2.8}$$

To deal with (FMP) and (CSP), we further suppose that

$$f \geq 0 \quad \text{on some } [0, \eta_0), \ \eta_0 > 0.$$

The validity of (FMP) and (CSP) is related to the next integrability requirement:

$$\frac{1}{K^{-1} \circ F} \in L^1(0^+). \tag{KO$_0$}$$

More precisely, (FMP) depends on the failure of (KO$_0$) while (CSP) on its validity. Regarding (SL), the relevant condition becomes an integrability at infinity, that to be expressed needs the further assumption

$$\frac{t\varphi'(t)}{l(t)} \notin L^1(\infty), \tag{2.9}$$

in order for K^{-1} to be defined on $\mathbb{R}_0^+$ (i.e. $K_\infty = \infty$). If we now suppose that

$$f \geq 0 \text{ on } \mathbb{R}^+,$$

then (SL) depends on the requirement

$$\frac{1}{K^{-1} \circ F} \in L^1(\infty). \tag{KO$_\infty$}$$

An important feature of (KO$_0$) and (KO$_\infty$) to notice is their independence on the underlying space and on the weight b. If $l \equiv 1$, K coincides with the function

$$H(t) = t\varphi(t) - \int_0^t \varphi(s)ds, \qquad t \geq 0, \tag{2.10}$$

that represents the pre-Legendre transform of

$$\Phi(t) = \int_0^t \varphi(s)ds,$$

and in this case we recover the necessary and sufficient conditions for (CSP), (FMP) and (SL) thoroughly investigated in [168, 192, 194, 196] on $\mathbb{R}^m$, see also the references therein. In the case of the p-Laplacian where $\varphi(t) = t^{p-1}$, and for $l \equiv 1$, (KO$_0$) and (KO$_\infty$) take, respectively, the well-known form

$$\frac{1}{F^{1/p}} \in L^1(0^+), \qquad \frac{1}{F^{1/p}} \in L^1(\infty). \tag{2.11}$$

In particular, for $p = 2$ the latter recovers the Keller–Osserman condition described by J.B. Keller and R. Osserman [135, 175] for the prototype case

$$\Delta u \geq f(u) \tag{2.12}$$

on $\mathbb{R}^m$. In [194, p. 125] and in [192] the reader can find a thorough account of the existing literature which concerns the relations between the validity (respectively, failure) of the first in (2.11) and that of, respectively, (CSP) an (FMP). For convenience, in what follows we name both (KO$_0$) and (KO$_\infty$) the *Keller–Osserman conditions*.

To our knowledge, the study of the relations between Keller–Osserman conditions and the geometry of M initiated with the influential paper [50] by S.Y. Cheng and S.T. Yau, for the semilinear example (2.12). In [50, Section 5], they proved that a complete manifold M satisfying

$$\text{Ric} \geq -(m-1)\kappa^2 \tag{2.13}$$

for some constant $\kappa > 0$, has the following property: if f matches the Keller–Osserman condition (2.6), then any solution of (2.12) satisfies

$$u^* < \infty, \qquad f(u^*) \leq 0. \tag{2.14}$$

Nowadays, it is known that (2.13) can be weakened, and much effort has been done to identify the sharp curvature conditions for estimates (2.14) to hold (cf. [182]). It is also known that (2.14) may fail, under the validity of the Keller–Osserman condition, without some control on the geometry at infinity of M. The way geometry relates to the Keller–Osserman conditions in order to give (SL) and (CSP) is one of the primary concerns of the present work, and will be expressed in terms of sharp curvature or volume growth bounds on M, and sharp estimates for b. In this respect, even when l is constant, in many instances such interplay is still partially unclear.

The bridge between geometry and the properties in Definition 2.6 is provided, at least in this book, by the validity of the weak and strong maximum principles at infinity, that we now define:

Definition 2.10 Assume (2.3) and fix b, l satisfying (2.5). We say that

- $(bl)^{-1}\Delta_\varphi$ satisfies the *weak maximum principle at infinity*, shortly, (WMP$_\infty$), if for each non-constant $u \in \text{Lip}_{\text{loc}}(M)$ such that $u^* = \sup_M u < \infty$, and for each $\eta < u^*$,

$$\inf_{\Omega_\eta} \left\{ \left(b(x)l(|\nabla u|) \right)^{-1} \Delta_\varphi u \right\} \leq 0,$$

where

$$\Omega_\eta = \{x \in M \, : \, u(x) > \eta\}$$

and the inequality has to be intended in the following sense: if u solves

$$\Delta_\varphi u \geq K b(x)l(|\nabla u|) \qquad \text{on } \Omega_\eta,$$

for some $K \in \mathbb{R}$, then necessarily $K \leq 0$.

- $(bl)^{-1}\Delta_\varphi$ satisfies the *strong maximum principle at infinity*, shortly, (SMP$_\infty$), if for each non-constant $u \in C^1(M)$ such that $u^* = \sup_M u < \infty$, and for each $\eta < u^*$, $\varepsilon > 0$, the set

$$\Omega_{\eta,\varepsilon} = \{x \in M : u(x) > \eta, \ |\nabla u(x)| < \varepsilon\} \tag{2.15}$$

is non-empty and

$$\inf_{\Omega_{\eta,\varepsilon}} \left\{ \left(b(x)l(|\nabla u|)\right)^{-1} \Delta_\varphi u \right\} \leq 0,$$

where, again, the inequality has to be intended in the way explained above.

Remark 2.11 Condition $\Omega_{\eta,\varepsilon} \neq \emptyset$ in (2.15) is not automatic: for example, consider the function $u(x) = \exp(-|x|)$ on $\mathbb{R}^m \backslash \{0\}$, for which $|\nabla u| \to 1$ when $u \to u^* = 1$. However, it is easy to see that $\Omega_{\eta,\varepsilon}$ is always non-empty if M is complete: by contradiction, if $|\nabla u| \geq \varepsilon$ on Ω_η, take any maximal flow line γ of $X = \nabla u/|\nabla u|$ starting from some $x \in \Omega_\eta$ (it might be locally non-unique since X is just continuous, but it exists by Peano theorem). Note that γ is defined on $\mathbb{R}^+$, since X is bounded and M is complete, and that $\gamma(\mathbb{R}^+) \subset \Omega_\eta$. Thus, integrating $(u \circ \gamma)' = |\nabla u| \geq \varepsilon$ on $\mathbb{R}^+$ we contradict $u^* < \infty$. This reasoning can also be seen as a consequence of I. Ekeland quasimaximum principle.

As we shall see in a short while, (WMP$_\infty$) and (SMP$_\infty$) hold under mild geometric assumptions, involving the Ricci curvature or the volume growth of geodesic balls. Moreover, (WMP$_\infty$) is equivalent to (L) for Lip$_{loc}$ solutions, for each f with $f(0) = 0$ and $f > 0$ on $\mathbb{R}^+$. Both principles relate to (KO$_0$) and (KO$_\infty$) to guarantee, respectively, (CSP) and (SL). To better describe our results and properly place them in the literature, we separately comment on each of the properties in Definitions 2.6 and 2.10.

2.3 Notation and Conventions

Hereafter, given two non-negative functions $h_1, h_2 : \mathbb{R} \to \mathbb{R}$, we write

$$h_1 \asymp h_2 \qquad \text{on an interval } (a, b) \subset \mathbb{R}$$

to indicate that there exists a constant $C \geq 1$ such that

$$C^{-1} h_1(t) \leq h_2(t) \leq C h_1(t) \qquad \forall t \in (a, b).$$

Given a complete manifold M, we denote with $r(x)$ the distance of x from a fixed subset $\mathcal{O} \subset M$ that we call an origin. The origin may be a point (in this case, we denote it by o)

or a relatively compact, open set with smooth boundary. It is known that r is smooth on an open, dense subset $D_{\mathcal{O}} \subset M \backslash \mathcal{O}$, and we denote as usual $\mathrm{cut}(\mathcal{O}) = M \backslash (D_{\mathcal{O}} \cup \mathcal{O})$ the cut-locus of $\mathcal{O}$, see Chap. 3 for more details. A geodesic ball of radius r centred at $\mathcal{O}$ will be denoted with B_r, and $|A|$ will mean either the Riemannian volume measure or the induced $(m-1)$-dimensional Hausdorff measure of a set A, according to the case and provided that there is no risk of confusion. For instance, $|B_r|$ and $|\partial B_r|$ denote, respectively, the volume of a geodesic ball B_r and the $(m-1)$-Hausdorff measure of ∂B_r. On $D_{\mathcal{O}}$, we define the radial sectional curvature $\mathrm{Sec}_{\mathrm{rad}}$ to be the restriction of the sectional curvature Sec to planes containing ∇r. Henceforth, with the notation

$$\mathrm{Sec}_{\mathrm{rad}} \leq G(r)$$

for some $G \in C(\mathbb{R}^+)$, we mean that

$$\mathrm{Sec}(X \wedge \nabla r)(x) \leq G(r(x))$$

for each $x \in D_{\mathcal{O}}$ and $X \perp \nabla r(x)$, $|X| = 1$, where $X \wedge \nabla r$ is the 2-plane spanned by X and ∇r.

2.4 The Finite Maximum Principle (FMP)

Beyond the basic requirements (2.3) and (2.5), assume also

$$\begin{cases} \varphi \in C^1(\mathbb{R}^+), & \varphi' > 0 \text{ on } \mathbb{R}^+, \\ \dfrac{t\varphi'(t)}{l(t)} \in L^1(0^+). \end{cases} \tag{2.16}$$

We construct F and K respectively as in (2.7) and (2.8). If $f > 0$ in a right neighbourhood of zero, the validity of (FMP) turns out to depend on the next non-integrability requirement:

$$\frac{1}{K^{-1} \circ F} \notin L^1(0^+). \tag{$\neg$KO$_0$}$$

If $l \equiv 1$, that is, K coincides with the function H in (2.10), and if f is non-decreasing and positive, in [192, 196] property (FMP) is shown to be equivalent to ($\neg$KO$_0$), see also Chapter 5 and Theorem 1.1.1 of [194]. We presently extend such a characterization to the case of a non-constant function l. The literature on the finite maximum principle for quasilinear inequalities is fairly intricate, with contributions from a number of different mathematicians. A detailed and commented account of previous works can be found in [194, p. 125] and in [192]. To introduce our main result, we begin with

Definition 2.12 Given a constant $C \geq 1$, we say that a function $h : \mathbb{R} \to \mathbb{R}$ is C-increasing on (a, b), $-\infty \leq a < b \leq \infty$, if

$$\forall t \in (a, b), \qquad \sup_{a < s < t} h(s) \leq Ch(t).$$

Clearly, h is 1-increasing if and only if it is non-decreasing, but on the other hand a C-increasing function is allowed to oscillate in a controlled way, so that, for example,

$$h(t) = t^2(2 + \sin t) \qquad \text{is 3-increasing on } [0, \infty).$$

Theorem 2.13 *Let M be a Riemannian manifold, and assume that φ, b, f, l satisfy (2.3), (2.5), and (2.16). Suppose further that*

- $f(0)l(0) = 0$;
- f *is non-negative and C-increasing on $(0, \eta_0)$, for some $\eta_0 > 0$;*
- l *is C-increasing on $(0, \xi_0)$, for some $\xi_0 > 0$.*

Then, (FMP) holds for non-negative solutions $u \in C^1(\Omega)$ of $(P_\leq)$ on a domain $\Omega \subset M$ if and only if either

$$f \equiv 0 \qquad \text{on } [0, \eta_0),$$

or

$$f > 0 \quad \text{on } (0, \eta_0), \quad \text{and} \quad \frac{1}{K^{-1} \circ F} \notin L^1(0^+).$$

Remark 2.14 For the sake of clarity, in [192] no differentiability of φ is needed: indeed, φ' does not appear in the definition of H, and the authors just require φ to be strictly increasing. However, the presence of a possibly only continuous function l forces us to increase the regularity of φ to be able to define K.

Example 2.15 Observe that Theorem 2.13 applies to the inequality

$$\Delta_p u \leq u^\omega |\nabla u|^q,$$

with $p > 1$, $\omega \geq 0$, $q \in [0, p)$, to guarantee that (FMP) holds if and only if

$$\omega + q \geq p - 1.$$

The proof of Theorem 2.13 follows the standard method used to prove Hopf type lemmas, that is, it relies on the construction of suitable radial solutions of $(P_\geq)$ defined on annuli (see [192, 194]). However, the study of the related ODE is, for nontrivial gradient terms l, considerably more involved than that in [194]. This calls for a detailed investigation of singular Dirichlet and mixed Dirichlet–Neumann problems for quasilinear ODEs, accomplished in Sect. 5. The results therein have independent interest, and are central in many of the main theorems of the present book.

For inequalities of the type

$$\Delta_{p,q} u \leq f(u),$$

where $\Delta_{p,q}$ is the operator in Example 2.4, very recently in [190] the authors succeeded to prove Theorem 2.13 *without requiring the C-monotonicity of* f. It is likely that the same is possible also for $(P_\leq)$ in our generality, and so we propose the following

Problem 1 Is it possible to prove Theorem 2.13 without requiring the C-monotonicity of f and l? Or, at least, keeping the C-monotonicity of just one of them?

2.5   Strong and Weak Maximum Principles at Infinity

We start describing the origin of properties (SMP$_\infty$) and (WMP$_\infty$), and for simplicity we restrict to the case $b \equiv 1, l \equiv 1$ and $\Delta_\varphi = \Delta$, the Laplace–Beltrami operator. In this case, when $u \in C^2(M)$, (SMP$_\infty$) and (WMP$_\infty$) can equivalently be restated as the existence of a sequence of points $\{x_k\}_{k\in\mathbb{N}} \subset M$ such that

$$u(x_k) > u^* - \frac{1}{k}, \qquad \Delta u(x_k) < \frac{1}{k}, \qquad |\nabla u|(x_k) < \frac{1}{k} \qquad (2.17)$$

for (SMP$_\infty$) to hold, and

$$u(x_k) > u^* - \frac{1}{k}, \qquad \Delta u(x_k) < \frac{1}{k}$$

for (WMP$_\infty$) to hold. The property is obvious if u attains its supremum, in particular if M is compact, since each x_k can be chosen to be equal to a maximum point of u. We can therefore argue that (SMP$_\infty$), (WMP$_\infty$) are ways to guarantee that M is, loosely speaking, "not too far from being compact", and in fact they effectively replace the lack of compactness of M in the investigation of geometric problems. The validity of (SMP$_\infty$) has first been studied in the pioneering papers by H. Omori [174] and S.T. Yau (cf. [239] and [50]), and for this reason is called the Omori–Yau principle: it proved to be a fundamental tool, and currently there is a huge number of results deriving from suitable applications of the principle. The interested reader is referred to [182] and [6] for a detailed account and for a thorough set of references. Omori in [174] realized that the validity of (SMP$_\infty$) is not granted on a generic Riemannian manifold, although it is sufficient that M enjoys very mild requirements. For example, by combining works of [31, 52, 182, 201], see [6, Thm. 2.4], Δ satisfies (SMP$_\infty$) whenever

$$\text{Ric}(\nabla r, \nabla r) \geq -G(r) \qquad \text{on } D_o, \tag{2.18}$$

$r(x)$ being the distance from a point o, and $G \in C^1(\mathbb{R})$ has the following properties:

$$G > 0, \quad G' \geq 0, \quad \frac{1}{\sqrt{G}} \notin L^1(\infty). \tag{2.19}$$

Clearly, in (2.18) and (2.19) what really matters is the growth of G at infinity. A borderline example is given, for instance, by

$$G(t) \asymp 1 + t^2 \qquad \text{on } \mathbb{R}^+.$$

This is a particular case of a criterion discovered in [182, 201], granting the validity of (SMP$_\infty$) provided that M supports a function satisfying

$$
\begin{aligned}
&w \in C(M) \cap C^2(M \backslash K) \ \text{ for some compact } K, \\
&w(x) \to +\infty \ \text{ as } x \ \text{ diverges in } M, \\
&|\nabla w| \leq \sqrt{G(w)}, \qquad \Delta w \leq \sqrt{G(w)} \qquad \text{on } M \backslash K,
\end{aligned}
\tag{2.20}
$$

where G meets the requirements in (2.19). Note that the second in (2.20) means that all lower level sets $\{w \leq \text{const}\}$ are compact. Indeed, it is sufficient that $w \in C(M)$ and solves the inequalities in (2.20) in the viscosity sense (cf. [154]): under the assumption (2.18), for instance, the function

$$w(x) = \log \int_0^{r(x)} \frac{ds}{\sqrt{G(s)}} \qquad \text{for } r \gg 1$$

satisfies (2.20) in the viscosity sense. For reasons that will be soon justified, we call w a *strong Khas'minskii potential*. To the best of our knowledge, this is essentially the only effective known condition, and w is often explicitly given not exclusively via curvature bounds like (2.18), but also by the geometrical nature of the problem at hand. This is the case, for instance, of immersed submanifolds, where w depends on extrinsic data, and of generic Ricci soliton structures, see [6].

When the operator is nonlinear and non-homogeneous, to guarantee (SMP$_\infty$) we need, instead of a single function w, an entire family of strong Khas'minskii potentials, see Sect. 8 below, [4] and Chapter 3 in [6]. For $b^{-1}\Delta_\varphi$, (SMP$_\infty$) has been studied in [182,

Sec. 6] and [191, Thm 1.1], and again a family of potentials of strong Khas'minskii type is exhibited to ensure (SMP$_\infty$) under an appropriate Ricci curvature bound. The construction of the potential in these papers is hand-made and appears not easily generalizable to cover the case of a non-constant l. Therefore, although our present strategy to prove (SMP$_\infty$) for $(bl)^{-1}\Delta_\varphi$ is still based on finding a strong Khas'minskii potential family, the construction of the latter relies on a different approach involving the study of the maximal domain of existence and the asymptotic behaviour of solutions of a singular two-points boundary ODE problem, see Sects. 5 and 8.1.

For the convenience of the reader, we summarize the assumptions of φ, l in the following:

$$
\begin{cases}
\varphi \in C(\mathbb{R}_0^+) \cap C^1(\mathbb{R}^+), \quad \varphi(0) = 0, \quad \varphi' > 0 \text{ on } \mathbb{R}^+; \\[2mm]
l \in C(\mathbb{R}_0^+), \quad l > 0 \text{ on } \mathbb{R}^+; \\[2mm]
\dfrac{t\varphi'(t)}{l(t)} \in L^1(0^+).
\end{cases}
\tag{2.21}
$$

We shall also require the growth conditions

$$
\begin{cases}
l(t) \geq C_1 \dfrac{\varphi(t)}{t^\chi} \quad \text{on } (0, 1], \text{ for some } C_1 > 0, \chi \geq 0; \\[3mm]
\varphi(t) \leq C_2 t^{p-1} \quad \text{on } [0, 1], \text{ for some } C_2 > 0, p > 1.
\end{cases}
\tag{2.22}
$$

Remark 2.16 Since l is continuous up to zero, if $\varphi(t) \asymp t^{p-1}$ near $t = 0$ the first condition in (2.22) forces the upper bound $\chi \leq p - 1$. For example, in the p-Laplacian case where $\varphi(t) = t^{p-1}$, chosen $l(t) = t^q$, the first in (2.22) holds if and only if $q \in [0, p - 1]$. Furthermore, to recover the case l constant the best choice of χ is

$$
\chi = p - 1;
$$

the choice $\chi = 0$ represents the borderline case of strong gradient dependence $l(t) \asymp \varphi(t)$ near $t = 0$. The latter often needs a special care to be treated.

We express our main result in terms of a sharp condition on the Ricci tensor.

Theorem 2.17 *Let M be a complete m-dimensional manifold such that, for some fixed origin $o \in M$, the distance $r(x)$ from o satisfies*

$$
\text{Ric}(\nabla r, \nabla r) \geq -(m - 1)\kappa^2 (1 + r^2)^{\alpha/2} \qquad \text{on } D_o,
\tag{2.23}
$$

for some $\kappa \geq 0$, $\alpha \geq -2$. Let l and φ satisfy (2.21) and (2.22). Consider $0 < b \in C(M)$ such that

$$b(x) \geq C\left(1 + r(x)\right)^{-\mu} \qquad on \ M,$$

for some constants $C > 0$, $\mu \in \mathbb{R}$. If

$$\mu \leq \chi - \frac{\alpha}{2} \quad and \ either \quad \begin{cases} \alpha \geq -2 \quad and \quad \chi > 0, \quad or \\ \alpha = -2, \quad \chi = 0 \quad and \quad \bar{\kappa} \leq \frac{p-1}{m-1}, \end{cases} \tag{2.24}$$

with $\bar{\kappa} = \frac{1}{2}\left(1 + \sqrt{1 + 4\kappa^2}\right)$, then $(bl)^{-1}\Delta_\varphi$ satisfies (SMP$_\infty$).

In particular, the Euclidean space $M = \mathbb{R}^m$ is recovered by choosing $\kappa = 0$, $\alpha = -2$, while, to deal with the hyperbolic space $\mathbb{H}^m$ of constant sectional curvature -1 we choose $\kappa = 1$, $\alpha = 0$. Even for these model manifolds, Theorem 2.17, in the above generality on b and l, is new. As an example, Corollary 8.7 in Sect. 8.1 expresses the result for the mean curvature operator both in $\mathbb{R}^m$ and in $\mathbb{H}^m$.

Next, we turn our attention to (WMP$_\infty$), introduced in [180] following the observation that, in many geometric applications, the gradient condition in (2.17) was unnecessary. It has various advantages with respect to (SMP$_\infty$): first, it can be stated for $u \in W^{1,p}_{\text{loc}}(M)$, $p \geq 1$, which is a natural regularity class for solutions of ($P_\geq$); second, the absence of the gradient bound allows to directly use the weak formulation together with refined integral estimates, to obtain sharp criteria for (WMP$_\infty$) that just depend on the volume growth of geodesic balls B_r, a requirement implied, but not equivalent, to (2.23). This approach will be described in more detail below.

Remark 2.18 It is important to observe that there exist manifolds satisfying (WMP$_\infty$) but not (SMP$_\infty$), hence the two principles are different. Counterexamples are very easy to construct in the setting of incomplete manifolds (indeed, $\mathbb{R}^m \backslash \{0\}$ satisfies (WMP$_\infty$) but not (SMP$_\infty$), see [182]), and a nice example in the complete case appeared recently in [32].

First, we introduce the following characterization improving on [155, 180, 184]. Despite the simplicity of the proof, the equivalences stressed below are particularly useful in geometric applications.

Proposition 2.19 *Let φ and b, f, l satisfy respectively (2.3) and (2.5). Then, the following properties are equivalent:*

(i) $(bl)^{-1}\Delta_\varphi$ satisfies (WMP$_\infty$);
(ii) (L) holds for Lip_{loc} solutions, for some (equivalently, every) f with

$$f(0) = 0, \qquad f > 0 \qquad on \ \mathbb{R}^+;$$

(iii) each solution $u \in \mathrm{Lip}_{\mathrm{loc}}(M)$ solving $(P_\geq)$ on M and bounded above satisfies
$f(u^*) \leq 0$.

It should be stressed that, by a generalization of work of R.Z. Khas'minskii [133] (see [107] for a nice exposition), (L) with the choice $f(t) = \lambda t$ and $\lambda > 0$ is related to the theory of the (minimal) Brownian motion on M, and indeed equivalent to the stochastic completeness of M, that is, the infinite lifetime of a.e. Brownian path. Exploiting this last equivalence, A. GrigorYan in [107, Thm. 9.1] found the weakest known geometric condition on a complete M for Δ to satisfy (L) with $f(t) = \lambda t$ and $\lambda > 0$, that is,

$$\frac{r}{\log |B_r|} \notin L^1(\infty).$$

However, the beautiful method of proof in [107] relies on the linearity of the Laplace–Beltrami operator. Hence, the search for similar volume conditions for general Δ_φ calls for different ideas, developed in a series of works [131,181,206,207] and refined in [157,182]. Our contributions are contained in Theorems 7.5 and 7.15 below.

Remark 2.20 A characterization similar to that of Proposition 2.19 holds for (L) when $f \equiv 0$ in a right neighbourhood of zero. In fact, by [184, Thm. A] (for Δ_p) and [182] (general Δ_φ), for these f 's property (L) is equivalent to the parabolicity of Δ_φ , see also [155]. For the p-Laplace operator, parabolicity is more often introduced via capacity estimates, see Sect. 4.1 below for details and references.

Remark 2.21 Khas'minskii introduced a sufficient condition for M to be stochastically complete in terms of the existence of w satisfying all of the properties in (2.20) but that on the gradient, with $G(t) = \lambda t$, $\lambda > 0$, see [107, 133]. This justifies the name strong Khas'minskii condition given to (2.20). It should be observed that, for a large class of operators including some geometrically relevant fully nonlinear ones, appropriate Khas'minskii conditions turn out to be *equivalent* to suitably defined maximum principles at infinity, see [155] and the recent [154].

To introduce a special case of our second main Theorem 7.5, observe that (2.23) implies, via the Bishop–Gromov comparison theorem, the following estimates:

$$\limsup_{r \to \infty} \frac{\log |B_r|}{r^{1+\alpha/2}} < \infty \qquad \text{if } \alpha > -2,$$

$$\limsup_{r \to \infty} \frac{\log |B_r|}{\log r} \leq (m-1)\bar{\kappa} + 1 \quad \text{if } \alpha > -2, \tag{2.25}$$

with

$$\bar{\kappa} = \frac{1}{2}\left(1 + \sqrt{1 + 4\kappa^2}\right).$$

Regarding our assumptions on φ and l, differently from (2.21) we now require the milder

$$\begin{cases} \varphi \in C(\mathbb{R}_0^+), & \varphi(0) = 0, & \varphi > 0 \text{ on } \mathbb{R}^+; \\[2mm] l \in C(\mathbb{R}_0^+), & l > 0 \text{ on } \mathbb{R}^+. \end{cases} \qquad (2.26)$$

We also need the next growth conditions, to be compared to those in (2.22).

$$\begin{cases} l(t) \geq C_1 \dfrac{\varphi(t)}{t^\chi} & \text{on } \mathbb{R}^+, \text{ for some } C_1 > 0, \chi \geq 0; \\[3mm] \varphi(t) \leq C_2 t^{p-1} & \text{on } [0, 1], \text{ for some } C_2 > 0, p > 1; \\[3mm] \varphi(t) \leq \bar{C}_2 t^{\bar{p}-1} & \text{on } [1, \infty), \text{ for some } \bar{C}_2 > 0, \bar{p} > 1. \end{cases} \qquad (2.27)$$

The use of different upper bounds for $\varphi(t)$ related to its behaviour near $t = 0$ and $t = \infty$ is crucial to obtain sharp results in the setting, for instance, of the mean curvature operator as well as of the (p, q)-Laplacian, that is $\Delta_{p,q} = \Delta_p + \Delta_q$. We are now ready to state

Theorem 2.22 *Let M be a complete m-dimensional manifold. Fix $\alpha \geq -2$ and suppose that*

$$\liminf_{r \to \infty} \frac{\log |B_r|}{r^{1+\alpha/2}} = V_\infty < \infty \qquad \text{if } \alpha > -2;$$

$$\liminf_{r \to \infty} \frac{\log |B_r|}{\log r} = V_\infty < \infty \qquad \text{if } \alpha = -2. \qquad (2.28)$$

Let φ and l satisfy (2.26) and (2.27), and consider $0 < b \in C(M)$ such that

$$b(x) \geq C\left(1 + r(x)\right)^{-\mu} \qquad \text{on } M,$$

for some constants $C > 0$, $\mu \in \mathbb{R}$. Suppose:

$$\mu \le \chi - \frac{\alpha}{2} \quad \text{and either} \quad \begin{cases} \alpha \ge -2, & \chi > 0, \quad \text{or} \\ \alpha \ge -2, & \chi = 0, \quad \mu < -\frac{\alpha}{2}, \quad \text{or} \\ \alpha > -2, & \chi = 0, \quad \mu = -\frac{\alpha}{2}, \quad V_\infty = 0, \quad \text{or} \\ \alpha = -2, & \chi = 0, \quad \mu = -\frac{\alpha}{2}, \quad V_\infty \le p. \end{cases} \quad (2.29)$$

Then, $(bl)^{-1}\Delta_\varphi$ satisfies (WMP$_\infty$).

Remark 2.23 As underlined in Remark 2.16, p and $\bar{p}$ are implicitly related to bounds on χ via (2.27). However, we feel remarkable that p, $\bar{p}$ do not appear in conditions (2.29), apart from the last borderline case. A detailed discussion follows the statement of Theorem 7.5 in Sect. 7.

Suitable counterexamples show the sharpness of Theorem 7.5, and consequently of Theorem 2.22, with respect to each parameter involved. In particular, the restrictions in (2.29) are sharp.

Conjecture 2.24 In the setting of Theorem 2.17, the full (SMP$_\infty$) holds if the range (2.24) is replaced by (2.29).

To better appreciate the range of applicability of Theorem 2.22, we state as a direct corollary the following extension of Do Carmo–Lawson's Theorem 1.3 in the minimal setting. The result is a particular case of Theorems 7.17 and 7.18 below.

Theorem 2.25 *Let M be a complete manifold, and consider the warped product $\bar{M} = \mathbb{R} \times_h M$, with warping function h satisfying either*

(i) h is strictly convex and $h^{-1} \in L^1(-\infty) \cap L^1(\infty)$, or
(ii) $h' > 0$ on $\mathbb{R}$, $h'(s) \ge C$ for $s \gg 1$ and $h^{-1} \in L^1(\infty)$.

If

$$\liminf_{r \to \infty} \frac{\log |B_r|}{r^2} < \infty,$$

then

under (i) the only entire minimal graph over M is the constant $u = s_0$, where s_0 is the unique minimum of h.
under (ii) there exists no entire minimal graph over M.

The corresponding statement for variable mean curvature will be given in Theorem 8.11, under the validity of (SMP$_\infty$). It should be noted that, besides bounded solutions, in Theorem 7.5 we can also consider solutions u of

$$\Delta_\varphi u \geq b(x) f(u) l(|\nabla u|) \qquad \text{on } \Omega_\eta = \{u > \eta\}$$

with a controlled growth at infinity. Indeed, under appropriate assumptions, the theorem guarantees both that $u^* < \infty$ and that $f(u^*) \leq 0$. The result is a significant improvement of [157, Thm 5.1] and [5, Thm. 2.1]; it applies, for instance, to differential inequalities with borderline gradient dependence of the type

$$\Delta_\varphi u \geq b(x) f(u) \varphi(|\nabla u|) \qquad \text{on } M,$$

to ensure that, under mild assumptions, any solution that grows polynomially is bounded from above and satisfies $f(u^*) \leq 0$. Recall that, by definition, u grows polynomially if there exists $\sigma \geq 0$ such that

$$|u(x)| = O\big(r(x)^\sigma\big) \qquad \text{as } r(x) \to \infty.$$

The reader that is interested in such borderline examples can see Corollary 7.9, as well as Theorem 7.12 in the particular setting of the mean curvature operator. Results in the spirit of Corollary 7.9 below, with a dependence on the gradient, appear in [86, 87] on Euclidean space $\mathbb{R}^m$, and will be compared with ours in Sect. 7. In a manifold setting, due to the possible lack of a polynomial bound for the growth of the volume of geodesic balls, the integral methods in [36, 63, 64, 86, 87] are, in most of the cases, not sufficient to get sharp conclusions. Indeed, even in the polynomial setting of Euclidean space, Theorem 7.5 complements and in some cases improves on the existing literature.

We report here the following application to entire vertical self-translators of the mean curvature flow:

Theorem 2.26 *Let $(M^m, \langle\,,\,\rangle)$ be a complete manifold and consider the product $\bar{M}^{m+1} = \mathbb{R} \times M$. Fix $0 \leq \sigma \leq 2$ and suppose that either*

$$\sigma < 2 \quad and \quad \liminf_{r \to \infty} \frac{\log |B_r|}{r^{2-\sigma}} < \infty, \qquad or$$

$$\sigma = 2 \quad and \quad \liminf_{r \to \infty} \frac{\log |B_r|}{\log r} < \infty. \tag{2.30}$$

Then, there exist no entire graph $\Sigma \subset \bar{M}$ *of* $v : M \to \mathbb{R}$ *which is a self-translator for the MCF with respect to the vertical direction* ∂_s *and satisfies*

$$|u(x)| = o(r(x)^\sigma) \qquad as \ r(x) \to \infty, \quad if \ \sigma > 0;$$

$$u^* < \infty, \qquad\qquad\qquad\qquad if \ \sigma = 0.$$

Remark 2.27 Specializing Theorem 2.26 in Euclidean space $\mathbb{R}^{m+1}$, there is no entire graph $v : \mathbb{R}^m \to \mathbb{R}$ over the horizontal $\mathbb{R}^m$ which is a self-translator in the vertical direction and satisfies $v = o(r^2)$ as $r \to \infty$. The result is sharp, since the bowl soliton in $\mathbb{R}^{m+1}$ (cf. [9] and [55, Lem. 2.2]) and the non-rotational manifolds in [233] for $m \geq 3$ are examples of entire (convex) graphs which translate vertically by MCF and have order of growth r^2.

Other applications for entire self-translators in $\mathbb{R}^{m+1}$ (not necessarily vertical) and for entire self-expanders will be given in Theorems 7.20 and 7.23, respectively.

We conclude with the next observations: in view of the volume estimates (2.25) that follow from the Ricci bound (2.23), Theorems 2.17 and 2.22 hold precisely for the same range of α, μ, χ (aside from some borderline cases covered by (2.29) but not by (2.24)). In view of this, the following problem seems interesting to us:

Problem 2 Prove or disprove by exhibiting a counterexample (in a complete manifold), the validity of (SMP$_\infty$) in the assumptions of Theorem 2.22.

2.6   The Strong Liouville Property (SL)

In the literature, the validity of (SL) has been mainly investigated by means of two different approaches: radialization techniques and refined comparison theorems [33, 157, 195, 198], or integral estimates, in the spirit of the work of E. Mitidieri and S.I. Pohozaev [162], see [63, 64, 86, 87, 220]. Assume the validity of (2.16) and (2.9), in order for the function K in (2.7) to realize a homeomorphism of $\mathbb{R}_0^+$ onto itself. Suppose that

$$f > 0 \qquad on \ (T, \infty), \quad for \ some \ T \geq 0,$$

and set

$$F(t) = \int_T^t f(s)ds. \tag{2.31}$$

Under these assumptions, the proof of the validity of (SL) via radialization techniques relies on the construction of suitable blowing-up radial supersolutions, explicitly related to the Keller–Osserman condition

$$\frac{1}{K^{-1} \circ F} \in L^1(\infty). \tag{KO$_\infty$}$$

As far as we know, (KO$_\infty$) first appeared for nontrivial l in the work of R. Redheffer [203] (Corollary 1 therein) for the inequality

$$\Delta u \geq f(u) l(|\nabla u|).$$

Since then, it has been systematically studied by various authors. Among them, for nontrivial l we quote

- [37] (for the 1-dimensional problem), [14, 91, 105] (when Δ_φ is the mean curvature operator) and [91, 95, 158] (when Δ_φ is the p-Laplacian);
- in a sub-Riemannian setting, [5, 33–35, 147].

For further generalizations to quasilinear inequalities, possibly with singular or degenerate weights, we refer to [65, 66, 92, 162].

We first discuss the necessity of (KO$_\infty$) for (SL), and recall that a point $o \in M$ is said to be a *pole* if the exponential map

$$\exp_o : \ T_o M \approx \mathbb{R}^m \to M$$

is a diffeomorphism. It can be proved that o is a pole for M if and only if the distance function $r(x) = \mathrm{dist}(x, o)$ is smooth outside of o, see [177] and the references therein. As a particular case of Theorem 10.3 below, we obtain the next theorem (recall Sect. 2.3 for the definition of the radial sectional curvature Sec$_{\mathrm{rad}}$).

Theorem 2.28 (Necessity of (KO$_\infty$)) *Let M^m be a complete Riemannian manifold with a pole $o \in M$, and assume that*

$$\mathrm{Sec}_{\mathrm{rad}} \leq \frac{1}{4r^2} \qquad on \ M \backslash \{o\}. \tag{2.32}$$

Let φ, b, f, l satisfy (2.21), (2.9), (2.5) and

$$f(0) = 0, \qquad f > 0 \ and \ C\text{-increasing on } \mathbb{R}^+.$$

Then, (KO$_\infty$) is necessary for the validity of (SL).

Inequality (2.32) is a mild requirement just needed to ensure that the model to be compared to M is complete and the volume of its geodesic spheres increases. It allows to apply Theorem 2.28 to all *Cartan–Hadamard manifolds* (cf. Definition 3.3), a class that includes both the Euclidean and the hyperbolic space. This might suggest that (2.32) be just a technical assumption (although seemingly not easy to remove) and thus, loosely speaking, that geometry does not affect implication (SL) $\Rightarrow$ (KO$_\infty$).

On the contrary, the sufficiency of (KO$_\infty$) heavily depends on the validity of maximum principles at infinity. To investigate the interplay, it is worth to consider (SL) as the combination of two properties:

– an L^∞-*estimate* for non-negative solutions of ($P_\geq$);
– property (L) for bounded, non-negative solutions of ($P_\geq$).

Note that (KO$_\infty$) plays a role in the first property, while, by Proposition 2.19, the second property is equivalent to (WMP$_\infty$) provided that $f(0) = 0$ and $f > 0$ on $\mathbb{R}^+$. As a first result, Theorem 10.5 below relates directly (SMP$_\infty$) to (SL), by showing that for some classes of operators, notably including the p-Laplacian with constant b, and for general f with $f > 0$ on $\mathbb{R}^+$,

$$(\text{KO}_\infty) \quad + \quad (\text{SMP}_\infty) \quad \Longrightarrow \quad (\text{SL}). \tag{2.33}$$

However, for more general operators, in particular for non-homogeneous ones, such a simple relation is currently unknown. Nevertheless, for large classes of functions φ, b, f, l, we can guarantee (SL) by coupling (KO$_\infty$) with the lower Ricci curvature bounds considered in Theorem 2.17, the latter being sharp for the validity of (SMP$_\infty$). This is the content of Theorems 10.19 and 10.20 below, dealing respectively with the case $\chi > 0$ and $\chi = 0$, that should be considered the main results of Sect. 10.2. We refer therein for the statements in full generality, and quote the following corollary for the mean curvature operator:

Theorem 2.29 *Let M^m be complete and assume that*

$$\text{Ric}(\nabla r, \nabla r) \geq -(m-1)\kappa^2(1+r^2)^{\alpha/2} \qquad on\ D_o,$$

for some constants $\kappa \geq 0$, $\alpha \geq -2$ and some origin o. Let b, f, l satisfy (2.5) and

$$b(x) \geq C_1(1+r(x))^{-\mu} \qquad on\ M,$$

$$f(0) = 0, \quad f > 0 \quad and\ C\text{-increasing on } \mathbb{R}^+,$$

$$l(t) \geq C_1 \frac{t^{1-\chi}}{\sqrt{1+t^2}} \qquad on\ \mathbb{R}^+,$$

for some constants $C, C_1 > 0$, $\mu \in \mathbb{R}$, $\chi \in (0, 1]$ with

$$\mu \leq \chi - \frac{\alpha}{2}.$$

Then, under the validity of the Keller–Osserman condition

$$F^{-\frac{1}{\chi+1}} \in L^1(\infty) \tag{2.34}$$

with F as in (2.31), (SL) holds for C^1 solutions of

$$\text{div}\left(\frac{\nabla u}{\sqrt{1 + |\nabla u|^2}}\right) \geq b(x) f(u) l(|\nabla u|) \qquad on \ M.$$

Suitable counterexamples will show the sharpness of (2.34), that seems to be new even in the Euclidean and hyperbolic space settings, corresponding, respectively, to $\alpha = -2$ and $\alpha = 0$.

Remark 2.30 If $\chi = 0$ and no Keller–Osserman condition is assumed, a Liouville theorem that well matches with the above result can be found in Theorem 7.12 below.

When $f(t)$ is a power of t, say $f(t) \asymp t^\omega$ in a neighbourhood of infinity, (2.34) becomes $\omega > \chi$. In this case, we will prove (SL) under a mere volume growth requirement. More precisely, we have the following

Theorem 2.31 *Assume that the conditions in Theorem 2.22 are satisfied, with the second and third of (2.27) replaced by*

$$\varphi(t) \leq C t^{p-1} \quad for \ t \in \mathbb{R}^+,$$

for some $p > 1$, $C > 0$. Let $f \in C(\mathbb{R})$ satisfy

$$f(t) \geq C_2 t^\omega \qquad for \ some \ C_2 > 0 \ and \ each \ t \gg 1.$$

If $\omega > \chi$, then any non-constant $u \in \text{Lip}_{\text{loc}}(M)$ solution of $(P_\geq)$ on M is bounded from above and satisfies $f(u^*) \leq 0$. In particular, if $f > 0$ on $\mathbb{R}^+$, (SL) holds for Lip_{loc} solutions.

The argument of the proof of Theorem 2.31, though close in spirit to that of Theorem 2.22, uses a different combination of integral estimates. Nevertheless, unlike [63, 64, 86, 87, 162] which treat similar results in $\mathbb{R}^m$, our method has again the advantage to work in settings where the volume growth of geodesic balls is not polynomial. This requires an iterative procedure originally due to [181, 182], of independent interest, but the appearance of the function l also requires a careful mixing with techniques in [87].

To describe the range of applicability of Theorem 2.31, we consider the capillarity equation

$$\text{div}\left(\frac{\nabla u}{\sqrt{1+|\nabla u|^2}}\right) = \kappa(x)u \qquad \text{on } M, \tag{2.35}$$

modelling a graphical interface in $M \times \mathbb{R}$ whose mean curvature is proportional to the height of the graph via the non-homogeneous coefficient $\kappa(x)$. Then, we have

Theorem 2.32 *Suppose that M is complete and that*

$$\kappa(x) \geq C\big(1 + r(x)\big)^{-\mu} \qquad \text{on } M, \tag{2.36}$$

for some constants $C > 0$ and $\mu < 2$. If there exists $\varepsilon > 0$ such that

$$\liminf_{r \to \infty} \frac{\log |B_r|}{r^{2-\varepsilon-\mu}} < \infty, \tag{2.37}$$

then the only solution of the capillarity equation (2.35) on M is $u \equiv 0$.

To the best of our knowledge, Theorem 10.39 seems to us to be the first result considering entire solutions of (2.35) in a manifold setting, in particular allowing the volume of geodesic balls to grow faster than polynomially. Nevertheless, it is of interest even for Euclidean space, guaranteeing $u \equiv 0$ whenever $\mu < 2$. To our knowledge, when $\kappa > 0$ is constant the vanishing of u on $\mathbb{R}^m$ solving (2.35) was first obtained in [168, 225] with no growth assumptions on u (cf. also [194, Thm. 8.1.3] for u growing polynomially). The methods in [168, 225] are different from one another; in particular, that in [225] has later been extended in [219] to more general inequalities, and Theorem 2.32 is shown to hold on $\mathbb{R}^m$ but only for $\mu < 1$. Recently, in [86] the authors were able to achieve the sharp bound $\mu < 2$ for solutions on $\mathbb{R}^m$. In Sect. 10.6, we will describe in more detail the relationship between our result and those in [86], and we will improve on [86, 168, 194, 219, 225] for a class of equations including (2.35). We stress that none of the methods therein easily adapts to manifolds just satisfying (2.37).

Remark 2.33 Theorem 2.32 has a curious and unexpected feature: although (2.35) does not contain a gradient term, the "artificial" inclusion of a suitable $l(|\nabla u|)$ in the right-hand side of (2.35) is the key to prove the corollary as a consequence of the Keller–Osserman condition $\omega > \chi$ in Theorem 2.31 (note that $\omega = 1$ for (2.35)). This is in striking contrast with previous results for equation

$$\text{div}\left(\frac{\nabla u}{\sqrt{1+|\nabla u|^2}}\right) = b(x)f(u), \tag{2.38}$$

in a manifold setting: for instance, to obtain the vanishing of u in (2.38) when $f(t)t \geq C|t|^{\omega+1}$ on $\mathbb{R}$, Theorem 4.8 in [182] needs inequality $\omega > 1$, which does not hold for (2.35). Loosely speaking, inserting a suitable gradient term enables us to weaken the requirement in the Keller–Osserman condition up to include the capillarity equation.

Observing that the volume growth conditions in Theorem 2.31 coincide with those in Theorem 2.22 for the validity of (WMP$_\infty$), one might wonder whether (2.33) can be weakened to

$$(\text{KO}_\infty) \quad + \quad (\text{WMP}_\infty) \quad \Longrightarrow \quad (\text{SL}). \tag{2.39}$$

In Example 10.11 below, we will show that (WMP$_\infty$) is *not* sufficient, and the full strength of (SMP$_\infty$) is needed. The counterexample is, however, on an incomplete manifold, and suggests the following

Problem 3 Prove or disprove: if M is a complete manifold, then the implication (2.39) holds at least for a subclass of operators Δ_φ and b, f, l.

2.7   The Compact Support Principle (CSP)

Unlike that on (L) and (SL), the literature on (CSP) is not so extensive. The subject initiated with the seminal paper by R. Redheffer [204], and received a renewed interest in the last 15 years starting from [196], see also [88, 109, 197] and the monograph [194]. However, all these works consider the problem in the setting of Euclidean space, and to our knowledge just [198, 208, 209] analyse the role played by the geometry of the underlying manifold. As we shall see, the link between geometry and (CSP) does not depend on the validity of a maximum principle at infinity: to explain which geometric conditions are to be expected, we first comment on the following result in [198, Thm. 1.1]:

Theorem 2.34 ([198]) *Let M be a complete manifold, and let r be the distance from a fixed origin o. Assume (2.3) and that φ is strictly increasing on $\mathbb{R}^+$. Let $f \in C(\mathbb{R})$ satisfy*

$$f(0) = 0, \qquad f > 0 \quad \text{and non-decreasing on some } [0, \eta_0), \ \eta_0 > 0. \tag{2.40}$$

Then, in order for (CSP) *to hold for* $(P_\geq)$ *with* $b(x) = 1$, $l(t) = 1$ *it is necessary that*

$$\frac{1}{H^{-1} \circ F} \in L^1(0^+), \tag{2.41}$$

where F *and* H *are defined in* (2.8) *and* (2.10). *Vice versa,* (2.41) *is also sufficient for* (CSP) *provided that*

$$\inf_M \Delta r > -\infty \tag{2.42}$$

holds in the weak sense.

As a matter of fact (cf. Proposition 3.10 in Sect. 3.2), (2.42) forces the origin o to be a pole for M, in particular r is smooth on $M \backslash \{o\}$. With the aid of the Hessian comparison theorem, (2.42) holds for a large class of manifolds including Cartan–Hadamard ones, such as the Euclidean and hyperbolic spaces. On the other hand, the topological restriction imposed by the existence of a pole is binding, and it would be desirable to remove it. However, already in [198] the authors realized that a condition like (2.42) or some other extra assumption needs necessarily to be included. Their example, reported below, is illustrative.

Example 2.35 Consider the radially symmetric model

$$M_g = (\mathbb{R}^m, \mathrm{d}s_g^2), \qquad \mathrm{d}s_g^2 = \mathrm{d}r^2 + g(r)^2 \mathrm{d}\theta^2,$$

where $\mathrm{d}\theta^2$ is the standard round metric of curvature 1 on the unit sphere, and $0 < g \in C^\infty(\mathbb{R}^+)$ satisfy $g(r) = r$ for $r \in [0, 1]$ and $g(r) = \exp\{-r^\alpha\}$ for $r \geq 2$, for some $\alpha > 2$. Clearly, M is a manifold with pole $o \in \mathbb{R}^m$. Then, for each $\omega \in (0, 1)$ the function

$$u(r) = r^{-\beta}, \qquad \beta \in \left(0, \frac{\alpha - 2}{1 - \omega}\right]$$

solves $\Delta u \geq Cu^\omega$ on the end $\Omega = M \backslash B_R$, for R large enough and for a suitable constant $C > 0$. Although (2.41) holds, u clearly contradicts (CSP). Note that in this case $\Delta r = -(m - 1)\alpha r^{\alpha - 1}$, hence (2.42) is violated.

A more elaborated example along these lines will be given in Sect. 2.7 below. The construction shows that, in sharp contrast with (SL) and (L), what matters in this case is that M should not possess ends shrinking too rapidly at infinity. This is the content of our first contribution to (CSP). We begin defining a weaker notion of a pole, to allow a nontrivial topology on M.

Definition 2.36 Let $\mathcal{O} \subset M$ be a relatively compact, open set with smooth boundary in M. We say that $\mathcal{O}$ is *a pole of* M if the normal exponential map $\exp_{\mathcal{O}} : T\mathcal{O}^{\perp} \to M\backslash\mathcal{O}$ realizes a diffeomorphism.

Here, $T\mathcal{O}^{\perp}$ is the subset of the normal bundle of $\partial\mathcal{O}$ consisting of vectors pointing outward from $\mathcal{O}$. The case of a point o being a pole can easily be recovered by choosing $\mathcal{O} = B_{\varepsilon}(o)$ for ε small enough. Let r be the distance function from $\mathcal{O}$, which is therefore smooth on $M\backslash\mathcal{O}$. Denote with $\text{II}_{-\nabla r}$ the second fundamental form of $\partial\mathcal{O}$ with respect to the inward unit normal $-\nabla r$, and let

$$B_R(\mathcal{O}) = \{x \in M : 0 < r(x) < R\}.$$

Beyond the standard requirements (2.3) and (2.5) we assume (2.16), in order for K to be defined, and the following condition corresponding to (2.40):

$$\begin{cases} f \text{ is positive and } C\text{-increasing on } (0, \eta_0), \text{ for some } \eta_0 > 0. \\ l \text{ is } C\text{-increasing and locally Lipschitz on } (0, \xi_0), \text{ for some } \xi_0 > 0. \\ f(0)l(0) = 0. \end{cases} \qquad (2.43)$$

Set F as in (2.8). We first address the necessity of the Keller–Osserman condition

$$\frac{1}{K^{-1} \circ F} \in L^1(0^+). \qquad (\text{KO}_0)$$

In analogy with Theorem 2.34, we see that there is no geometric obstruction, at least on manifolds with a pole.

Theorem 2.37 (Necessity of (KO$_0$)) *Let M be a manifold with a pole $\mathcal{O}$. Assume (2.3), (2.5), (2.16) and (2.43). Then, (KO$_0$) is necessary for the validity of (CSP) for (P$_\geq$).*

The proof relies on the construction of a suitable radial solution of a Dirichlet problem at infinity for singular ODEs, of independent interest. As for the sufficiency part, geometry enters into play, and the statement is considerably more elaborated. We state the following corollary of our main result, Theorem 9.13 in Sect. 9.

Theorem 2.38 *Let $(M, \langle\,,\,\rangle)$ be a manifold with a pole $\mathcal{O}$, whose radial sectional curvature satisfies*

$$\text{Sec}_{\text{rad}} \leq -\kappa^2(1+r)^{\alpha} \qquad \text{on } M\backslash\mathcal{O}, \qquad (2.44)$$

for some $\kappa \geq 0$, $\alpha \geq -2$. Suppose

$$II_{-\nabla r} \geq -C_{\alpha,\kappa}\langle\,,\rangle \qquad on\ T\partial\mathcal{O}, \tag{2.45}$$

with

$$C_{\alpha,\kappa} = \begin{cases} \kappa & if\ \alpha \geq 0\ or\ \kappa = 0, \\ \left[\dfrac{\alpha+\sqrt{\alpha^2+16\kappa^2}}{4}\right] & otherwise. \end{cases}$$

Consider φ, b, f, l satisfying (2.3), (2.5), (2.16), and (2.43). Fix $\chi, \mu \in \mathbb{R}$ with

$$\chi > 0, \qquad \mu \leq \chi - \frac{\alpha}{2} \tag{2.46}$$

and assume that

$$\begin{aligned} l(t) &\asymp t^{1-\chi}\varphi'(t) & for\ t \in (0,1), \\ b(x) &\geq C_1\big(1+r(x)\big)^{-\mu} & for\ r(x) \geq r_0, \end{aligned} \tag{2.47}$$

for some constants $r_0, C_1 > 0$. If there exists a constant $c_F \geq 1$ such that

$$F(t)^{\frac{\chi}{\chi+1}} \leq c_F f(t) \qquad for\ each\ t \in (0, \eta_0), \tag{2.48}$$

then,

$$(CSP)\ holds\ for\ (P_\geq) \qquad \Longleftrightarrow \qquad (KO_0).$$

Remark 2.39 By (2.47), (KO_0) is equivalent to

$$F^{-\frac{1}{\chi+1}} \in L^1(0^+).$$

Note also that the C-increasing property of l implies that $t^{1-\chi}l(t)$ is C-increasing too, possibly with a different C, and this forces a bound on the vanishing of l near $t = 0$. For instance, if

$$\varphi'(t) \asymp t^{p-2}, \qquad f(t) \asymp t^\omega \qquad for\ t \in (0, t_0),$$

for some $p > 1$, $\omega > 0$, then (2.47) holds with l C-increasing if and only if $\chi \leq p - 1$, (2.48) is satisfied whenever $\omega \leq \chi$, and (KO_0) is equivalent to $\omega < \chi$.

The bound (2.45) means, roughly speaking, that $\partial\mathcal{O}$ should not be too concave in the inward direction. In particular, if $\kappa = 0$, (2.45) requires $\mathcal{O}$ to have a convex boundary. By choosing $\mathcal{O} = B_\varepsilon(o)$, the theorem applies to Cartan–Hadamard manifolds.

Example 2.40 Another example to which Theorem 2.38 applies is that of hyperbolic manifolds with finite volume. It is known by the thick-thin decomposition (see Theorem D.3.3 and Proposition D.3.12 in [17]) that a manifold M^m with sectional curvature -1 and finite volume decomposes as the disjoint union $\mathcal{O} \cup \Omega_1 \cup \cdots \cup \Omega_s$, where $\mathcal{O}$ is a smooth, relatively compact open set and, for each j, Ω_j is a non-compact cusp end isometric to the warped product $\mathbb{R}_0^+ \times_{e^{-r}} N^{m-1}$, for some compact flat manifold (N, g_N), with metric $dr^2 + e^{-2r} g_N$. Therefore, $\mathcal{O}$ is a pole of M, r is the distance from $\mathcal{O}$ and a direct computation gives

$$\mathrm{II}_{-\nabla r} = -\langle\,,\,\rangle \qquad \text{on } T\partial\mathcal{O},$$

precisely the borderline case in (2.45).

Remark 2.41 Observe that the bound $\mu \le \chi - \frac{\alpha}{2}$ in (2.46) is the same as that in (2.24) and (2.29) for the validity, respectively, of (SMP$_\infty$) and (WMP$_\infty$). Example 9.4 below shows that it cannot be weakened (see the range (9.18) of the parameter there).

Because of the presence of nontrivial b, l, the proof of Theorem 2.38 is technically considerably more demanding than that of Theorem 2.34, and calls for a few extra-assumptions guaranteeing certain mild "homogeneity properties", which account for conditions (2.47). However, the underlying principle is the same and relies on the explicit construction of a radial, compactly supported, C^1 solution of

$$\begin{cases} \Delta_\varphi w \le b(x) f(w) l(|\nabla w|) & \text{on } M\backslash B_R(\mathcal{O}), \\ w \ge 0 & \text{on } M\backslash B_R(\mathcal{O}), \\ w \equiv 0 & \text{on } M\backslash B_{R_1}(\mathcal{O}), \end{cases} \qquad (2.49)$$

for some $R_1 > R$, via a direct use of (KO$_0$). We provide two variants of the construction, that work under a mildly different set of assumptions: one of them is quite involved and closely related to that in [208] (which seems to be the only reference investigating (CSP) with a gradient term l), while the other is new and considerably simpler.

Although sharp, Theorem 2.38 still requires the presence of a pole in order to apply the Laplacian comparison theorem from below and deduce, from the combination of (2.44) and (2.45), the lower bound

$$\Delta r \ge -Cr^{\frac{\alpha}{2}} \qquad \text{for } r \ge 1,$$

and some constant $C > 0$, which is the weighted version of (2.42). For the relevant case of the p-Laplace operator, we introduce a different radialization method, that we believe to be of independent interest. It is based on smooth functions replacing the distance from $\mathcal{O}$, and called for this reason *fake distances*. The method does not require the existence of a pole. Suppose that M is a complete manifold satisfying

$$\mathrm{Ric} \geq -(m-1)\kappa^2\langle\,,\,\rangle \qquad \text{on } M, \tag{2.50}$$

for some $\kappa \geq 0$, and let $v_\kappa(r)$ be the volume of a geodesic sphere of radius r in the space form of constant sectional curvature $-\kappa^2$ (i.e. $\mathbb{R}^m$ for $\kappa = 0$ or the hyperbolic space $\mathbb{H}^m_\kappa$ of curvature $-\kappa^2$ for $\kappa > 0$). We restrict here to the case

$$p \in (1, m],$$

the complementary case $p > m$ being slightly different and discussed in Sect. 4.2. Assume

$$v_\kappa^{-\frac{1}{p-1}} \in L^1(\infty)$$

(which always holds if $\kappa > 0$, while it is equivalent to $p < m$ if $\kappa = 0$). Assume that Δ_p is non-parabolic on M, equivalently, that for each fixed origin o there exists a positive Green kernel $\mathcal{G}(x, o)$ for Δ_p with pole at o, that is, a solution of

$$\begin{cases} \Delta_p \mathcal{G} = -\delta_o & \text{distributionally,} \\ \mathcal{G}(x, o) > 0 & \text{for } x \in M \backslash \{o\}, \end{cases}$$

where δ_o is the Dirac's distribution at o. In view of the comparison theory for kernels (cf. Chap. 6), in this case we can take $\mathcal{G}$ to be the minimal, positive Green kernel on M, and define the fake distance $\varrho : M \to \mathbb{R}$ implicitly via the equation

$$\mathcal{G}(x, o) = \int_{\varrho(x)}^\infty v_\kappa(s)^{-\frac{1}{p-1}}\,\mathrm{d}s.$$

In the literature, the fake distance modelled on the Green kernel of the Laplace–Beltrami operator has been used with great success to study the geometry in the large of manifolds with non-negative Ricci curvature, see for instance the works of Cheeger–Colding [46–49], Colding–Minicozzi [57] and Colding [56], together with the references therein. On the contrary, in a quasilinear setting and especially for the purposes of the present book, its use seems to be new. Differentiating the identity defining ϱ, we get

$$\Delta_p \varrho = \frac{v_\kappa'(\varrho)}{v_\kappa(\varrho)} |\nabla \varrho|^p.$$

Consequently, for each diffeomorphism $\psi : \mathbb{R} \to \mathbb{R}$ we obtain

$$\Delta_p \psi(\varrho) = \left[v_\kappa^{-1} \left(v_\kappa |\psi'|^{p-2} \psi' \right)' \right] (\varrho) |\nabla \varrho|^p . \tag{2.51}$$

Since $v_\kappa^{-1} \left(v_\kappa |\psi'|^{p-2} \psi' \right)'$ is the expression of the p-Laplacian of a radial function in the model of curvature $-\kappa^2$, (2.51) enables us to construct solutions of (2.49) by radializing with respect to ϱ instead of r. The striking advantage is that ϱ is smooth and p-subharmonic, hence a bound of the type in (2.42) is automatically satisfied. However, to be able to conclude the validity of (CSP), we need to control $|\nabla \varrho|$ from above and to guarantee that ϱ be an exhaustion function, equivalently, we need the vanishing of $\mathcal{G}$ as x diverges. Both problems have been tackled in the very recent [156], that has been conceived at the same time of the present monograph. Indeed, our results in Chap. 4 are based on those in [156]. Concerning the gradient bound, the main Theorem 4.15 below establishes the sharp estimate

$$|\nabla \varrho| \leq 1 \qquad \text{on } M \backslash \{o\}.$$

On the other hand, sufficient conditions for the vanishing of $\mathcal{G}$ follow, for instance, from the validity of Sobolev inequalities on M. We apply these ideas to obtain a characterization of (CSP) on a class of manifolds satisfying (2.50), which is a particular case of Theorem 9.24 and is somehow complementary to Theorem 2.38. Unfortunately, a technical point forces us to make a further assumption on the geometry of the Green kernel of Δ_p. Although the requirement is rather weak, we are aware of no result that guarantees its validity in a general setting.

Definition 2.42 We say that the weak Sard property ($\mathcal{WS}$) holds if, for some origin o, there exists a sequence $s_j \to 0$ such that the upper level sets $U_{s_j} = \{\mathcal{G} > s_j\}$ of the Green kernel with pole at o have the exterior ball condition:

$$\forall x \in \partial U_{s_j}, \text{ there exists a ball } B \subset M \backslash U_{s_j} \text{ such that } x \in \partial B.$$

In view of Sard's and the implicit function Theorems, ($\mathcal{WS}$) is satisfied whenever $\mathcal{G} \in C^m(M \backslash \{o\})$, which is clearly the case if $p = 2$. Refined versions of Sard's Theorem for Sobolev functions can be found in [12, 70, 90]. In a few special cases, sharp results about the structure of the set of critical points of a p-harmonic function were given, notably in $\mathbb{R}^2$ [129, 143] and for the p-capacity potential of convex rings in $\mathbb{R}^m$ [142]. However, it seems reasonable to hope that ($\mathcal{WS}$) holds without geometric conditions on M at least in the range $p \in (1, 2]$, which is considered in our next result. We stress that the result is new and of interest even in the semilinear case $p = 2$, for which ($\mathcal{WS}$) is automatic.

Theorem 2.43 *Let $(M, \langle \, , \, \rangle)$ be a complete m-dimensional Riemannian manifold with*

$$\mathrm{Ric} \geq -(m-1)\kappa^2 \langle \, , \, \rangle \qquad on \ M, \tag{2.52}$$

and assume that, for some $p \in (1, 2]$ and $v > p$, the Sobolev inequality

$$\left(\int |\psi|^{\frac{vp}{v-p}} \right)^{\frac{v-p}{v}} \leq S_{p,v} \int |\nabla\psi|^p \tag{2.53}$$

holds for each $\psi \in \mathrm{Lip}_c(M)$. Fix $\chi \in (0, p-1]$ and

$$if \ \chi \in (0, p-1) \ and \ p \in (1, 2), \ assume \ property \ (\mathcal{WS}).$$

Let $f \in C(\mathbb{R})$ satisfy

$$f(0) = 0, \qquad f > 0 \ on \ (0, \eta_0), \qquad f \ is \ C\text{-increasing on } (0, \eta_0),$$

for some $\eta_0 > 0$. Define F as in (2.8), and furthermore suppose

$$F(t)^{\frac{\chi}{\chi+1}} \leq c_F f(t) \qquad on \ (0, \eta_0),$$

for some $c_F > 0$. Let Ω be an end of M. Then, (CSP) holds for solutions of

$$\begin{cases} \Delta_p u \geq f(u)|\nabla u|^{p-1-\chi} & on \ \Omega \\ u \geq 0, \qquad \lim\limits_{x\in\Omega, \, x\to\infty} u(x) = 0 \end{cases}$$

if and only if

$$F^{-\frac{1}{\chi+1}} \in L^1(0^+).$$

Example 2.44 Conditions for the validity of (2.53) with $v = m$ will be given in Sect. 4.4, see Examples 4.20, 4.21, 4.22, and 4.23. In particular, we stress that (2.53) holds with $v = m$ if M is (complete and) minimally immersed in a Cartan–Hadamard ambient space N. By Gauss equations, in this setting Theorem 2.43 can be applied provided that the minimal immersion $M \to N$ has bounded second fundamental tensor in order to satisfy (2.52).

We stress that Theorem 9.24 below allows a dependence on the function b, which is required to be bounded from below in terms of a decaying function of ϱ. However, to bound ϱ from *below* with the more manageable r, one needs to know an effective decay estimate for $\mathcal{G}$ at infinity. A particularly neat case is when Ric ≥ 0 and

$$\lim_{r \to \infty} \frac{|B_r|}{r^m} > 0,$$

namely, M has maximal volume growth compatible with the condition on Ric. If $m \geq 3$, it is known that M supports an isoperimetric inequality (cf. Example 4.21), thus our estimates for $\mathcal{G}$ and $|\nabla \varrho|$ guarantee that, for some constant $C \geq 1$,

$$\varrho \leq r \leq C\varrho \qquad \text{on } M.$$

The reader is referred to Theorem 9.26 below for the precise statement of the result.

Observe that in Theorem 2.43 we require $p \in (1, 2]$, that is, that the p-Laplacian be non-degenerate. This condition is technical, and is necessary to apply the comparison theorems that are currently available in the literature. For this reason, we feel interesting to investigate the following

Problem 4 Prove or disprove the validity of Theorem 2.43 (or, more generally, Theorems 9.24 and 9.26 below) in the full range $p \in (1, \infty)$.

When l is constant a further fake distance, recently constructed in [20], allows to improve on Theorems 2.43 and 9.26 when u solves

$$\Delta u \geq (1+r)^{-\mu} f(u),$$

by reducing the geometric conditions to the only

$$\text{Ric} \geq -(m-1)\kappa^2 (1+r^2)^{\alpha/2} \langle \, , \, \rangle,$$

for some $\kappa > 0$ and $\alpha \in (-2, 2]$. The threshold $\alpha = 2$ is sharp and related to a probabilistic requirement called the Feller property, cf. [13, 178]. More details on this issue are given at page 214.

2.8 More General Inequalities

The techniques discussed in the present book are also effective to investigate more general inequalities of the type

$$\Delta_\varphi u \geq b(x) f(u) l(|\nabla u|) - \bar{b}(x) \bar{f}(u) \bar{l}(|\nabla u|). \tag{2.54}$$

This class includes (1.12), describing a geodesic graph with a prescribed mean curvature that is neither minimal nor a MCF soliton, and the inequality

$$\Delta_p u \geq f(u) - c|\nabla u|^q, \qquad \text{with } q > 0, c > 0,$$

that, in the last section of the previous chapter, is discussed in connection to a stochastic control problem (cf. [140, 199]). Existence and nonexistence of entire solutions have been investigated in a series of papers, notably

- [104, 139], where the authors consider solutions of

$$\Delta u \pm \bar{b}(x)|\nabla u|^q = b(x)u^\gamma \qquad \text{on } \mathbb{R}^m, \tag{2.55}$$

  for $q, \gamma \in \mathbb{R}^+$ and non-negative $b, \bar{b}$ that are allowed to vanish in a controlled (yet very general) way. See also improvements in [227].
- [94, 157], that concern quasilinear analogues of (2.55) when the driving operator is, respectively, a weighted p-Laplacian and a general φ-Laplacian in a manifold setting. Although related, their techniques differ from those in [104, 139].
- [89], that considers a fully nonlinear version of (2.55) of the type

$$\mathcal{M}[u] \geq f(u) \pm g(|\nabla u|),$$

  where the driving operator $\mathcal{M}$ is uniformly elliptic.

Apart from some special cases, the existence-nonexistence problem for such inequalities is far from being completely understood, and many interesting questions are still unanswered even in the Euclidean setting. For instance, none of the references considers the interplay of the weights and nonlinearities in the generality of (2.54). By a way of example, with the aid of (SMP$_\infty$), in Corollary 8.8 we establish sharp estimates for solutions of the differential inequality

$$\Delta_p u \geq b(x)f(u)|\nabla u|^q - \bar{b}(x)\bar{f}(u)|\nabla u|^{\bar{q}}$$

that are bounded from above.

Preliminaries from Riemannian Geometry

3

We briefly recall some facts from Riemannian Geometry, mostly to fix notation and conventions. Our main source for the present chapter is P. Petersen's book [177]. Let $(M^m, \langle\,,\,\rangle)$ be a connected Riemannian manifold. We denote with ∇ the Levi–Civita connection induced by $\langle\,,\,\rangle$, and with R the $(4, 0)$ curvature tensor of ∇, with the usual sign agreement:

$$R(X, Y, Z, W) \doteq \langle \nabla_Z \nabla_W Y - \nabla_W \nabla_Z Y - \nabla_{[Z,W]} Y, X \rangle.$$

We also denote with Sec the sectional curvature of M, defined as usual on 2-planes $X \wedge Y \in \Lambda^2(TM)$ by the formula

$$\mathrm{Sec}(X \wedge Y) = \frac{R(X, Y, X, Y)}{|X|^2 |Y|^2 - \langle X, Y \rangle^2}.$$

The Ricci tensor is obtained by tracing the curvature tensor in the second and fourth indices:

$$\mathrm{Ric}(X, Y) = \sum_i R(X, e_i, Y, e_i),$$

where, at a given point x, $\{e_i\}$ is an orthonormal basis for $T_x M$. In particular,

$$\mathrm{Ric}(X, X) = \sum_\alpha \mathrm{Sec}(X \wedge e_\alpha) \qquad \forall X \in T_x M, \ |X| = 1,$$

© The Author(s), under exclusive license to Springer Nature Switzerland AG 2021
B. Bianchini et al., *Geometric Analysis of Quasilinear Inequalities on Complete Manifolds*, Frontiers in Mathematics, https://doi.org/10.1007/978-3-030-62704-1_3

where $\{e_\alpha\}$, $2 \le \alpha \le m$ is an orthonormal basis for $X^\perp$. The distance induced by $\langle \, , \, \rangle$ via the formula

$$\text{dist}(x, y) = \inf \left\{ \int_0^1 |\gamma'(t)| dt \; : \; \gamma : [0, 1] \to M \text{ curve from } x \text{ to } y \right\}$$

turns M into a metric space, and we say that M is complete provided that (M, dist) is so as a metric space. The metric ball of radius r centred at a fixed origin $o \in M$ will be denoted with $B_r(o)$, or simply with B_r when no possible confusion about the chosen origin arises. If $x, y \in M$ are sufficiently close, $\text{dist}(x, y)$ is realized by a unique curve that solves $\nabla_{\gamma'} \gamma' = 0$, called a geodesic. For given $o \in M$, the exponential map is defined as follows:

$$\exp_o \; : \; U \subset T_o M \to M, \qquad v \mapsto \gamma_v(1),$$

where γ_v is the unique geodesic issuing from o with velocity v. The identity $\gamma_{tv}(1) = \gamma_v(t)$ for $t \in [0, 1]$ shows that U is geodesically starshaped with respect to o, that is, $\exp_o v \in U$ implies $\exp_o(tv) \in U$ for every $t \in [0, 1]$. Also, $\exp_o$ is smooth on U. By the Hopf–Rinow theorem, M is complete if and only if for some (equivalently, every) $o \in M$ the exponential map is defined on the entire $T_o M$, namely, if and only if geodesics issuing at any given origin can be extended on the entire $[0, \infty)$. If M is complete, every pair of points x, y can be joined by a geodesic realizing $\text{dist}(x, y)$, hereafter called a segment. We let $\mathscr{D}_o$ be the maximal set where $\exp_o$ is a diffeomorphism, and D_o be its image. It turns out that D_o is an open, dense subset of M which is geodesically starshaped with respect to o. Furthermore, the distance function $r(x) \doteq \text{dist}(x, o)$ is smooth on $D_o \backslash \{o\}$. Since $\exp_o : \mathscr{D}_o \to D_o$ is a chart, the set D_o is called the maximal domain of normal coordinates. Its complement $\text{cut}(o) \doteq M \backslash D_o$ is a closed subset of measure zero, called the cut-locus of o. All of these facts can be found in [177, Sec. 5.7].

Definition 3.1 Let M be a complete Riemannian manifold. A point $o \in M$ is said to be a pole for M if $\exp_o : T_o M \to M$ is a diffeomorphism, namely, if $\text{cut}(o) = \emptyset$.

Remark 3.2 If o is a pole for M, observe that M is diffeomorphic to $\mathbb{R}^m$ and that the distance from o is smooth on $M \backslash \{o\}$.

Relevant examples of manifolds for which every origin is a pole include Cartan–Hadamard ones, according to the following

Definition 3.3 $(M, \langle \, , \, \rangle)$ is said to be a Cartan–Hadamard manifold if M is complete, simply connected and satisfies Sec ≤ 0.

As anticipated in the Introduction, we shall also be interested in considering the distance from a relatively compact, smooth open subset. To this aim, hereafter an origin $\mathcal{O} \subset M$ will be either a single point, or a relatively compact, open subset with smooth boundary. In the second case, assume from the very beginning that M is complete for convenience. Let $T\mathcal{O}^\perp$ be the normal bundle of $\partial\mathcal{O}$ restricted to directions that are either vanishing or outward pointing, and denote with $\mathscr{D}_\mathcal{O}$, $D_\mathcal{O}$ the maximal domains where the normal exponential map

$$\exp_\mathcal{O} \; : \; \mathscr{D}_\mathcal{O} \subset T\mathcal{O}^\perp \longrightarrow D_\mathcal{O} \subset M \backslash \mathcal{O}$$
$$v \qquad\qquad \longmapsto \qquad \gamma_v(1)$$

is a diffeomorphism. As in the case of a point, $D_\mathcal{O}$ is open in $M \backslash \mathcal{O}$, starshaped with respect to normal geodesics issuing from $\partial\mathcal{O}$, and $r(x) = \mathrm{dist}(x, \mathcal{O})$ is smooth on $D_\mathcal{O}$ and up to $\partial\mathcal{O}$. Note moreover that the Hessian of r satisfies

$$\nabla^2 r = \mathrm{II}_{-\nabla r} \qquad \text{on } \partial\mathcal{O},$$

where $\mathrm{II}_{-\nabla r}$ is the second fundamental form of $\partial\mathcal{O}$ in the inward direction $-\nabla r$. The cut-locus

$$\mathrm{cut}(\mathcal{O}) \doteq M \backslash (\mathcal{O} \cup D_\mathcal{O})$$

is still a closed set of measure zero. If it is empty, we say that $\mathcal{O}$ is a pole for M. As introduced in Sect. 2.3, on $D_\mathcal{O}$ we define the radial sectional curvature $\mathrm{Sec}_{\mathrm{rad}}$ to be the restriction of the sectional curvature Sec to planes containing ∇r and we write

$$\mathrm{Sec}_{\mathrm{rad}} \leq G(r)$$

for some $G \in C(\mathbb{R}^+)$, whenever

$$\mathrm{Sec}(X \wedge \nabla r)(x) \leq G\big(r(x)\big)$$

for each $x \in D_\mathcal{O}$ and $X \perp \nabla r(x)$, $|X| = 1$.

3.1   Model Manifolds

Comparison theory in Riemannian geometry allows to deduce the behaviour of relevant geometric quantities on M depending on the distance r from a fixed origin, from the knowledge of the corresponding ones on a simpler, rotationally symmetric model example. The conditions needed to compare the two are bounds on the curvatures of M in terms of those of the model. Classical comparison theorems, as described for instance in Chapters 6 and 7 of [177] or in Chapter 2 of [185], involve radially symmetric manifolds M_g defined as being topologically $\mathbb{R}^m$ with a metric given, in polar coordinates

$$(r, \theta) \in \mathbb{R}^+ \times \mathbb{S}^{m-1} \qquad \text{on } \mathbb{R}^m \backslash \{0\}$$

as follows:

$$\mathrm{d}s_g^2 = \mathrm{d}r^2 + g(r)^2 \mathrm{d}\theta^2,$$

with $\mathrm{d}\theta^2$ the round metric of curvature 1 on $\mathbb{S}^{m-1}$. If no conditions are posed on g, even for $g \in C^\infty(\mathbb{R}_0^+)$ the metric might be singular at the puncture 0, obtained as the limit $r \to 0$. However, the metric is C^2 on the entire M_g (that suffices for all of our purposes) provided that

$$g \in C^2(\mathbb{R}_0^+), \qquad g(0) = 0, \qquad g'(0) = 1, \tag{3.1}$$

and r is in fact the geodesic distance from the origin 0. Indeed, $\mathrm{d}s_g^2$ is a smooth metric if and only if $g \in C^\infty(\mathbb{R}_0^+)$ and satisfies the further condition $g^{(2j)}(0) = 0$ for every $j \geq 0$. Hereafter, $(M_g, \mathrm{d}s_g^2)$ will be called a *model*. We will mostly assume that $g > 0$ on $\mathbb{R}^+$, although the case in which g is positive only on some maximal $(0, R) \subset \mathbb{R}^+$ is of interest. Important examples include the space forms of constant curvature in Riemannian geometry:

– the sphere of curvature κ^2 for some $\kappa \in \mathbb{R}^+$, where

$$g(r) = \frac{\sin(\kappa r)}{\kappa}$$

on the chart $(0, \pi/\kappa) \times \mathbb{S}^{m-1}$ with the north and south poles removed;
– the Euclidean space, where $g(r) = r$;
– the hyperbolic space of curvature $-\kappa^2$ for some $\kappa \in \mathbb{R}^+$, for which

$$g(r) = \frac{\sinh(\kappa r)}{\kappa}.$$

In the next chapters, particularly, in the study of (CSP), we need to consider the distance r from a relatively compact, open subset $\mathcal{O} \subset M$, not necessarily coinciding with a point. Consequently, we will need to compare M to models whose defining function g does not satisfy (3.1) at $r = 0$, but rather $g(0) = 1$, $g'(0) = \lambda$ for some $\lambda \in \mathbb{R}$.

A direct computation shows that the radial sectional curvature $\mathrm{Sec}_{\mathrm{rad}}$ and the Ricci curvature Ric in the radial direction ∇r satisfy

$$\mathrm{Sec}(X \wedge \nabla r) = -\frac{g''(r)}{g(r)} \doteq -G(r) \qquad \text{for every } 0 \neq X \in T_{(r,\theta)}M_g, \ X \perp \nabla r;$$

$$\mathrm{Ric}(\nabla r, \nabla r) = -(m-1)\frac{g''(r)}{g(r)} \doteq -(m-1)G(r).$$

In particular, a model M_g can also be constructed given the radial sectional curvature $G \in C(\mathbb{R}_0^+)$, by letting $g \in C^2(\mathbb{R}_0^+)$ be the solution of

$$\begin{cases} g'' - Gg = 0 & \text{on } \mathbb{R}^+ \\ g(0) = 0, \quad g'(0) = 1 \end{cases}$$

(resp. with $g(0) = 1$, $g'(0) = \lambda$). In this case, M_g is thought to be defined on the maximal interval $(0, R)$ where $g > 0$.

Remark 3.4 As shown in [22, Prop. 1.21], if $g(0) = 0$, $g'(0) = 1$ a sufficient condition to guarantee that $g > 0$ and $g' > 0$ on $\mathbb{R}^+$ is

$$t \int_t^{\infty} G_-(s)ds \leq \frac{1}{4} \qquad \forall t \in \mathbb{R}^+ \tag{3.2}$$

with $G_- = -\min\{G, 0\}$. This condition allows for model manifolds of positive curvature that open at infinity like paraboloids, and strengthens a classical condition due to A. Kneser, cf. [74]. For instance, (3.2) holds if $G(r) \geq -(4r^2)^{-1}$ on $\mathbb{R}^+$.

The Hessian and Laplacian of r on M_g have the following expression:

$$\nabla^2 r = \frac{g'(r)}{g(r)}\left(ds_g^2 - dr \otimes dr\right),$$

$$\Delta r = (m-1)\frac{g'(r)}{g(r)}.$$

In particular, if $\overline{M}_g$ has a boundary $\{r = 0\}$, its second fundamental form $\mathrm{II}_{-\nabla r}$ and its (unnormalized) mean curvature $H_{-\nabla r}$ in the direction $-\nabla r$ satisfy

$$\mathrm{II}_{-\nabla r} = \frac{g'(0)}{g(0)}ds_g^2, \qquad H_{-\nabla r} = (m-1)\frac{g'(0)}{g(0)}.$$

We also define

$$v_g(r) = \omega_{m-1} g(r)^{m-1}, \qquad V_g(r) = \int_0^r v_g(t)\,dt,$$

where $\omega_{m-1} = |\mathbb{S}^{m-1}|$. Note that $v_g(r)$ and $V_g(r)$ are, respectively, the volume of a geodesic sphere and ball centred at $\{r = 0\}$ in M_g.

3.2   Comparison Theory for the Distance Function

Although a thorough and well-organized account of comparison theorems can already be found in [22, 177, 185] for the distance to a point, the less standard case when the model has a boundary requires some adaptation that we feel better to made explicit for the ease of reference. For this reason, we also provide (sketchy) proofs. The starting point of comparison theory is the Riccati equation satisfied by the second fundamental form of the level sets of r, namely, by $\nabla^2 r$ restricted to $\nabla r^\perp$. For each $x \in D_{\mathcal{O}}$, let $\gamma : [0, r(x)] \to D_{\mathcal{O}}$ be the unique unit speed, minimizing geodesic normal to $\mathcal{O}$, starting from $\partial\mathcal{O}$ and ending at x. Fix a parallel, orthonormal basis $\{\gamma', E_2, \ldots, E_m\}$ along γ and note that $\{E_\alpha(0)\}_{\alpha \geq 2}$ span $T_{\gamma(0)}\partial\mathcal{O}$. Differentiating twice the identity $|\nabla r|^2 = 1$ and using Schwarz lemma together with the Ricci commutation rules

$$v_{jik} = v_{jki} + v_l R_{ljik} \qquad \forall\, v \text{ of class } C^3,$$

where R_{ljik} are the components of the curvature tensor R in an orthonormal basis, we get

$$0 = \tfrac{1}{2}(r_i r_i)_{jk} = r_i r_{ijk} + r_{ij} r_{ik}$$

$$= r_i r_{jik} + r_{ij} r_{ik}$$

$$= r_i r_{jki} + r_i r_l R_{ljik} + r_{ij} r_{ik}.$$

Contracting with respect to $\{E_\alpha\}$, the matrix-valued function

$$B : [0, r(x)] \to \mathrm{Sym}^2(\mathbb{R}^{m-1})$$

$$B_{\alpha\beta}(t) = \nabla^2 r\big(E_\alpha(t), E_\beta(t)\big),$$

representing the second fundamental form of $\{r = t\}$ at $\gamma(t)$, solves the matrix Riccati equation

$$B' + B^2 + R_\gamma = 0, \tag{3.3}$$

where

$$(R_\gamma)_{\alpha\beta}(t) = R(\nabla r, E_\alpha(t), \nabla r, E_\beta(t)),$$

with initial condition

$$B(0)_{\alpha\beta} = \mathrm{II}_{-\nabla r}\left(E_\alpha(0), E_\beta(0)\right).$$

We begin with the matrix Riccati comparison theorem, as stated in [82], see also [22, Thm. 1.14]).

Theorem 3.5 *Let* $R_1, R_2 : [0, T] \to \mathrm{Sym}^2(\mathbb{R}^{m-1})$ *be continuous, and let* $B_1, B_2 :$ $(0, T] \to \mathrm{Sym}^2(\mathbb{R}^{m-1})$ *solve*

$$\begin{cases} B_1' + B_1^2 + R_1 \leq 0 & \text{on } (0, T], \\ B_2' + B_2^2 + R_2 \geq 0 & \text{on } (0, T], \end{cases}$$

with initial condition $(B_1 - B_2)(0^+) \leq 0$. *If* $R_1 \geq R_2$ *on* $[0, T]$, *then*

$$B_1 \leq B_2 \qquad \text{on } (0, T],$$

and $\dim \ker(B_2 - B_1)$ *is non-increasing. In particular, if* $B_1(t_0) = B_2(t_0)$ *for some* t_0, *then* $B_1 \equiv B_2$ *on* $[0, t_0]$.

The Hessian comparison theorem is a direct corollary.

Theorem 3.6 (Hessian Comparison from Below) *Let* $(M^m, \langle\,,\,\rangle)$ *be a complete Riemannian manifold, define* $\mathcal{O}, D_\mathcal{O}, r$ *as above and suppose that*

$$\mathrm{Sec}_{\mathrm{rad}} \leq -G(r) \qquad \text{on } D_\mathcal{O}, \tag{3.4}$$

for some $G \in C^2(\mathbb{R}_0^+)$. *Fix* $\underline{\lambda} \in \mathbb{R}$ *such that*

$$\inf_{\partial\mathcal{O}} \mathrm{II}_{-\nabla r} \geq \underline{\lambda}, \tag{3.5}$$

consider a solution g of

$$\begin{cases} g'' - Gg \leq 0 & \text{on } \mathbb{R}^+ \\ g(0) = 1, \quad g'(0^+) \leq \underline{\lambda}, \end{cases} \tag{3.6}$$

and let $[0, R)$ be maximal interval where $g > 0$. Then,

$$\nabla^2 r \geq \frac{g'(r)}{g(r)}\left(\langle\,,\,\rangle - dr \otimes dr\right) \qquad on \ D_{\mathcal{O}} \cap B_R(\mathcal{O}). \tag{3.7}$$

Proof Let $x \in D_{\mathcal{O}}$ and let γ, B, R_γ be as above. Clearly, by (3.4)

$$R_\gamma \geq G(r)I_{m-1}, \qquad B(0) \geq \underline{\lambda}I_{m-1}.$$

Since the function $\bar{B} = g'/g\,I_{m-1}$ solves

$$\begin{cases} \bar{B}' + \bar{B}^2 - G(t) \leq 0 \qquad \text{on } (0, R), \\ \bar{B}(0^+) \leq \underline{\lambda}I_{m-1}, \end{cases}$$

by Riccati comparison we get $B \geq \bar{B}$ on $[0, \min\{r(x), R\})$. In other words,

$$\nabla^2 r \geq \frac{g'(r)}{g(r)}\langle\,,\,\rangle \qquad \text{on } \nabla r^\perp.$$

Taking into account that $\nabla^2 r(\nabla r, \cdot) = 0$ (this can be seen by differentiating $|\nabla r|^2 = 1$), estimate (3.7) follows at once. $\qquad\qquad\square$

The Laplacian comparison from below simply follows by taking traces in (3.7):

Theorem 3.7 (Laplacian Comparison from Below) *Let $(M^m, \langle\,,\,\rangle)$ be a complete manifold, let $\mathcal{O}$, $D_{\mathcal{O}}$, r be as above and suppose that*

$$\mathrm{Sec}_{\mathrm{rad}} \leq -G(r) \qquad on \ D_{\mathcal{O}},$$

for some $G \in C^2(\mathbb{R}_0^+)$. Fix $\underline{\lambda} \in \mathbb{R}$ such that

$$\inf_{\partial\mathcal{O}} \mathrm{II}_{-\nabla r} \geq \underline{\lambda}.$$

If g solves (3.6) and is positive on $[0, R)$, then

$$\Delta r \geq (m - 1)\frac{g'(r)}{g(r)} \qquad on \ D_{\mathcal{O}} \cap B_R(\mathcal{O}).$$

The Hessian comparison from above is obtained by reversing all the inequalities in (3.4), (3.5) (that is, assume $\sup_{\partial\mathcal{O}} \mathrm{II}_{-\nabla r} \leq \bar{\lambda}$), (3.6) and (3.7). As a matter of fact, in this case one can also prove that $D_{\mathcal{O}} \subset B_R(\mathcal{O})$ and that (3.7) (with the reversed sign) holds on all of $M\backslash\mathcal{O}$ in the support sense (Calabi sense, see [177]). The Laplacian comparison from above, on the other hand, requires a milder curvature requirement and an initial estimate just involving the unnormalized mean curvature $H_{-\nabla r}$ of $\partial \mathcal{O}$.

Theorem 3.8 (Laplacian Comparison from Above) *Let* $(M^m, \langle\,,\,\rangle)$ *be a complete manifold, let* $\mathcal{O}, D_\mathcal{O}, r$ *be as above and suppose that*

$$\mathrm{Ric}(\nabla r, \nabla r) \geq -(m-1)G(r) \qquad \textit{on } D_\mathcal{O}, \tag{3.8}$$

for some $G \in C^2(\mathbb{R}_0^+)$. *Fix* $\bar{\lambda} \in \mathbb{R}$ *such that*

$$\sup_{\partial \mathcal{O}} H_{-\nabla r} \leq (m-1)\bar{\lambda}, \tag{3.9}$$

consider a solution g of

$$\begin{cases} g'' - Gg \geq 0 & \textit{on } \mathbb{R}^+ \\ g(0) = 1, \quad g'(0^+) \geq \bar{\lambda}, \end{cases} \tag{3.10}$$

and let $[0, R)$ *be the maximal interval where* $g > 0$. *Then,* $D_\mathcal{O} \subset B_R(\mathcal{O})$ *and*

$$\Delta r \leq (m-1)\frac{g'(r)}{g(r)} \tag{3.11}$$

holds pointwise on $D_\mathcal{O}$ *and weakly on* $M \backslash \mathcal{O}$.

Proof Taking traces in (3.3) and applying Newton's inequality $\mathrm{tr}(B^2) \geq \dfrac{(\mathrm{tr}(B))^2}{m-1}$ one deduces that the function

$$u(t) = \frac{\mathrm{tr}B(t)}{m-1} = \frac{\Delta r(\gamma(t))}{m-1}$$

solves

$$u' + u^2 + \mathrm{Ric}(\gamma', \gamma') \leq 0, \qquad u(0) = \frac{1}{m-1}H_{-\nabla r}(\gamma(0)) \leq \bar{\lambda}.$$

On the other hand, $\bar{u} = g'/g$ satisfy

$$\bar{u}' + \bar{u}^2 - G \geq 0, \qquad \bar{u}(0^+) \geq \bar{\lambda}.$$

Riccati comparison now applied to $B_1 = u I_{m-1}$ and $B_2 = \bar{u} I_{m-1}$ implies $u \leq \bar{u}$ on $[0, \min\{R, r(x)\})$, whence (3.11) holds on $D_{\mathcal{O}} \cap B_R$. However, $\bar{u} \to -\infty$ as $t \to R^-$, so necessarily u is unbounded from below as $t \to R^-$, which implies $r(x) < \bar{R}$. Hence, $D_{\mathcal{O}} \subset B_R$. The weak inequality can be proved as in [185, Lem. 2.5] (see also [22, Thm. 1.19]).

□

Example 3.9 The initial condition satisfied by $g(r)$ is crucial for the validity of the Hessian and Laplacian comparison theorems, as illustrated by the following example. Fix $\delta \geq 1$ and consider the model M_δ with metric

$$ds_\delta^2 = dt^2 + g_\delta(t)^2 d\theta^2, \qquad \text{where} \quad \begin{cases} g_\delta \in C^2(\mathbb{R}_0^+) & g_\delta > 0 \quad \text{on } \mathbb{R}^+ \\ g_\delta(t) = t & \text{if } t \leq 1/2 \\ g_\delta(t) = \exp\{-t^\delta\} & \text{if } t \geq 1. \end{cases}$$

Define $\mathcal{O} = \{t < 1\}$ and $G_\delta = g_\delta''(t)/g_\delta(t)$. Note that $r = t - 1$ is the distance from $\mathcal{O}$, and that on $M \backslash \mathcal{O}$,

$$\mathrm{II}_{-\nabla r} = -\delta ds_\delta^2, \qquad \Delta_\delta r = -\delta(m-1)(1+r)^{\delta-1}$$

$$\mathrm{Ric}_\delta(\nabla_\delta r, \nabla_\delta r) = -(m-1)\delta\big[- (\delta - 1)(1+r)^{\delta-2} + \delta(1+r)^{2\delta-2}\big].$$

Observe that $\mathrm{Ric}_\delta(\nabla_\delta r, \nabla_\delta r)$ is a decreasing function of δ, but also $\Delta_\delta r$ is so. This is, however, not in contradiction with Theorem 3.8. Indeed, to apply the latter with $\delta < \bar{\delta}$, $M = M_\delta$ and $G = G_{\bar{\delta}}$, condition (3.8) would be satisfied, while (3.9) requires $-\delta \leq \lambda$. In this way, the function $g_{\bar{\delta}}$ does not solve (3.10) because $g_{\bar{\delta}}'(0) = -\bar{\delta} < \lambda$.

For most manifolds the trace of inequality (3.7), that is,

$$\Delta r \geq (m-1)\frac{g'(r)}{g(r)}$$

does not hold weakly on $M \backslash \mathcal{O}$ even if $g'(r)/g(r)$ is well defined on $M \backslash \mathcal{O}$, that is, if $R = \infty$. To motivate this claim, and to introduce our next result, we shall have a closer look at the fine structure of the cut-locus. A point x belongs to $\mathrm{cut}(\mathcal{O})$ if and only if either of the following two possibilities occurs:

− $\exp_{\mathcal{O}}$ is not a diffeomorphism at x. These points are called focal for $\mathcal{O}$ (if $\mathcal{O} = o$ is a point, they are said to be conjugate to o);
− x is joined to $\mathcal{O}$ by at least two segments.

The two possibilities do not mutually exclude. The set of non-focal points $x \in \mathrm{cut}(\mathcal{O})$ where exactly 2 minimizing geodesics meet is named the *normal cut-locus*. The distribution Δr, acting on $\phi \in C_c^\infty(M)$ via the formula

$$< \Delta r, \phi > \doteq \int_M r \Delta \phi = - \int \langle \nabla r, \nabla \phi \rangle,$$

has been investigated in [151], in the case $\mathcal{O}$ is a point. Their argument, however, extends verbatim to smooth, relatively compact open sets $\mathcal{O}$, see also [150] and Section 3.9 of [10]. The singular part of the distribution Δr with respect to the Riemannian volume measure dx acts as a negative Radon measure concentrated on the cut-locus $\mathrm{cut}(\mathcal{O})$, and can be written as

$$\Delta r = (\Delta r)_{AC} dx - |\nabla^+ r - \nabla^- r| \mathcal{H}^{m-1} \llcorner \mathrm{cut}(\mathcal{O}), \tag{3.12}$$

where $\mathcal{H}^{m-1}$ is the $(m-1)$-dimensional Hausdorff measure, and

- $(\Delta r)_{AC}$ coincides with the $L_{\mathrm{loc}}^1(M \backslash \mathcal{O})$ function given by Δr outside $\mathrm{cut}(\mathcal{O})$;
- $|\nabla^+ r - \nabla^- r|$ is a function defined on the normal cut-locus and $\nabla^+ r(x)$ and $\nabla^- r(x)$ are the tangent vectors of the two segments from $\partial \mathcal{O}$ to x.

As a consequence of the work of various authors (the reader can find an account in Section 1.1 of [22]), the normal cut-locus is a smooth $(m-1)$-dimensional manifold, possibly with many connected components, and the complement of the normal cut-locus has Hausdorff dimension at most $(m-2)$. In particular, the expression for the singular part of (3.12) is meaningful, being defined $\mathcal{H}^{m-1}$ almost everywhere on $\mathrm{cut}(\mathcal{O})$. Furthermore, the normal cut-locus is dense in the set of non-focal cut-points. Therefore,

$$\Delta r = (\Delta r)_{AC} dx \qquad \Longleftrightarrow \qquad \mathrm{cut}(\mathcal{O}) \text{ consists only of focal points.}$$

In particular, in the assumptions of Theorem 3.7, if $R = \infty$ the inequality

$$\Delta r \geq (m-1) \frac{g'(r)}{g(r)}$$

holds weakly on $M \backslash \mathcal{O}$ provided that $\mathrm{cut}(\mathcal{O})$ consists only of focal points.

In the Introduction, and in particular in Theorem 2.34, we claimed that (2.42) implies that o be a pole of M. We prove this statement in the next

Proposition 3.10 *In the above notation, suppose that the negative part* $(\Delta r)_-$ *of the measure* Δr *satisfies*

$$(\Delta r)_- \in L^\infty_{\text{loc}}(M \setminus \mathcal{O}). \tag{3.13}$$

Then, $\mathcal{O}$ *is a pole of* M.

Proof Inequality (3.11) coming from the Laplacian comparison from above implies that the positive part $(\Delta r)_+ \in L^\infty_{\text{loc}}(M \setminus \mathcal{O})$. Because of (3.13), Δr is absolutely continuous and represented by a locally bounded function and thus, by (3.12) and the discussion above, $\text{cut}(\mathcal{O})$ has just conjugate points. On the other hand, if $x_0 \in \text{cut}(\mathcal{O})$ is conjugate to $\mathcal{O}$ and denoting with $g(r, \theta)$ the determinant of $\langle\,,\,\rangle$ in normal coordinates $(r, \theta) \in \mathscr{D}_\mathcal{O} \subset \mathbb{R}^+_0 \times \partial \mathcal{O}$ for $D_\mathcal{O}$, by the identity

$$\Delta r = \frac{1}{2} \partial_r \log g(r, \theta)$$

we see that $\Delta r(y) \to -\infty$ as $y \in D_\mathcal{O},\ y \to x$. Hence, Δr is not bounded in a neighbourhood of x, a contradiction which shows that $\text{cut}(\mathcal{O})$ is in fact empty, equivalently, that $\mathcal{O}$ is a pole. $\qquad \square$

The volume comparison theorems can be deduced by integration. For $R > 0$, we define

$$\partial B_R(\mathcal{O}) = \Big\{ x \in M : r(x) = R \Big\}, \qquad B_R(\mathcal{O}) = \Big\{ x \in M : r(x) \in (0, R) \Big\}.$$

Note that $\mathcal{O} \not\subseteq B_R(\mathcal{O})$, and thus $|B_R(\mathcal{O})| \to 0$ as $R \to 0$. The proof of the next result follows verbatim the version in [185, Thm. 2.14] (see also [22, Thm. 1.24]).

Theorem 3.11 (Volume Comparison) *Let* $(M^m, \langle\,,\,\rangle)$ *be a complete Riemannian manifold, and define* $\mathcal{O}, D_\mathcal{O}, r$ *as above.*

(1) In the assumptions of Theorem 3.7, the functions

$$\frac{|\partial B_r(\mathcal{O})|}{v_g(r)}, \qquad \frac{|B_r(\mathcal{O}) \setminus B_{r_0}(\mathcal{O})|}{V_g(r) - V_g(r_0)}$$

are non-decreasing in r *provided that* $r_0 \le r < R$ *and* $B_r(\mathcal{O}) \subset D_\mathcal{O}$. *In particular, there exists* $C > 0$ *such that for all such* r

$$|\partial B_r(\mathcal{O})| \ge C v_g(r),$$

$$|B_r(\mathcal{O}) \setminus B_{r_0}(\mathcal{O})| \ge C\big(V_g(r) - V_g(r_0)\big).$$

(2) *In the assumptions of Theorem 3.8, the functions*

$$\frac{|\partial B_r(\mathcal{O})|}{v_g(r)}, \qquad \frac{|B_r(\mathcal{O})\setminus B_{r_0}(\mathcal{O})|}{V_g(r) - V_g(r_0)}$$

are non-increasing in r for each r $\geq r_0$ (a.e. r for the first one). In particular, there exists $C > 0$ such that for all such r

$$|\partial B_r(\mathcal{O})| \leq C v_g(r),$$

$$|B_r(\mathcal{O})\setminus B_{r_0}(\mathcal{O})| \leq C\big(V_g(r) - V_g(r_0)\big).$$

The comparison theorem from above, in (2), is due to Bishop–Gromov, see [185, Sec.2] for references. In the particular case

$$\mathrm{Ric}(\nabla r, \nabla r) \geq -(m-1)\kappa^2\big(1 + r^2\big)^{\alpha/2},$$

for some $\kappa \geq 0$ and $\alpha \geq -2$, and when $\mathcal{O}$ is a point, detailed computations of the asymptotic behaviour of a suitable solution g of

$$\begin{cases} g'' - \kappa^2(1+t^2)^{\alpha/2} \geq 0 & \text{on } \mathbb{R}^+, \\ g(0) = 0, \qquad g'(0) \geq 1 \end{cases}$$

can be found in [185, Prop. 2.1]: more precisely, the above inequality admits a solution g with

$$g(r) \asymp \begin{cases} \exp\left\{\frac{2\kappa}{2+\alpha}(1+r)^{1+\frac{\alpha}{2}}\right\} & \text{if } \alpha \geq 0 \\ r^{-\frac{\alpha}{4}}\exp\left\{\frac{2\kappa}{2+\alpha}r^{1+\frac{\alpha}{2}}\right\} & \text{if } \alpha \in (-2, 0) \\ r^{\bar{\kappa}}, \quad \bar{\kappa} = \frac{1+\sqrt{1+4\kappa^2}}{2} & \text{if } \alpha = -2 \end{cases} \qquad (3.14)$$

in a neighbourhood of infinity. In particular, setting $v_g(r)$ as above, from

$$\log \int_{r_0}^r v_g \sim \begin{cases} \dfrac{2\kappa(m-1)}{2+\alpha}r^{1+\frac{\alpha}{2}} & \text{if } \alpha > -2; \\ \dfrac{2\kappa(m-1)}{2+\alpha}r^{1+\frac{\alpha}{2}} - \dfrac{\alpha(m-1)}{4}\log r & \text{if } \alpha \in (-2, 0); \\ \big[(m-1)\bar{\kappa} + 1\big]\log r & \text{if } \alpha = -2 \end{cases}$$

and from Theorem 3.11 we deduce

$$\limsup_{r \to \infty} \frac{\log |B_r|}{r^{1+\alpha/2}} < \infty \qquad\qquad \text{if } \alpha > -2,$$

$$\limsup_{r \to \infty} \frac{\log |B_r|}{\log r} \le (m-1)\bar{\kappa} + 1 \quad \text{if } \alpha = -2.$$

We conclude by extending the examples in (3.14) to a larger class of solutions of (3.6) and (3.10), that enables us to include more general initial conditions. When $\mathcal{O}$ reduces to a point, further examples can be found in the appendix of [24]. The proof of the next lemma is by direct computation.

Lemma 3.12 *Let $G \in C^1(\mathbb{R}^+) \cap C(\mathbb{R}_0^+)$ be non-negative, set*

$$\theta_* = \inf_{\mathbb{R}^+} \frac{G'}{2G^{3/2}}, \qquad \theta^* = \sup_{\mathbb{R}^+} \frac{G'}{2G^{3/2}}, \qquad D_\pm(t) = \frac{1}{2}\left(-t \pm \sqrt{t^2 + 4}\right).$$

For constants $C > 0$, $D \in \mathbb{R}$ consider the function

$$g(t) = 1 + C\left\{\exp\left(D\int_0^t \sqrt{G(s)}\,ds\right) - 1\right\}.$$

Then, for a fixed $\lambda \in \mathbb{R}$,

(1) g solves

$$\begin{cases} g'' - Gg \ge 0 & \text{on } \mathbb{R}^+ \\ g(0) = 1, & g'(0) \ge \lambda \end{cases}$$

provided that

$$C \ge 1, \qquad CD\sqrt{G(0)} \ge \lambda, \qquad D \in \left(-\infty, D_-(\theta^*)\right] \cup \left[D_+(\theta_*), \infty\right);$$

(2) g solves

$$\begin{cases} g'' - Gg \le 0 & \text{on } \mathbb{R}^+ \\ g(0) = 1, & g'(0) \le \lambda \end{cases}$$

provided that

$$C \in (0, 1], \qquad CD\sqrt{G(0)} \le \lambda, \qquad D \in \left[D_-(\theta_*), D_+(\theta^*)\right].$$

In both (1) and (2), if θ^* or θ_* are infinite, then $D_\pm(\theta_*)$, $D_\pm(\theta^*)$ are intended in the limit sense and are excluded from the range of D.

Radialization and Fake Distances

4

The proof of some of our main results, for instance the (CSP), relies on the construction of a suitable radial solution of $(P_\geq)$ or $(P_\leq)$ to be compared with a given one. For convenience, hereafter we extend φ to an odd function on all of $\mathbb{R}$ by setting

$$\varphi(s) = -\varphi(-s) \qquad \text{for each } s < 0. \tag{4.1}$$

Suppose that $w \in C^1(\mathbb{R}_0^+)$ satisfies $\varphi(w') \in C^1(\mathbb{R}_0^+)$. If $u(x) = w(r(x))$, where $r(x)$ is the distance from a fixed origin $\mathcal{O}$ (a point, or a relatively compact open set with smooth boundary), then

$$\Delta_\varphi u = \operatorname{div}\left(\frac{\varphi(|w'|)}{|w'|} w' \nabla r\right) = \operatorname{div}\big(\varphi(w')\nabla r\big) = \big(\varphi(w')\big)' + \varphi(w')\Delta r,$$

therefore u solves, say, $(P_\geq)$ if and only if

$$\big(\varphi(w')\big)' + \varphi(w')\Delta r \geq b(x)f(w)l(|w'|). \tag{4.2}$$

Take $0 < g \in C^2(\mathbb{R}_0^+)$ and a model manifold M_g with metric

$$ds_g^2 = dr^2 + g(r)^2 d\theta^2,$$

and let $v_g(r) = \omega_{m-1} g(r)^{m-1}$ be the volume of geodesic spheres at the pole of M_g. By comparison theory for Δr, given a radial bound for the function b of the type $b(x) \geq \beta(r(x))$, for some positive $\beta \in C(\mathbb{R}_0^+)$, to find solutions of (4.2) one is first lead to solve

$$\big[v_g \varphi(w')\big]' = v_g \beta f(w) l(|w'|) \tag{4.3}$$

© The Author(s), under exclusive license to Springer Nature Switzerland AG 2021
B. Bianchini et al., *Geometric Analysis of Quasilinear Inequalities on Complete Manifolds*, Frontiers in Mathematics, https://doi.org/10.1007/978-3-030-62704-1_4

on an interval of $\mathbb{R}^+$. Furthermore, a solution of (4.3) gives rise to a solution of (4.2) provided the sign of w' matches appropriately with the inequalities coming from the comparison theorems for Δr, and with the sign of $f(w)$. Therefore, the monotonicity of w becomes relevant. We will devote the next chapter to the study of the ODE (4.3).

The investigation of the compact support principle along these lines needs the use of comparison theorems from below, that requires r to be smooth, equivalently, $\mathcal{O}$ to be a pole. As outlined in the Introduction, in this chapter we develop a different radialization procedure that uses a "fake distance" ϱ modelled on the operator Δ_φ, in the particular case of the p-Laplacian

$$\Delta_p u = \operatorname{div}(|\nabla u|^{p-2}\nabla u).$$

The goal of this section is to find conditions for the properness of ϱ and for the gradient estimate $|\nabla\varrho| \leq 1$ to hold. Our treatment closely follows [156].

4.1 Basic Facts on Nonlinear Potential Theory

We begin by recalling some terminology and basic results in nonlinear potential theory. The reader can find a thorough treatment in [107] for $p = 2$, while for $p \neq 2$ we suggest the book [116] and the papers [120, 121, 186, 228, 229].

Given $K \subset \Omega$, K compact and Ω open, the p-capacity of the condenser (K, Ω) is by definition

$$\operatorname{cap}_p(K, \Omega) = \inf\left\{\int_\Omega |\nabla\psi|^p \; : \; \psi \in \operatorname{Lip}_c(\Omega), \; \psi \geq 1 \text{ on } K\right\}.$$

If K is the closure of a smooth domain, and if Ω is smooth, the infimum coincides with the energy $\|\nabla u\|_p^p$ of the unique solution of

$$\begin{cases} \Delta_p u = 0 & \text{on } \Omega\backslash K, \\ u = 0 \text{ on } \partial\Omega, & u = 1 \text{ on } \partial K, \end{cases}$$

extended with $u \equiv 1$ on K. The solution u is called the p-capacity potential of (K, Ω). By the strong maximum principle, $0 < u < 1$ on $\Omega\backslash K$. If Ω has non-compact closure, or if it has a rough boundary (in the sense that some point of $\partial\Omega$ does not satisfy Wiener's test in [148]), by exhausting Ω with a family of smooth open sets Ω_j satisfying

$$K \Subset \Omega_j \Subset \Omega_{j+1} \Subset \Omega \quad \text{for each } j \geq 1, \qquad \bigcup_{j=1}^{\infty} \Omega_j = \Omega,$$

and using elliptic estimates, the sequence $\{u_j\}$ of the p-capacity potentials of (K, Ω_j) converges to a limit function $u : \Omega \rightarrow (0, 1]$, still called the p-capacity potential of (K, Ω), which is independent of the chosen exhaustion and satisfies

$$\begin{cases} \Delta_p u = 0 & \text{on } \Omega \backslash K, \\ u = 1 & \text{on } K, \quad 0 < u \leq 1 \quad \text{on } \Omega \backslash K. \end{cases}$$

Furthermore, $\mathrm{cap}_p(K, \Omega) = \|\nabla u\|_p^p$. For a proof and more details, see [186].

Remark 4.1 (Regularity) By the regularity theory in [226], we recall that a solution $u \in W_{\mathrm{loc}}^{1,p}(\Omega)$ of $\Delta_p u = 0$ belongs to $C_{\mathrm{loc}}^{1,\alpha}(\Omega)$ for some $\alpha \in (0, 1)$. Therefore, Schauder's theory implies that u is C^∞ away from the (closed) singular set $\{|\nabla u| = 0\}$.

Definition 4.2 We say that Δ_p is *non-parabolic on* Ω if some $K \subset \Omega$ compact with non-empty interior satisfies $\mathrm{cap}_p(K, \Omega) > 0$, that is, the p-capacity potential u of (K, Ω) is not identically 1.

If Δ_p is non-parabolic on Ω, it can be proved that $\mathrm{cap}_p(K, \Omega) > 0$ holds indeed for *every* compact set K with non-empty interior. Also, from [120, 121] (cf. also [228, 229]), the non-parabolicity of Δ_p is equivalent to the existence, for each fixed $o \in \Omega$, of a positive Green kernel $\mathcal{G}$ with pole at o, that is, to the existence of a positive, distributional solution of $\Delta_p \mathcal{G} = -\delta_o$ on Ω:

$$\int_\Omega |\nabla \mathcal{G}(x)|^{p-2} \langle \nabla \mathcal{G}(x), \nabla \psi(x) \rangle dx = \psi(o) \qquad \forall \psi \in C_c^1(\Omega). \tag{4.4}$$

A kernel $\mathcal{G}$ was constructed in [120, 121] by exhausting Ω with an increasing family $\{\Omega_j\}$ of smooth domains, and taking the limit of the Green kernels $\mathcal{G}_j$ with pole at o and Dirichlet boundary conditions on $\partial \Omega_j$. The existence of each $\mathcal{G}_j$ was shown in [120, Thm. 3.19] for $p \in (1, m]$, and in [121] for $p > m$ (condition $p \leq m$ in [120] has to be assumed since the kernel $\mathcal{G}_j$ is required to diverge as $x \rightarrow o$, cf. [121, p. 656]). The convergence of $\mathcal{G}_j$ to a finite limit and its equivalence to the non-parabolicity of Δ_p on Ω can be found in [120, Thm. 3.27]. As we shall see in a moment, a comparison theorem holds for Green kernels, in particular $\mathcal{G}$ is independent of the chosen exhaustion and qualifies as the minimal, positive Green kernel of Ω. Hereafter, we shall shortly say that $\mathcal{G}$ is the Green kernel of Ω.

Model Manifolds

Fix a model M_g such that

$$g \in C^2(\mathbb{R}_0^+), \quad g > 0 \quad \text{on } \mathbb{R}^+, \quad g(0) = 0, \quad g'(0) = 1.$$

Then, the Green kernel of a geodesic ball $B_R^g \subset M_g$ with pole at the origin is

$$\mathscr{G}_R^g(r) \doteq \int_r^R v_g(s)^{-\frac{1}{p-1}}\,\mathrm{d}s. \tag{4.5}$$

For instance, if $M_g = \mathbb{R}^m$, then $g(r) = r$ and (4.5) becomes

$$\mathscr{G}_R^g(r) = \begin{cases} \omega_{m-1}^{-\frac{1}{p-1}} \dfrac{p-1}{m-p}\left[r^{-\frac{m-p}{p-1}} - R^{-\frac{m-p}{p-1}}\right] & \text{if } p \neq m \\[2ex] \omega_{m-1}^{-\frac{1}{m-1}} \log(R/r) & \text{if } p = m. \end{cases}$$

From (4.5), Δ_p is non-parabolic on M_g if and only if the limit of $\mathscr{G}_R^g(r)$ as $R \to \infty$ is finite, namely, if and only if

$$v_g^{-\frac{1}{p-1}} \in L^1(\infty).$$

In this case,

$$\mathscr{G}^g(r) = \int_r^\infty v_g(s)^{-\frac{1}{p-1}}\,\mathrm{d}s, \qquad r \in \mathbb{R}^+,$$

is the minimal positive Green kernel of Δ_p on M_g with pole at the origin. Observe that, for every $0 < s < R$, the p-capacity potential of the condenser (B_s^g, B_R^g) in M_g is

$$u(r, \theta) = \frac{\mathscr{G}_R^g(r)}{\mathscr{G}_R^g(s)},$$

in particular,

$$\mathrm{cap}_p\left(B_s^g, B_R^g\right) = \int_{B_R^g \setminus B_s^g} |\nabla u|^p = \mathscr{G}_R^g(s)^{1-p}. \tag{4.6}$$

Note also that $\mathscr{G}_R^g(0^+) < \infty$ if and only if $p > m$, and in this case the origin 0 of M_g has positive capacity $\mathscr{G}_R^g(0)^{1-p}$ in the ball B_R^g.

Uniqueness and Comparison Theory for $\mathcal{G}$

To our knowledge, the question whether a positive Green kernel constructed by exhaustion is unique (in particular, independent of the exhaustion) on every Riemannian manifold has been settled only recently in [156]. Uniqueness follows from a comparison theorem, for which one needs to know precisely the behaviour of $\mathcal{G}$ in neighbourhood of the puncture o. If $p > m$, every kernel $\mathcal{G}$ constructed by exhaustion admits a continuous extension to $x = o$ with a positive, finite value (see [116, Thm. 6.33] and [121]), and $\mathcal{G} \in W^{1,p}_{loc}(\Omega)$, in particular $\mathcal{G}$ is Hölder continuous. A short proof of all these facts can be found in the Appendix of [156]. On the other hand, if $p \leq m$, the local behaviour of a solution $\mathcal{G}$ of $\Delta_p \mathcal{G} = -\delta_o$ on Ω has been investigated by J. Serrin in [216, Thm 12]. Striking refinements were later obtained in [136, 232] in the Euclidean setting, see in particular pp. 243-251 in [232], and their arguments have been adapted to manifolds in [156]. The following theorem summarizes the properties of $\mathcal{G}$ that we need. Set

$$\mu(r) = \begin{cases} \omega_{m-1}^{-\frac{1}{p-1}} \left(\dfrac{p-1}{m-p} \right) r^{-\frac{m-p}{p-1}} & \text{if } p \neq m \\[3ex] \omega_{m-1}^{-\frac{1}{m-1}} (-\log r) & \text{if } p = m. \end{cases}$$

Note that $\mu(|x|)$ is a solution of $\Delta_p u = -\delta_0$ on $\mathbb{R}^m$ and it is the Green kernel of $\mathbb{R}^m$ for $p < m$. We then have

Theorem 4.3 ([156], Thms. 2.5 and 2.7) *Let $\mathcal{G}$ be a positive Green kernel for Δ_p constructed by exhaustion on an open set $\Omega \subset M^m$ containing o.*

- *If $p \leq m$, $\mathcal{G}$ is smooth in a punctured neighbourhood of o and, as $x \to o$,*

$$(1) \quad \mathcal{G} \sim \mu(r),$$

$$(2) \quad |\nabla \mathcal{G} - \mu'(r)\nabla r| = o(\mu'(r)).$$

- *If $p > m$, then $\mathcal{G} \in W^{1,p}_{loc}(\Omega) \cap C^{\frac{p-m}{p-1}}_{loc}(\Omega)$, and*

$$\mathcal{G}(x) < \mathcal{G}(o) = \left[\operatorname{cap}_p(\{o\}, \Omega) \right]^{-\frac{1}{p-1}} \qquad \forall x \in \Omega \backslash \{o\}. \tag{4.7}$$

Remark 4.4 The construction of $\mathcal{G}$ for $p > m$ will be necessary in a moment, so we briefly sketch it. More details can be found in the Appendix of [156]. For $\varepsilon > 0$ small, let $u_\varepsilon \in W^{1,p}_{loc}(\Omega)$ be the p-capacity potential of $(B_\varepsilon(o), \Omega)$, extended with 1 on $B_\varepsilon(o)$. By comparison, $u_\varepsilon \downarrow u$ for some bounded solution u of $\Delta_p u = 0$ on $\Omega \backslash \{o\}$. To prove that u does not vanish identically, fix R small enough to satisfy $B_R(o) \Subset \Omega \cap \mathscr{D}_o$, and consider the function

$$w(x) = \frac{\mathscr{G}^g_R(r(x))}{\mathscr{G}^g_R(0)},$$

where, having chosen $\kappa \in \mathbb{R}^+$ such that $\mathrm{Ricc} \geq -(m-1)\kappa^2$ on $B_R(o)$, $\mathscr{G}_R^g$ is the Green kernel of the hyperbolic space of curvature $-\kappa^2$. Note that the definition of w makes sense since $\mathscr{G}_R^g$ extends continuously at 0. By the Laplacian comparison theorem and since $\mathscr{G}_R^g$ is decreasing,

$$\Delta_p w = \frac{1}{\mathscr{G}^g(0)^{p-1}} \left[(p-1)|\mathscr{G}_R^g(r)'|^{p-2}\mathscr{G}_R^g(r)'' - |\mathscr{G}_R^g(r)'|^{p-1}\Delta r \right]$$

$$\geq \frac{1}{\mathscr{G}^g(0)^{p-1}} |\mathscr{G}_R^g(r)'|^{p-2} \left[(p-1)\mathscr{G}_R^g(r)'' + \frac{v_g(r)'}{v_g(r)}\mathscr{G}_R^g(r)' \right] = 0 \qquad (4.8)$$

pointwise on the punctured ball $B_R(o)\backslash\{o\}$, with $w(x) \to 1$ as $x \to o$ and $w = 0$ on ∂B. By comparison, $u \geq w$ and thus $u \to 1$ as $x \to o$. Passing to the limit the identity $\mathrm{cap}_p(B_\varepsilon(o), \Omega) = \|\nabla u_\varepsilon\|_p^p$ we deduce $\mathrm{cap}_p(\{o\}, \Omega) = \|\nabla u\|_p^p$. To conclude, integrating $\Delta_p u$ against a test function $\varphi \in C_c^\infty(\Omega)$ that is constant in a small neighbourhood of o, we deduce $\Delta_p u = -c\delta_o$, for some suitable $c > 0$. To find c, choose $\{\varphi_\delta\} \subset C_c^\infty(M)$ such that $\varphi_\delta \to u$ as $\delta \to 0$ in $W^{1,p}(\Omega) \subset C_{\mathrm{loc}}^\alpha(\Omega)$. Testing with φ_δ the weak definition of $\Delta_p u = -c\delta_o$ and letting $\delta \to 0$, we get

$$\mathrm{cap}_p(\{o\}, \Omega) = \int_\Omega |\nabla u|^p = cu(o) = c.$$

The kernel $\mathcal{G}$ is therefore

$$\mathcal{G}(x) = u(x)\left[\mathrm{cap}_p(\{o\}, \Omega)\right]^{-\frac{1}{p-1}},$$

and (4.7) holds by the strong maximum principle.

With the aid of Theorem 4.3, one easily deduces the next corollary that can be found in [136, Thm. 2.1]) when $M = \mathbb{R}^m$, and in [156, Cor. 2.8].

Corollary 4.5 *Let $p > 1$.*

(i) *If $\mathcal{G}_1, \mathcal{G}_2 \in W_{\mathrm{loc}}^{1,p}(\Omega\backslash\{o\})$ satisfy*

$$\begin{cases} \Delta_p\mathcal{G}_1 \geq \Delta_p\mathcal{G}_2 & \text{on } \Omega \Subset M \\ \mathcal{G}_1 \leq \mathcal{G}_2 & \text{on } \partial\Omega, \\ \displaystyle\limsup_{x \to o} \frac{\mathcal{G}_1(x)}{\mathcal{G}_2(x)} \leq 1, \end{cases}$$

then, $\mathcal{G}_1 \leq \mathcal{G}_2$ on Ω.

(ii) *If* $\Omega_1 \subset \Omega_2 \Subset M$ *are open domains containing* o, *and* $\mathcal{G}_j$ *be a Green kernel for* Δ_p *constructed by exhaustion on* Ω_j, $j \in \{1, 2\}$, *then* $\mathcal{G}_1 \leq \mathcal{G}_2$. *In particular, the Green kernel of an open set, if it exists, is unique.*

Proof

(i) Suppose that

$$\Omega_\varepsilon = \Omega \cap \{\mathcal{G}_1 > (1 + \varepsilon)\mathcal{G}_2\} \neq \emptyset$$

for some $\varepsilon > 0$. Since, in our assumptions, $\Omega_\varepsilon \Subset \Omega \backslash \{o\}$, we infer $\mathcal{G}_1, \mathcal{G}_2 \in W^{1,p}(\Omega_\varepsilon)$ and thus $\mathcal{G}_1 \leq (1 + \varepsilon)\mathcal{G}_2$ by classical comparison results (cf. [194, Thm. 3.4.1]), contradicting the definition of Ω_ε. Therefore, $\Omega_\varepsilon = \emptyset$, and we let $\varepsilon \to 0$ to conclude.

(ii) We prove that the kernel $\mathcal{G}'$ of a smooth domain $\Omega' \Subset \Omega_1$ satisfies $\mathcal{G}' \leq \mathcal{G}_2$ on Ω', and the thesis follows by letting $\mathcal{G}' \uparrow \mathcal{G}_1$. If $p > m$, it is enough to apply again standard comparison, since both $\mathcal{G}', \mathcal{G}_2 \in W^{1,p}(\Omega')$ and $\Delta_p \mathcal{G}' = \Delta_p \mathcal{G}_2$ on Ω', $\mathcal{G}' = 0 \leq \mathcal{G}_2$ on $\partial \Omega'$. If $p \leq m$, Theorem 4.3 guarantees that all of the assumptions in item (i) are satisfied, which is enough to conclude.   □

Under suitable curvature conditions, the kernels of Δ_p on M and M_g can be compared to each other, as described in the following

Proposition 4.6 *Let* $(M^m, \langle\,,\,\rangle)$ *be a complete manifold satisfying*

$$\text{Ric}(\nabla r, \nabla r) \geq -(m - 1)G(r) \qquad \text{on } M \backslash \text{cut}(o),$$

for some $0 \leq G \in C(\mathbb{R}_0^+)$, *where* $r(x)$ *is the distance from a point* o. *If* Δ_p *is non-parabolic on* M *for some* $p \in (1, \infty)$, *then it is non-parabolic also on* M_g, *the model of radial sectional curvature* $-G(r)$. *Moreover, the kernel* $\mathcal{G}$ *of* Δ_p *on* M *with pole at* o *satisfies the following inequalities:*

(i) *If* $p \leq m$, *then*

$$\mathcal{G}(x) \geq \mathcal{G}^g(r(x)) = \int_{r(x)}^{\infty} v_g(s)^{-\frac{1}{p-1}} \, ds \qquad \text{on } M \backslash \{o\}. \tag{4.9}$$

(ii) *If* $p > m$, *then*

$$\mathcal{G}(x) \geq \frac{1}{c_o} \mathcal{G}^g(r(x)) \qquad \text{on } M \backslash \{o\},$$

where

$$c_o \doteq \left[\frac{\mathrm{cap}_p(\{o\}, M)}{\mathrm{cap}_p(\{0\}, M_g)} \right]^{\frac{1}{p-1}} \in (0, 1].$$

Proof By performing the same computations as in (4.8), the Laplacian comparison theorem from above guarantees that, for every $R > 0$, the transplanted function

$$\bar{\mathcal{G}}(x) = \mathscr{G}_R^g(r(x))$$

satisfies $\Delta_p \bar{\mathcal{G}} \geq 0$ weakly on $B_R(o) \setminus \{o\}$. In fact, integrating against $0 \leq \psi \in C_c^\infty(B_R)$ we obtain

$$\int_{B_R \setminus B_\varepsilon} |\nabla \bar{\mathcal{G}}|^{p-2} \langle \nabla \bar{\mathcal{G}}, \nabla \psi \rangle = - \int_{\partial B_\varepsilon} \psi |\nabla \bar{\mathcal{G}}|^{p-2} \langle \nabla \bar{\mathcal{G}}, \nabla r \rangle - \int_{B_R \setminus B_\varepsilon} \psi \Delta_p \bar{\mathcal{G}}$$

$$\leq - \int_{\partial B_\varepsilon} \psi |\nabla \bar{\mathcal{G}}|^{p-2} \langle \nabla \bar{\mathcal{G}}, \nabla r \rangle$$

$$= \int_{\partial B_\varepsilon} \frac{\psi}{v_g(\varepsilon)} \rightarrow \psi(o) \qquad \text{as } \varepsilon \rightarrow 0,$$

thus $\Delta_p \bar{\mathcal{G}} \geq -\delta_o$ on $B_R(o)$. Moreover, by Theorem 4.3,

if $p \leq m$, then $\bar{\mathcal{G}} \sim \mu(r)$ as $x \rightarrow o$;
if $p > m$, then $\bar{\mathcal{G}} \in W^{1,p}(B_R(o))$ and the identity $\bar{\mathcal{G}}(o) = c_o \mathcal{G}(o)$ holds.

Define for convenience $c_o = 1$ for $p \leq m$. It follows from Corollary 4.5 that

$$c_o \mathcal{G}(x) \geq \mathscr{G}_R^g(r(x)) \qquad \text{on } B_R(o),$$

for every $R > 0$ and $x \in M \setminus \{o\}$. Letting $R \rightarrow \infty$ we deduce (4.9), that in particular implies the non-parabolicity of Δ_p on M_g.

To show that $c_o \leq 1$ when $p > m$, observe that $\bar{u}_{R,\varepsilon} = \mathscr{G}_R^g(r)/\mathscr{G}_R^g(\varepsilon)$, extended with 1 on $B_\varepsilon(o)$, is an admissible test function for $\mathrm{cap}(\{o\}, M)$. Using Bishop–Gromov comparison theorem, letting $\varepsilon \rightarrow 0$ and $R \rightarrow \infty$, and using (4.6), we get

$$\mathrm{cap}_p(\{o\}, M) \leq \int_{B_R} |\nabla \bar{u}_{R,\varepsilon}|^p = \mathscr{G}_R^g(\varepsilon)^{-p} \int_\varepsilon^R \frac{|\partial B_t|}{v_g(t)^{\frac{p}{p-1}}} \mathrm{d}t \leq \mathscr{G}_R^g(\varepsilon)^{1-p}$$

$$\rightarrow \mathscr{G}^g(0)^{1-p} = \mathrm{cap}_p(\{0\}, M_g).$$

$\square$

4.2   The Fake Distance and Its Basic Properties

We shall assume the following:

$(\mathscr{H}_p)$ $(M^m, \langle\,,\,\rangle)$ is complete, non-compact and, for a fixed origin $o \in M$, writing $r(x) = \mathrm{dist}(x, o)$ it holds

$$\mathrm{Ric} \geq -(m-1)G(r)\langle\,,\,\rangle \qquad \text{on } M,$$

for some $0 \leq G \in C(\mathbb{R}_0^+)$. Moreover, Δ_p is non-parabolic on M.

In view of Proposition 4.6, Δ_p is non-parabolic also on M_g, the model of radial sectional curvature $-G(r)$. Also, note that g, and therefore v_g, are monotone increasing and diverging as $r \to \infty$.

Definition 4.7 Let M satisfy $(\mathscr{H}_p)$, for some $p > 1$ and origin $o \in M$, and let $\mathcal{G}$ be the Green kernel with pole at $x = o$. The *fake distance* $\varrho : M\backslash\{o\} \to \mathbb{R}_0^+$ is implicitly defined as follows:

– If $p \leq m$ we set $\mathcal{G}(x) = \mathscr{G}^g(\varrho(x))$, that is,

$$\mathcal{G}(x) = \int_{\varrho(x)}^{\infty} v_g(s)^{-\frac{1}{p-1}} \, ds \qquad \text{on } M\backslash\{o\}. \tag{4.10}$$

– If $p > m$ we set

$$c_o \mathcal{G}(x) = \int_{\varrho(x)}^{\infty} v_g(s)^{-\frac{1}{p-1}} \, ds \qquad \text{on } M\backslash\{o\},$$

with

$$c_o = \left[\frac{\mathrm{cap}_p(\{o\}, M)}{\mathrm{cap}_p(\{0\}, M_g)} \right]^{\frac{1}{p-1}}.$$

Remark 4.8 Observe that ϱ is well defined and locally in $C^{1,\alpha}$ on $M\backslash\{o\}$, positive there and can be extended by continuity with $\varrho(o) = 0$. In fact, for $p \leq m$ this follows from Theorem 4.3, while if $p > m$ it follows from (4.7) and (4.6), which, in particular, imply the identity $c_o = \mathscr{G}^g(0)/\mathcal{G}(o)$.

If $M = M_g$, then ϱ coincides with r, the distance function from o. On the other hand, the following proposition holds:

Proposition 4.9 *Let M satisfy $(\mathcal{H}_p)$ for some $p > 1$, let ϱ be the fake distance associated to the kernel $\mathcal{G}$ of Δ_p with pole at o. Then, $\varrho \leq r$ on M.*

Proof For $p \leq m$, define for convenience $c_o = 1$. Then, it follows by Proposition 4.6 and the definition of ϱ that both if $p \leq m$ and if $p > m$,

$$\mathcal{G}^g\big(\varrho(x)\big) = c_o \mathcal{G}(x) \geq \mathcal{G}^g\big(r(x)\big) \qquad \text{on } M\backslash\{o\},$$

and the conclusion follows since $\mathcal{G}^g$ is decreasing. $\square$

Differentiating (4.10) we obtain

$$
\begin{aligned}
|\nabla \varrho| &= v_g(\varrho)^{\frac{1}{p-1}} |\nabla \mathcal{G}| = \left[v_g^{\frac{1}{p-1}} \int_\varrho^\infty \frac{ds}{v_g(s)^{\frac{1}{p-1}}} \right] |\nabla \log \mathcal{G}| \\
&= \frac{|\nabla \log \mathcal{G}|}{\big|\log(\mathcal{G}^g)'(\varrho)\big|}.
\end{aligned}
\tag{4.11}
$$

and

$$\Delta_p \varrho = \frac{v_g'(\varrho)}{v_g(\varrho)} |\nabla \varrho|^p \qquad \text{weakly on } M\backslash\{o\}, \tag{4.12}$$

hence, for each $\psi \in C^2(\mathbb{R})$ with $\psi' \neq 0$ everywhere we obtain

$$
\begin{aligned}
\Delta_p[\psi(\varrho)] &= |\psi'(\varrho)|^{p-2} \psi'(\varrho) \left[(p-1)\psi''(\varrho) + \frac{v_g'(\varrho)}{v_g(\varrho)} \psi'(\varrho) \right] |\nabla \varrho|^p \\
&= \left[v_g^{-1} \Big(v_g |\psi'|^{p-2}\psi' \Big)' \right] (\varrho) |\nabla \varrho|^p.
\end{aligned}
\tag{4.13}
$$

As we have already observed,

$$v_g^{-1} \Big(v_g |\psi'|^{p-2}\psi' \Big)'$$

is the expression of the p-Laplacian of the radial function ψ in the model M_g, making it possible to radialize with respect to ϱ. However, in order for this procedure to be effective we need to control the L^∞-norm of $|\nabla \varrho|$ and to guarantee properness of ϱ. The latter, by (4.10), is equivalent to the property $\mathcal{G}(x) \to 0$ as $r(x) \to \infty$.

Relations with Hardy Weights

We focus our attention on estimating $|\nabla \varrho|$. Under the validity of $(\mathcal{H}_p)$ with G non-increasing, the goal is to obtain the sharp upper bound

$$|\nabla \varrho| \leq 1,$$

that in particular, by (4.11), implies

$$|\nabla \log \mathcal{G}(x)| \leq \left| \log(\mathcal{G}^g)'(\varrho(x)) \right| \qquad \text{on } M\setminus\{o\}. \tag{4.14}$$

In the linear case $p = 2$, and on non-parabolic manifolds with $\mathrm{Ric} \geq 0$, the inequality has been obtained by T. Colding in [56, Thm. 3.1], see also [58] for improvements. The function $|\log(\mathcal{G}^g)'(\varrho)|$ relates, via comparison theory, to weights in Hardy-type inequalities on M, and it thus appears in geometrical problems where stationary Schrödinger type operators (linear or nonlinear) are considered. In fact, if Δ_p is non-parabolic on M, then the following Hardy inequality holds:

$$\left(\frac{p-1}{p}\right)^p \int |\nabla \log \mathcal{G}|^p |\phi|^p \leq |\nabla \phi|^p \qquad \forall \phi \in \mathrm{Lip}_c(M).$$

(cf. [25]), and for this reason we call $|\nabla \log \mathcal{G}|^p$ a Hardy weight. In particular, $|(\log \mathcal{G}^g)'(r)|^p$ is a Hardy weight for M_g. In view of (4.14), $|\nabla \varrho| \leq 1$ can be interpreted as a comparison result for Hardy weights, that seems however unable to allow to replace $|\nabla \log \mathcal{G}|^p$ with a more manageable function constructed from the model M_g. Nevertheless, comparison theory guarantees that the function $|(\log \mathcal{G}^g)'(r)|^p$, transplanted to M, is a Hardy weight provided that $\mathrm{cut}(o) = \emptyset$ and that the radial sectional curvature of M is bounded from *above* in terms of that of M_g (see Section 5 in [25]). For a systematic study of Hardy weights and their role in geometric problems in the linear case $p = 2$ we refer the reader to [22–24].

To prove our desired gradient estimates, we shall first collect some useful properties of the weight

$$|(\log \mathcal{G}^g)'(r)|^p$$

on M_g. The first is its monotonicity for a suitable G.

Lemma 4.10 ([156], Lem. 2.12) *Let $0 \leq G \in C(\mathbb{R}_0^+)$ be non-increasing. Then, v_g'/v_g and $|(\log \mathscr{G}^g)'|$ have negative derivatives on $\mathbb{R}^+$.*

Proof The behaviour of v_g, hence of $\mathscr{G}^g$, at zero guarantees that the sets

$$\{t \; : \; (v_g'/v_g)'(t) < 0\}, \qquad \{t \; : \; |(\log \mathscr{G}^g)'|'(t) < 0\}$$

are both non-empty. Assume, by contradiction, that $(v_g'/v_g)'(t_0) = 0$ for some $t_0 \in \mathbb{R}^+$. Then, the Riccati equation

$$\left(\frac{g'}{g}\right)' + \left(\frac{g'}{g}\right)^2 = G$$

implies the equality $(v_g'/v_g)(t_0) = (m-1)\kappa$, where we have set $\kappa \doteq \sqrt{G(t_0)}$. Because G is non-increasing, by Sturm comparison on $(0, t_0)$ with the model of curvature $-\kappa^2$ and volume v_κ we deduce that $v_g'/v_g \geq v_\kappa'/v_\kappa$ on $(0, t_0]$. However, $(v_\kappa'/v_\kappa)(t) = (m-1)\kappa \coth(\kappa t) > (m-1)\kappa$ for each $t \in \mathbb{R}^+$, contradiction. To show the second part of the statement, set for convenience $\chi(t) = |(\log \mathscr{G}^g)'(t)|$. Differentiating and using the definition of $\mathscr{G}^g$ yields

$$\chi' = \chi \left[\chi - \frac{1}{p-1} \frac{v_g'}{v_g}\right].$$

Suppose that $\chi'(t_0) = 0$ for some $t_0 \in \mathbb{R}^+$; since v_g'/v_g has negative derivative on $\mathbb{R}^+$, then by inspecting the ODE we deduce that $\chi' > 0$ on (t_0, ∞). Therefore, fixing $t_1 > t_0$ there exists $\varepsilon > 0$ such that

$$\chi - \frac{1}{p-1} \frac{v_g'}{v_g} \geq \varepsilon \chi \qquad \text{on } [t_1, \infty).$$

By comparison, χ lies above the solution $\bar{\chi}$ to $\bar{\chi}' = \varepsilon \bar{\chi}^2$ on $[t_1, \infty)$. However, $\bar{\chi}$ explodes in finite time, contradiction. $\square$

The second lemma is a comparison theorem for Hardy weights on model manifolds.

Lemma 4.11 ([22], Prop. 4.12) *Suppose that h/g is increasing on $\mathbb{R}^+$. Then,*

$$|(\log \mathscr{G}^h)'(r)| \geq |(\log \mathscr{G}^g)'(r)| \qquad \forall r \in \mathbb{R}^+.$$

Proof Since both $\mathscr{G}^h$ and $\mathscr{G}^g$ are C^1 and decreasing on $\mathbb{R}^+$, the inequality is equivalent to

$$\int_R^r (\log \mathscr{G}^h)'(t)\,dt \le \int_R^r (\log \mathscr{G}^g)'(t)\,dt \qquad \forall\, 0 < R < r,$$

that is, to the inequality

$$\frac{\int_r^\infty v_h^{-\frac{1}{p-1}}}{\int_R^\infty v_h^{-\frac{1}{p-1}}} \le \frac{\int_r^\infty v_g^{-\frac{1}{p-1}}}{\int_R^\infty v_g^{-\frac{1}{p-1}}}.$$

Set $\chi_h = v_h^{-\frac{1}{p-1}}$ and $\chi_g = v_g^{-\frac{1}{p-1}}$. By assumptions, χ_h/χ_g is non-increasing on $\mathbb{R}^+$, thus

$$\left[\int_R^\infty \chi_h\right]\left[\int_r^\infty \chi_g\right] = \left[\int_R^r \chi_h + \int_r^\infty \chi_h\right]\left[\int_r^\infty \chi_g\right]$$

$$= \left[\int_R^r \frac{\chi_h}{\chi_g}\chi_g + \int_r^\infty \chi_h\right]\left[\int_r^\infty \chi_g\right]$$

$$\ge \left[\frac{\chi_h(r)}{\chi_g(r)}\int_R^r \chi_g + \int_r^\infty \chi_h\right]\left[\int_r^\infty \chi_g\right]$$

$$= \left[\int_R^r \chi_g\right]\left[\frac{\chi_h(r)}{\chi_g(r)}\int_r^\infty \chi_g\right] + \int_r^\infty \chi_h \int_r^\infty \chi_g$$

$$\ge \left[\int_R^r \chi_g\right]\left[\int_r^\infty \frac{\chi_h}{\chi_g}\chi_g\right] + \int_r^\infty \chi_h \int_r^\infty \chi_g$$

$$= \left[\int_R^\infty \chi_g\right]\left[\int_r^\infty \chi_h\right],$$

as claimed. □

4.3 Gradient Estimate

Inequality $|\nabla \varrho| \le 1$ will be the consequence of a new Böchner formula for $|\nabla \varrho|$, recently discovered in [156] and inspired by previous work in [56] in the linear case. For $X \in TM$, $X \ne 0$ consider the linearization of the p-Laplacian

$$A(X) \ : \ TM \to TM,$$

$$A(X) = |X|^{p-2}\left(\mathrm{Id} + (p-2)\left\langle \cdot, \frac{X}{|X|}\right\rangle \frac{X}{|X|}\right).$$

The endomorphism $A(X)$ has two distinct eigenvalues: $(p-1)|X|^{p-2}$ in the direction of X, and $|X|^{p-2}$ in the orthogonal complement. Define also $\langle\ ,\ \rangle_B$ as the $(2, 0)$-version of $A(X)^{-1/2}$, and note that $\langle\ ,\ \rangle_B$ is a metric for each $X \neq 0$. Norms and traces with respect to $\langle\ ,\ \rangle_B$ will be denoted with $|\cdot|_B, \mathrm{tr}_B$. Setting $\nu = X/|X|$ and considering an orthonormal frame $\{e_i, \nu\}$, $1 \leq i \leq m - 1$ for $\langle\ ,\ \rangle$ with dual coframe $\{\theta^j, \theta^\nu\}$, for every covariant 2-tensor C we can write

$$\langle\ ,\ \rangle_B = |X|^{-\frac{p-2}{2}}\left\{(p-1)^{-1/2}\theta^\nu \otimes \theta^\nu + \sum_j \theta^j \otimes \theta^j\right\}$$

$$\mathrm{tr}_B C = |X|^{\frac{p-2}{2}}\left\{\sqrt{p-1}C_{\nu\nu} + \sum_j C_{jj}\right\}$$

$$|C|_B^2 = |X|^{p-2}\left\{(p-1)C_{\nu\nu}^2 + p\sum_j C_{\nu j}^2 + \sum_{i,j} C_{ij}^2\right\}.$$

The Böchner formula was established in [156, Prop. 2.18], and we refer to it for a complete proof. Similar, but different, identities were obtained in [58] and in [1] in the linear case, and in [100, Thm. 3.4], [2] for generic p in the Euclidean setting.

Proposition 4.12 ([156], Prop. 2.18) *Let u be a positive solution of $\Delta_p u = 0$ in an open set $\Omega \subset M$. Fix a model M_g with radial sectional curvature $-G(r)$ and such that Δ_p is non-parabolic on M_g, and define ϱ according to*

$$u(x) = \mathscr{G}^g\big(\varrho(x)\big) = \int_{\varrho(x)}^\infty v_g(s)^{-\frac{1}{p-1}}\,\mathrm{d}s.$$

If $p > m$, also suppose that $u < \mathscr{G}^g(0)$ on Ω. Set

$$\mu = -\frac{mp - 3p + 2}{p - 1}, \qquad F(t) = \int_0^t g(s)^{\frac{1}{\sqrt{p-1}}}\,\mathrm{d}s. \qquad (4.15)$$

Then, on $\{|\nabla\varrho| > 0\}$ and denoting with $\nu = \nabla\varrho/|\nabla\varrho|$, we have

$$\frac{1}{2}g^{-\mu}\mathrm{div}\left(g^\mu A(\nabla\varrho)\nabla|\nabla\varrho|^2\right) \geq$$

$$(F')^{-p}\left|\nabla^2 F - \frac{\mathrm{tr}_B\nabla^2 F}{m}\langle\ ,\ \rangle_B\right|_B^2$$

$$+ \frac{1}{m}\left[(p-1)^{1/2} - (p-1)\right]^2 |\nabla\varrho|^{p-2}[\nabla^2\varrho(\nu, \nu)^2]$$

$$+ |\nabla\varrho|^p\left[\mathrm{Ric}(\nu, \nu) + (m-1)G|\nabla\varrho|^2\right], \qquad (4.16)$$

where, with a slight abuse of notation, $F = F(\varrho)$, $B = B(\nabla F)$ and G, g are evaluated at ϱ.

To explain why an L^∞ bound for $|\nabla \varrho|$ might be expected from (4.16), assume for ease of presentation that $G = \kappa^2 > 0$ is constant, and that u is defined on the entire M. Setting $u = |\nabla \varrho|^2$, (4.16) can be written as

$$\mathscr{L}u \geq (m-1)\kappa^2 u^{\frac{p}{2}}(u-1),$$

where $\mathscr{L}$ is a linear operator. The nonlinearity

$$f(u) = (m-1)\kappa^2 u^{\frac{p}{2}}(u-1)$$

satisfies the Keller–Osserman condition

$$\frac{1}{\sqrt{F}} \in L^1(\infty), \qquad \text{with } F(t) = \int_0^t f(s)ds,$$

and thus, in view of the discussion in Chaps. 1 and 2, it is reasonable to hope that u satisfies

$$u^* < \infty, \qquad f(u^*) \leq 0,$$

at least if the eigenvalues of the operator $\mathscr{L}$ are sufficiently well behaved. The two properties would lead to $\sup_M |\nabla \varrho|^2 = u^* \leq 1$, as we wanted to prove. In this respect, the proof of the next key Lemma resembles that of Theorem 10.33 below, which is itself inspired by [182, Thm. 4.8].

Lemma 4.13 ([156], Lem. 2.20) *Let $0 \leq G \in C(\mathbb{R}_0^+)$ be non-increasing, consider a model M_g with radial sectional curvature $-G(r)$, and assume that Δ_p is non-parabolic on M_g. Let u be a positive solution of $\Delta_p u = 0$ in an open set $\Omega \subset M$, possibly the entire M, and define ϱ according to*

$$u(x) = \mathscr{G}^g(\varrho(x)) = \int_{\varrho(x)}^\infty v_g(s)^{-\frac{1}{p-1}} ds.$$

When $p > m$, also suppose that $u < \mathscr{G}^g(0)$ on Ω. If

$$\mathrm{Ric} \geq -(m-1)G(\varrho) \qquad \text{on } \Omega, \tag{4.17}$$

then

$$\sup_{\Omega} |\nabla \varrho| \leq \max \left\{ 1, \limsup_{x \to \partial \Omega} |\nabla \varrho(x)| \right\}, \tag{4.18}$$

where we set

$$\limsup_{x \to \partial \Omega} |\nabla \varrho(x)| \doteq \inf \left\{ \sup_{\Omega \setminus V} |\nabla \varrho| \ : \ V \ open, \ \overline{V} \subsetneq \Omega \right\}.$$

In particular, if $\partial \Omega = \emptyset$ then $|\nabla \varrho| \leq 1$.

Remark 4.14 The bound (4.17) holds in the following relevant cases:

1) if

$$\mathrm{Ric} \geq -(m-1)\kappa^2 \qquad \text{on } \Omega,$$

for some constant $\kappa \geq 0$, and we choose $G(t) = \kappa^2$. If $\kappa = 0$, we further assume that $p < m$ in order for Δ_p to be non-parabolic on M_g;

2) if M satisfies $(\mathscr{H}_p)$ for some $p > 1$ and a non-increasing G, $\Omega \subset M \setminus \{o\}$ and u is the restriction to Ω of the Green kernel of M with pole at o. Indeed, by Proposition 4.9, the fake distance ϱ associated to u satisfies $\varrho \leq r$ and therefore

$$\mathrm{Ric} \geq -(m-1)G(r) \geq -(m-1)G(\varrho) \qquad \text{on } M.$$

Proof Suppose (4.18) fails. Then, $\limsup_{x \in \partial \Omega} |\nabla \varrho(x)|^2 < \infty$ and we can pick $\delta_0 > 0$ such that, for each $\delta \in [\delta_0, \sup_M |\nabla \varrho|^2 - 1)$, the set

$$U_\delta = \left\{ |\nabla \varrho|^2 > 1 + \delta \right\}$$

is non-empty and $\overline{U}_\delta \subset \Omega$. We remark that $|\nabla \varrho| \in C^\infty(\overline{U}_\delta)$, by the regularity of u on the complement of its stationary points. We first examine the case $G_* \doteq \inf G > 0$. Inserting (4.17) into (4.16) shows that the following inequality holds on U_δ:

$$\frac{1}{2} g^{-\mu} \mathrm{div}\left(g^\mu A(\nabla \varrho) \nabla |\nabla \varrho|^2 \right) \geq |\nabla \varrho|^p (m-1)G(\varrho) \left[|\nabla \varrho|^2 - 1 \right]$$

$$\geq (m-1)G_* |\nabla \varrho|^{p+2} \left[\tfrac{\delta}{\delta+1} \right] \tag{4.19}$$

$$\geq c_0 |\nabla \varrho|^{p+2},$$

where $c_0 = (m-1)\delta_0/(1+\delta_0)G_*$. For $R \geq 1$ pick $\psi \in C_c^2(B_{2R}(o))$ and $\lambda \in C^1(\mathbb{R})$ satisfying

$$0 \leq \psi \leq 1 \text{ on } M, \qquad \psi \equiv 1 \text{ on } B_R(o), \qquad |\nabla\psi| \leq \frac{8}{R}\psi^{1/2}$$

$$0 \leq \lambda \leq 1 \text{ on } \mathbb{R}, \qquad \mathrm{supp}(\lambda) = [1+2\delta, \infty), \qquad \lambda' \geq 0 \text{ on } \mathbb{R}.$$

For $\eta, \alpha \geq 1$ to be chosen later, we use the test function

$$\varphi = \lambda(|\nabla\varrho|^2)\psi(x)^\eta|\nabla\varrho|^\alpha \in C_c^1(U_\delta).$$

in the weak definition of (4.19). Writing $A = A(\nabla\varrho)$, $\lambda = \lambda(|\nabla\varrho|^2)$ we get

$$\frac{\alpha}{2}\int \langle A\nabla|\nabla\varrho|^2, \lambda|\nabla\varrho|^{\alpha-1}\psi^\eta\nabla|\nabla\varrho|\rangle g(\varrho)^\mu + c_0\int \lambda\psi^\eta|\nabla\varrho|^{\alpha+p+2}g(\varrho)^\mu$$

$$\leq -\frac{\eta}{2}\int \psi^{\eta-1}\lambda|\nabla\varrho|^\alpha\langle A\nabla|\nabla\varrho|^2, \nabla\psi\rangle g(\varrho)^\mu$$

$$\qquad (4.20)$$

$$\qquad -\frac{1}{2}\int \lambda'\psi^\eta\langle A\nabla|\nabla\varrho|^2, \nabla|\nabla\varrho|^2\rangle g(\varrho)^\mu$$

$$\leq -\frac{\eta}{2}\int \psi^{\eta-1}\lambda|\nabla\varrho|^\alpha\langle A\nabla|\nabla\varrho|^2, \nabla\psi\rangle g(\varrho)^\mu,$$

where, in the last inequality, we used $\lambda' \geq 0$ and the non-negativity of A. From the definition of A,

$$\langle A\nabla|\nabla\varrho|^2, \nabla|\nabla\varrho|\rangle = \frac{1}{2|\nabla\varrho|}\langle A\nabla|\nabla\varrho|^2, \nabla|\nabla\varrho|^2\rangle$$

$$\geq \frac{1}{2}\min\{1, p-1\}|\nabla\varrho|^{p-3}\big|\nabla|\nabla\varrho|^2\big|^2$$

while, by Cauchy–Schwarz inequality,

$$\langle A\nabla|\nabla\varrho|^2, \nabla\psi\rangle \leq \left\{\langle A\nabla|\nabla\varrho|^2, \nabla|\nabla\varrho|^2\rangle\right\}^{1/2}\left\{\langle A\nabla\psi, \nabla\psi\rangle\right\}^{1/2}$$

$$\leq \max\{1, p-1\}|\nabla\varrho|^{p-2}\big|\nabla|\nabla\varrho|^2\big||\nabla\psi|$$

$$\leq \frac{8\max\{1, p-1\}}{R}|\nabla\varrho|^{p-2}\big|\nabla|\nabla\varrho|^2\big|\psi^{1/2}.$$

Substituting into (4.20) we obtain

$$\frac{\alpha}{4} \min\{1, p-1\} \int \psi^\eta \lambda |\nabla \varrho|^{p+\alpha-4} |\nabla|\nabla\varrho|^2|^2 g(\varrho)^\mu$$

$$+ c_0 \int \lambda \psi^\eta |\nabla\varrho|^{\alpha+p+2} g(\varrho)^\mu \qquad (4.21)$$

$$\leq \frac{4\eta \max\{1, p-1\}}{R} \int \psi^{\eta-\frac{1}{2}} \lambda |\nabla\varrho|^{\alpha+p-2} |\nabla|\nabla\varrho|^2| g(\varrho)^\mu.$$

By Young's inequality, for $\tau > 0$

$$2\psi^{\eta-\frac{1}{2}} \lambda |\nabla\varrho|^{\alpha+p-2} |\nabla|\nabla\varrho|^2| \leq \tau \psi^\eta \lambda |\nabla\varrho|^{\alpha+p-4} |\nabla|\nabla\varrho|^2|^2 + \frac{1}{\tau} \psi^{\eta-1} \lambda |\nabla\varrho|^{\alpha+p},$$

whence, choosing

$$\tau = \frac{\alpha R \min\{1, p-1\}}{8\eta \max\{1, p-1\}}$$

and inserting into (4.21), we deduce the existence of a constant c_p, depending only on p, such that

$$c_0 \int \lambda \psi^\eta |\nabla\varrho|^{\alpha+p+2} g(\varrho)^\mu \leq c_p \frac{\eta^2}{\alpha R^2} \int \lambda \psi^{\eta-1} |\nabla\varrho|^{\alpha+p} g(\varrho)^\mu. \qquad (4.22)$$

We next apply Young's inequality again with exponents

$$q = \frac{p+\alpha+2}{p+\alpha}, \qquad q' = \frac{p+\alpha+2}{2}$$

and a free positive parameter $\bar{\tau}$ to obtain

$$\psi^{\eta-1} |\nabla\varrho|^{p+\alpha} \leq \frac{\bar{\tau}^q}{q} \psi^\eta |\nabla\varrho|^{p+\alpha+2} + \frac{1}{q'\bar{\tau}^{q'}} \psi^{\eta-q'}.$$

We choose $\eta = 2q' = p+\alpha+2$ and $\bar{\tau}$ such that

$$c_p \frac{\eta^2}{\alpha R^2} \frac{\bar{\tau}^q}{q} = \frac{c_0}{2},$$

so that, inserting into (4.22) and rearranging, we deduce that there exists a constant $c_1 = c_1(c_0, c_p)$ such that

$$\int \lambda \psi^\eta |\nabla \varrho|^{\alpha+p+2} g(\varrho)^\mu$$

$$\leq \frac{c_0}{p+\alpha} \left[\frac{2c_p}{c_0} \frac{\eta^2}{\alpha R^2} \frac{p+\alpha}{p+\alpha+2} \right]^{\frac{p+\alpha+2}{2}} \int \lambda \psi^{\eta/2} g(\varrho)^\mu \qquad (4.23)$$

$$\leq \left[\frac{c_1(p+\alpha+2)}{R^2} \right]^{\frac{p+\alpha+2}{2}} \int \lambda \psi^{\eta/2} g(\varrho)^\mu.$$

Set

$$I(R) = \int_{B_R} \lambda g(\varrho)^\mu.$$

Taking into account the definition of ψ and the fact that $|\nabla \varrho|^2 \geq 1 + 2\delta$ on the support of $\lambda \psi$, (4.23) yields

$$I(R) \leq \left[\frac{c_1(p+\alpha+2)}{R^2(1+2\delta)} \right]^{\frac{p+\alpha+2}{2}} I(2R).$$

Choosing α in such a way that

$$\frac{c_1(p+\alpha+2)}{R^2(1+2\delta)} = \frac{1}{e},$$

we get

$$I(R) \leq e^{-\frac{p+\alpha+2}{2}} I(2R) = e^{-\frac{R^2(1+2\delta)}{2c_1 e}} I(2R).$$

Iterating and taking logarithms as in [182, Lem. 4.7] shows that there exists $S > 0$ independent of R, δ such that for each $R > 2R_0$,

$$\frac{\log I(R)}{R^2} \geq \frac{\log I(R_0)}{R^2} + S \frac{(1+2\delta)}{c_1}. \qquad (4.24)$$

To conclude, we estimate $I(R)$ from above. Since $G \in L^\infty(\mathbb{R}^+)$ and $\inf_{\mathbb{R}^+} G > 0$, we pick $\kappa, \bar{\kappa} > 0$ such that $\kappa^2 \le G \le \bar{\kappa}^2$. By Sturm comparison,

$$\frac{\sinh(\kappa t)}{\kappa} \le g(t) \le \frac{\sinh(\bar{\kappa} t)}{\bar{\kappa}} \qquad \text{on } \mathbb{R}^+.$$

Consequently, both for positive and negative μ, by the Bishop–Gromov volume comparison theorem and since $\lambda \le 1$ there exist constants $b_j = b_j(\mu, m, \kappa, \bar{\kappa})$ such that

$$I(R) \le e^{b_1 R}|B_R| \le e^{b_2 R}.$$

Taking limits in (4.24) for $R \to \infty$ we then deduce $0 \ge S\frac{(1+2\delta)}{c_1}$, contradiction.

To examine the case $G_* = 0$, fix $c > 0$ and consider a model of curvature $-G(t) - c$. Let $v_{g,c}, \mathscr{G}_c^g$ denote, respectively, the volume of geodesic spheres and the Green kernel centred at the origin of the model. Note that

$$\mathscr{G}_c^g \uparrow \mathscr{G}^g \qquad \text{in } C_{\text{loc}}^1(\mathbb{R}^+) \quad \text{as } c \downarrow 0.$$

Suppose first that $p \le m$. In this case, $\mathscr{G}_c^g(0) = \infty$ and thus the fake distance ϱ_c associated to u and $\mathscr{G}_c^g$ is defined on the entire Ω. Moreover, $\varrho_c \to \varrho$ locally uniformly in $C^1(\Omega)$ and monotonically from below as $c \downarrow 0$, which implies

$$|\nabla\varrho(x)| = \lim_{c \to 0} |\nabla\varrho_c(x)| \qquad \forall x \in \Omega.$$

We claim that

$$v_{g,c}(\varrho_c) \le v_g(\varrho) \qquad \text{on } \Omega. \tag{4.25}$$

We postpone for a moment its proof, and conclude the argument. Since

$$|\nabla\varrho_c| = |\nabla u| v_{g,c}(\varrho_c)^{\frac{1}{p-1}} \le |\nabla u| v_g(\varrho)^{\frac{1}{p-1}} = |\nabla\varrho|,$$

applying the first part of the proof to ϱ_c we deduce, for each $x \in \Omega$,

$$\begin{aligned} |\nabla\varrho(x)| = \lim_{c \to 0} |\nabla\varrho_c(x)| &\le \liminf_{c \to 0} \max\left\{1, \limsup_{y \to \partial\Omega} |\nabla\varrho_c(y)|\right\} \\ &\le \max\left\{1, \limsup_{y \to \partial\Omega} |\nabla\varrho(y)|\right\}, \end{aligned}$$

as required. To prove (4.25), by Sturm comparison $v_{g,c}/v_g$ is increasing on $\mathbb{R}^+$, thus Lemma 4.11 implies the inequality $|(\log \mathscr{G}_c^g)'| \geq |(\log \mathscr{G}^g)'|$ on $\mathbb{R}^+$, that rewrites as

$$\frac{v_{g,c}^{-\frac{1}{p-1}}}{\mathscr{G}_c^g}(t) \geq \frac{v_g^{-\frac{1}{p-1}}}{\mathscr{G}^g}(t) \qquad \forall t \in \mathbb{R}^+.$$

We evaluate at $t = \varrho_c$, and use $\varrho_c \leq \varrho$ together with the monotonicity of $|(\log \mathscr{G}^g)'|$ that follows from Lemma 4.10, to deduce

$$\frac{v_{g,c}(\varrho_c)^{-\frac{1}{p-1}}}{\mathscr{G}_c^g(\varrho_c)} \geq \frac{v_g(\varrho_c)^{-\frac{1}{p-1}}}{\mathscr{G}^g(\varrho_c)} \geq \frac{v_g(\varrho)^{-\frac{1}{p-1}}}{\mathscr{G}^g(\varrho)}.$$

Inequality (4.25) follows since $\mathscr{G}_c^g(\varrho_c) = \mathscr{G}^g(\varrho) = u$.

Next, we assume $p > m$. In this case, ϱ_c is defined on

$$\Omega_c \doteq \left\{ x \in \Omega \ : \ u(x) < \mathscr{G}_c^g(0) \right\} \subset \Omega.$$

However, since ϱ_c vanishes on $\partial\Omega_c \cap \Omega$, it holds

$$|\nabla \varrho_c(x)| = |\nabla u(x)| v_{g,c}(\varrho_c(x))^{\frac{1}{p-1}} \to 0 \qquad \text{as } x \to \partial\Omega_c \cap \Omega,$$

and therefore

$$\limsup_{\Omega_c \ni y \to \partial\Omega_c} |\nabla \varrho(y)| = \limsup_{\Omega_c \ni y \to \partial\Omega} |\nabla \varrho(y)| \leq \limsup_{y \to \partial\Omega} |\nabla \varrho(y)|.$$

We conclude by noticing that $\Omega_c \uparrow \Omega$, so for fixed $x \in \Omega$ we can choose c such that $x \in \Omega_c$. By the above,

$$|\nabla \varrho(x)| \leq \max\left\{ 1, \limsup_{\Omega_c \ni y \to \partial\Omega_c} |\nabla \varrho(y)| \right\} \leq \max\left\{ 1, \limsup_{y \to \partial\Omega} |\nabla \varrho(y)| \right\},$$

concluding the proof.   □

Theorem 4.15 *Suppose that M satisfies $(\mathscr{H}_p)$ for some $p > 1$ and*

$$G(t) \geq 0, \qquad G(t) \text{ non-increasing on } \mathbb{R}^+.$$

Then, having defined ϱ as in Definition 4.7,

(i) $|\nabla\varrho| \leq 1$ *on* $M\backslash\{o\}$.
(ii) Equality $|\nabla\varrho(x)| = 1$ *holds for some* $x \in M\backslash\{o\}$ *if and only if* $\varrho = r$ *and* M *is the radially symmetric model* M_g.

Proof

(i) We split the argument according to whether $p \leq m$ or $p > m$.
  The case $p \leq m$.
  Inequality $\varrho \leq r$ holds because of Proposition 4.9, hence (2) in Theorem 4.3 guarantees

$$|\nabla\varrho(x)| = v_g(\varrho(x))^{\frac{1}{p-1}}|\nabla\mathcal{G}(x)| \leq v_g(r(x))^{\frac{1}{p-1}}|\nabla\mathcal{G}(x)| \to 1$$

as $x \to o$. Thus, $\limsup_{x\to o}|\nabla\varrho(x)| \leq 1$. In view of Remark 4.14, we are in the position to apply Lemma 4.13 with the choice $\Omega = M\backslash\{o\}$ and conclude *(i)*.
  The case $p > m$.
  Fix

$$c > c_o = \left[\frac{\text{cap}_p(\{o\}, M)}{\text{cap}_p(\{0\}, M_g)}\right]^{\frac{1}{p-1}}$$

and define ϱ_c according to

$$c\mathcal{G}(x) = \int_{\varrho_c(x)}^{\infty} \frac{ds}{v_g(s)^{\frac{1}{p-1}}}.$$

Then, ϱ_c is well defined on

$$\Omega_c = \left\{x \in M \ : \ c\mathcal{G}(x) < c_o\mathcal{G}(o)\right\}.$$

Note that $o \notin \overline{\Omega}_c$, $\Omega_c \uparrow M\backslash\{o\}$ as $c \downarrow c_o$, and $\varrho_c \uparrow \varrho$ pointwise. In particular, by Proposition 4.9, $\varrho_c \leq r$ for each c and the monotonicity of G implies (4.17) on Ω_c. From $\mathcal{G} \in C^1(\overline{\Omega}_c)$ and $\varrho_c = 0$ on $\partial\Omega_c$ we deduce

$$|\nabla\varrho_c(x)| = |\nabla\mathcal{G}(x)|v_g(\varrho_c(x))^{\frac{1}{p-1}} \to 0 \qquad \text{as } x \to \partial\Omega_c.$$

We can therefore apply Lemma 4.13 to $u = c\mathcal{G}(x)$ and $\Omega = \Omega_c$ to deduce $|\nabla\varrho_c| \leq 1$ on Ω_c. The limit ϱ is therefore 1-Lipschitz, hence $|\nabla\varrho| \leq 1$ on $M\backslash\{o\}$.

To show (ii), because of (4.16) the function $u = 1 - |\nabla \varrho|^2 \geq 0$ solves

$$
\begin{aligned}
\tfrac{1}{2} g^{-\mu} \mathrm{div}\big(g^{\mu} A(\nabla \varrho) \nabla u\big) &\leq -|\nabla \varrho|^p \Big[\mathrm{Ric}(v, v) + (m-1)G(\varrho)|\nabla \varrho|^2 \Big] \\
&\leq -(m-1)|\nabla \varrho|^p \Big[-G(r) + G(\varrho)|\nabla \varrho|^2 \Big] \\
&\leq -(m-1)|\nabla \varrho|^p \Big[-G(r) + G(r)|\nabla \varrho|^2 \Big] \\
&= (m-1)G(r)|\nabla \varrho|^p u.
\end{aligned}
$$

If u vanishes at some point, by the strong minimum principle $u \equiv 0$ on M, that is, $|\nabla \varrho| \equiv 1$. In this case, again by (4.16) we deduce

$$
\nabla^2 F = \frac{1}{m} \mathrm{Tr}_B(\nabla^2 F)\langle \,, \,\rangle_B \qquad \text{on } M, \tag{4.26}
$$

with F as in (4.15). Consider the flow Φ_t of $\nabla \varrho$. Since $|\nabla \varrho| = 1$, the integral curves of the flow are unit speed geodesics. Because of the completeness of M, for each $x \in M \setminus \{o\}$ the geodesic $\Phi_t(x)$ is defined on the maximal interval $(-\varrho(x), \infty)$, and $\lim_{t \to -\varrho(x)} \Phi_t(x) = o$, being o the unique zero of ϱ. Hence, Φ_t is a unit speed geodesic issuing from o to x. If $x \in M \setminus \mathrm{cut}(o)$, it therefore holds $\varrho(x) = r(x)$, and by continuity $\varrho = r$ on M. The function r is thus C^1 outside of o, and this implies $\mathrm{cut}(o) = \emptyset$, that is, o is a pole of M. Indeed, the distance function r is not differentiable at any point $y \in \mathrm{cut}(o)$ joined to o by at least two minimizing geodesics. Such points are dense in $\mathrm{cut}(o)$ by [26, 236], so $\mathrm{cut}(o) = \emptyset$ whenever r is everywhere differentiable outside of o. Rewriting (4.26) in terms of $\nabla^2 \varrho = \nabla^2 r$ we deduce

$$
\nabla^2 r = \frac{g'(r)}{g(r)} \Big(\langle \,, \,\rangle - dr \otimes dr \Big) \qquad \text{on } M \setminus \{o\}.
$$

Writing $\langle \,, \,\rangle$ in polar coordinates as $dr^2 + b_{\alpha\beta}(r, \theta) d\theta^{\alpha} d\theta^{\beta}$, with $\{\theta^{\alpha}\}$ coordinates on $\mathbb{S}^{m-1}$, we get

$$
\partial_r b_{\alpha\beta} = \big(L_{\partial r} \langle \,, \,\rangle\big)_{\alpha\beta} = 2(\nabla^2 r)_{\alpha\beta} = 2 \frac{g'(r)}{g(r)} b_{\alpha\beta}.
$$

Integrating and using the asymptotic relation

$$
b_{\alpha\beta}(r, \theta) \sim r^2 (d\theta^2)_{\alpha\beta}
$$

with $d\theta^2$, as usual, the metric on the unit sphere $\mathbb{S}^{m-1}$, we deduce

$$
b_{\alpha\beta} = g(r)^2 (d\theta^2)_{\alpha\beta},
$$

thus $\langle \,, \,\rangle = dr^2 + g(r)^2 d\theta^2$ and M is isometric to M_g. $\qquad \square$

4.4   Properness of ϱ

The properness of ϱ, equivalently, the fact that $\mathcal{G}(x) \to 0$ as $r(x) \to \infty$, is a nontrivial fact intimately related to the geometry of M at infinity. The first class that we are going to investigate is that of manifolds with Ric ≥ 0. We introduce the following bound due to I. Holopainen, that extended, to the nonlinear setting, a previous estimate in [146, Thm. 5.2] valid for the Laplace–Beltrami operator.

Theorem 4.16 (Prop. 5.10 of [122]) *Let M be complete, and suppose that* Ric ≥ 0. *Denote by B_r a geodesic ball of radius r centred at some fixed origin o, and with $\partial M(r)$ the portion of ∂B_r which is the boundary of an unbounded connected component of $M \backslash B_r$. Fix $p \in (1, \infty)$. Then, Δ_p is non-parabolic on M if and only if*

$$\left(\frac{s}{|B_r|} \right)^{\frac{1}{p-1}} \in L^1(\infty). \tag{4.27}$$

In this case there exists a constant $C \geq 1$ independent of r such that

$$\frac{1}{C} \int_{2r}^{\infty} \left(\frac{s}{|B_s|} \right)^{\frac{1}{p-1}} ds \leq \mathcal{G}(x) \leq C \int_{2r}^{\infty} \left(\frac{s}{|B_s|} \right)^{\frac{1}{p-1}} ds \tag{4.28}$$

for each $x \in \partial M(r)$.

As a consequence, we deduce

Corollary 4.17 *Let $p \in (1, \infty)$. If M is complete with* Ric ≥ 0, *and Δ_p is non-parabolic, then the Green kernel $\mathcal{G}$ of Δ_p with pole at $o \in M$ satisfies $\mathcal{G}(x) \to 0$ as $r(x) \to \infty$.*

Proof Suppose that there exists $c > 0$ and a sequence $\{x_j\}$ with $r_j = r(x_j) \to \infty$ such that $\mathcal{G}(x_j) \geq c$ for each j. By (4.28), up to removing a finite number of x_j and relabelling, x_j necessarily belongs to the boundary of a compact connected component of $M \backslash B_{r_j}$. For $r \leq r_j$, let U_r be the connected component of $M \backslash B_r$ containing x_j, and define

$$I_j = \left\{ r \in (0, r_j] \ : \ U_r \text{ is compact} \right\}, \qquad \bar{r}_j = \inf(I_j).$$

Note that $I_j \neq \emptyset$. For $r \in I_j$, since $\mathcal{G}$ is p-harmonic on U_r the maximum principle gives $\mathcal{G}(x_j) \leq \max_{\partial U_r} \mathcal{G}$, and by continuity $\mathcal{G}(x_j) \leq \max_{\partial U_{\bar{r}_j}} \mathcal{G}$. Choose $\hat{r}_j \in (\bar{r}_j - 1, \bar{r}_j)$ in order to satisfy $\mathcal{G}(x_j) \leq \max_{\partial U_{\hat{r}_j}} \mathcal{G} + c/2$. Applying (4.28) to points of $\partial U_{\hat{r}_j}$ we have

$$c \leq \mathcal{G}(x_j) \leq \max_{\partial U_{\hat{r}_j}} \mathcal{G} + \frac{c}{2} \leq \frac{c}{2} + C \int_{2\hat{r}_j}^{\infty} \left(\frac{s}{|B_s|} \right)^{\frac{1}{p-1}} ds.$$

This implies that $\{\hat{r}_j\}$, and hence $\{\bar{r}_j\}$, is bounded. Fix $R > \sup_j \bar{r}_j$. Since $r(x_j) \to \infty$ and using that, by construction, each x_j belongs to a bounded connected component of $M \backslash B_R$, we deduce that $M \backslash B_R$ should necessarily have infinitely many connected components, a contradiction. $\qquad\square$

Remark 4.18 The above proof does not provide an explicit rate of decay of $\mathcal{G}(x)$ at infinity in terms of bounds on the geometry of M, the problem being to control the diameter of the *bounded* connected components of $M \backslash B_r$. However, it can easily be modified to guarantee a uniform decay of the type

$$\mathcal{G}(x) \le C \int_{r(x)}^{\infty} \left(\frac{s}{|B_s|}\right)^{\frac{1}{p-1}} ds \qquad (4.29)$$

provided that each bounded connected component of $M \backslash B_r$ lies inside an *unbounded* connected component of $M \backslash B_{\mu r}$, for some constant $\mu = \mu(m) \in (0, 1)$. To the best of our knowledge, whether or not such property holds on every complete manifold with $\mathrm{Ric} \ge 0$ is still an open problem. A sufficient condition for this to hold has been identified by V. Minerbe in [161]: it is sufficient that $\mathrm{Ric} \ge 0$ and that, for some origin o, there exist $C_1 > 0, b > 1$ and a point $o \in M$ such that

$$\forall t \ge s > 0, \qquad \frac{|B_t|}{|B_s|} \ge C_1 \left(\frac{t}{s}\right)^b, \qquad (4.30)$$

where balls are centred at o. In this case, the constant μ only depends on m, b, C_1. More details can also be found in Section 3 of [156].

If the Ricci tensor is negative somewhere, some additional condition must be placed on M in order to guarantee that $\mathcal{G}(x) \to 0$ as $r(x) \to \infty$. We will be interested in manifolds supporting a L^p Sobolev inequality of the type

$$\left(\int |\psi|^{\frac{vp}{v-p}}\right)^{\frac{v-p}{v}} \le \mathscr{S}_{p,v} \int |\nabla \psi|^p \qquad \forall \psi \in \mathrm{Lip}_c(M), \qquad (4.31)$$

for some constants $\mathscr{S} > 0, v > p$. We recall that if (4.31) holds for $p = 1$, that is, if

$$\left(\int |\psi|^{\frac{v}{v-1}}\right)^{\frac{v-1}{v}} \le \mathscr{S}_{1,v} \int |\nabla \psi| \qquad \forall \psi \in \mathrm{Lip}_c(M) \qquad (4.32)$$

holds for some $\mathscr{S}_{1,v} > 0$, then applying (4.32) to the test function

$$|\psi|^{\frac{p(v-1)}{v-p}} \qquad \text{for } p \in (1, v)$$

and using Hölder inequality, one deduces the validity of (4.31) with constant

$$\mathscr{S}_{p,v} = \left[\frac{\mathscr{S}_{1,v}\, p(v-1)}{v-p}\right]^p \to \mathscr{S}_{1,v} \qquad \text{as } p \to 1. \tag{4.33}$$

Remark 4.19 Because of [38, Prop. 2.5] and [186], if (4.31) holds on $M \backslash K$ for some compact set K, then (4.31) holds on the entire M, possibly with a different constant.

Examples of manifolds supporting (4.31) with $v = m$ include the following classes:

Example 4.20 Let $M^m \to N^n$ be a complete, minimal immersion into a Cartan–Hadamard space (cf. Definition 3.3). By [119], the L^1 Sobolev inequality

$$\left(\int |\psi|^{\frac{m}{m-1}}\right)^{\frac{m-1}{m}} \le \mathscr{S}_{1,m} \int |\nabla\psi| \qquad \forall\, \psi \in \mathrm{Lip}_c(M) \tag{4.34}$$

holds on M.

Example 4.21 If $\mathrm{Ric} \ge 0$ on M, and $m \ge 3$, it is known that

$$M \text{ enjoys (4.34)} \qquad \Longleftrightarrow \qquad \lim_{r\to\infty} \frac{|B_r|}{r^m} > 0,$$

that is, M has maximal volume growth. Implication $\Leftarrow$ holds by [214, Thm. 3.3.8], while $\Rightarrow$ holds irrespectively of a bound on the Ricci tensor, see [38] and [185, Lem. 7.15].

Example 4.22 Let M^m be a complete manifold satisfying

(i) $\mathrm{Ric} \ge -(m-1)\kappa^2\langle\,,\,\rangle$ for some constant $\kappa > 0$, and

$$\inf_{x\in M} |B_1(x)| = v > 0; \tag{4.35}$$

(ii) for some $\mathscr{P}_p \in \mathbb{R}^+$, the Poincaré inequality

$$\int |\psi|^p \le \mathscr{P}_p \int |\nabla\psi|^p \qquad \forall\, \psi \in \mathrm{Lip}_c(M). \tag{4.36}$$

Note that a striking estimate in [60, Prop. 14] guarantees the validity of (4.35) whenever $\mathrm{inj}(M) > 0$, where $\mathrm{inj}(M)$ is the injectivity radius of M. By work of N. Varopoulos (see [112], Thm. 3.2), because of (i) M enjoys the L^1 Sobolev inequality with potential

$$\left(\int |\psi|^{\frac{m}{m-1}}\right)^{\frac{m-1}{m}} \le \mathscr{S}_{1,m} \int \left[|\nabla\psi| + |\psi|\right] \qquad \forall\, \psi \in \mathrm{Lip}_c(M),$$

for some $\mathscr{S}_{1,m}$ depending on (m, κ, υ). Using again as a test function $|\psi|^{\frac{p(m-1)}{m-p}}$, by Hölder inequality and rearranging we get

$$\left(\int |\psi|^{\frac{mp}{m-p}}\right)^{\frac{m-p}{m}} \leq \mathscr{S}_{p,m} \int \left[|\nabla \psi|^p + |\psi|^p\right] \qquad \forall \psi \in \mathrm{Lip}_c(M),$$

for some $\mathscr{S}_{p,m}$ depending on (m, κ, υ, p). Assumption (ii) then guarantees (4.31) with $\nu = m$.

Example 4.23 The notion of rough isometry between metric spaces was introduced by M. Kanai in [130], and proved to be very powerful in connection with the validity of Sobolev inequalities. Two metric spaces (M, d_M) and (N, d_N) are said to be *roughly isometric* if there exist $\varphi : M \to N$ such that

- $B_\varepsilon(\varphi(M)) = N$ for some $\varepsilon > 0$;
- for some constants $C_1 \geq 1, C_2 \geq 0$,

$$C_1^{-1}\mathrm{d}_M(x, y) - C_2 \leq \mathrm{d}_N(\varphi(x), \varphi(y)) \leq C_1\mathrm{d}_M(x, y) + C_2 \qquad \forall x, y \in M.$$

Kanai proved in [130, Thm. 4.1] that if M and N are roughly isometric manifolds of the same dimension, both satisfying

(iii) $\quad \mathrm{Ric} \geq -(m-1)\kappa^2\langle\,,\,\rangle, \qquad \mathrm{inj}(M) > 0,$

for some constant $\kappa > 0$, then

$$(4.34) \text{ holds on } M \qquad \Longleftrightarrow \qquad (4.34) \text{ holds on } N.$$

In particular, a manifold M^m satisfying (iii) and roughly isometric to $\mathbb{R}^m$ enjoys (4.34), and therefore (4.31) with $\nu = m$. Under the same assumptions, Δ_p is parabolic for each $p \geq m$, see [121, Thm. 3.16], and thus (4.31) fails for any $m \leq p < \nu$.

We are ready to prove our main result of this section, that provides a sharp decay estimate for $\mathcal{G}$ whenever an L^p Sobolev inequality holds. For $p = 2$, an analogous decay estimate for $\mathcal{G}$ was obtained by L. Ni in [170], by integrating a corresponding estimate for the heat kernel. However, his technique does not extend to $p \neq 2$. The theorem is of interest even for relatively compact, smooth sets Ω.

Theorem 4.24 *Let $\Omega \subset M$ be a connected open set, and denote with r the distance from a fixed origin $o \in \Omega$. Assume that Ω supports the Sobolev inequality (4.31) for some $p \in (1, m), \nu > p, \mathscr{S}_{p,\nu} > 0$ and every $\psi \in \mathrm{Lip}_c(\Omega)$.*

Then, Δ_p is non-parabolic on Ω, and the kernel $\mathcal{G}(x)$ of Δ_p with pole at o satisfies

$$\mathcal{G}(x) \leq C_{p,\nu}^{\frac{1}{p-1}} r(x)^{-\frac{\nu-p}{p-1}}, \qquad \forall x \in \Omega \backslash \{o\}, \tag{4.37}$$

with

$$C_{p,\nu} = \mathscr{S}_{p,\nu}^{\frac{\nu}{p}} \left[2^{\nu} p (1 + p)^p \left(\frac{p}{p-1} \right)^{p-1} \right]^{\frac{\nu-p}{p}}$$

In particular, $\mathcal{G}(x) \to 0$ as $r(x) \to \infty$ in Ω.

Remark 4.25 If M satisfies the L^1 Sobolev inequality (4.32), then $S_{p,\nu}$ can be chosen as in (4.33). Consequently, the constant $C_{p,\nu}$ in (4.37) remains bounded as $p \to 1$. This is a crucial step in the application of the above estimates to the construction of the Inverse Mean Curvature Flow by means of p-Laplace approximation. The interested reader is referred to [156] and to the references therein for details.

Remark 4.26 The non-parabolicity of Δ_p under the validity of a Sobolev inequality is, indeed, a corollary of a general result for non-negative operators of type $\Delta_p + V(x)$ known as the ground-state alternative, see [187] and [25, Thm. 4.1] for further insight.

Proof We begin by proving that Δ_p is non-parabolic on Ω, a rather standard fact that can be found in [186] for general p, and [185, Lemma 7.13] for $p = 2$. If Δ_p were parabolic, by the characterization of parabolicity via capacity there would exist a sequence $\{\phi_j\} \subset \text{Lip}_c(\Omega)$ such that

$$\int |\nabla \phi_j|^p \to 0, \qquad \phi_j = 1 \qquad \text{on a fixed open domain } K \Subset \Omega.$$

Plugging ϕ_j in (4.31) and letting $j \to \infty$ we obtain $|K| = 0$, contradiction.

We prove the bound when $\mathcal{G}$ is the kernel of Δ_p on a smooth, relatively compact set $\Omega' \Subset \Omega$ containing o, since the general case follows by taking an exhaustion of Ω and the limit of the corresponding kernels.

Let $x \in \Omega$, and assume that $\Omega' \cap \partial B_{r(x)} \neq \emptyset$, otherwise there is nothing to prove. Let $\psi \in C^1(\Omega')$ and consider (4.4) with test function $\psi \eta(\mathcal{G}) \in \text{Lip}_c(\Omega')$, where

$$\eta(s) = \begin{cases} 0 & \text{on } [0, \ell - \varepsilon], \\ \varepsilon^{-1}(s - \ell + \varepsilon) & \text{on } (\ell - \varepsilon, \ell), \\ 1 & \text{otherwise.} \end{cases}$$

Letting $\varepsilon \to 0$ and using the coarea's formula we deduce that

$$\psi(o) = \int_{\{\mathcal{G}>\ell\}} |\nabla\mathcal{G}|^{p-2}\langle\nabla\mathcal{G}, \nabla\psi\rangle + \int_{\{\mathcal{G}=\ell\}} |\nabla\mathcal{G}|^{p-1}\psi \tag{4.38}$$

holds for almost every $\ell \in \mathbb{R}^+$. Similarly, the identity

$$0 = \int_{\{\mathcal{G}<\ell\}} |\nabla\mathcal{G}|^{p-2}\langle\nabla\mathcal{G}, \nabla\psi\rangle - \int_{\{\mathcal{G}=\ell\}} |\nabla\mathcal{G}|^{p-1}\psi \tag{4.39}$$

holds for every $\psi \in C^1(\Omega'\backslash\{o\})$ and a.e. ℓ. Fix a constant $\theta \in (0, 1)$ to be specified later, and choose ℓ satisfying

$$\sup_{\Omega'\cap\partial B_{(1-\theta)r(x)}} \mathcal{G} < \ell \tag{4.40}$$

and such that (4.38), (4.39) hold with, respectively, $\psi \equiv 1$ and $\psi = \mathcal{G}$. In particular,

$$\int_{\{\mathcal{G}\leq\ell\}} |\nabla\mathcal{G}|^p = \ell \int_{\{\mathcal{G}=\ell\}} |\nabla\mathcal{G}|^{p-1} = \ell. \tag{4.41}$$

Since $\mathcal{G}$ is p-harmonic on $\Omega' \backslash \{o\}$ and it diverges at the pole o (by Theorem 4.3, because $p < m$), the strong maximum principle implies that its superlevel sets are connected open sets containing o. In particular, using (4.40),

$$\{\mathcal{G} > \ell\} \subset B_{(1-\theta)r(x)}.$$

Furthermore, the strong maximum principle also implies that $\sup_{\partial B_r\cap\Omega} \mathcal{G}$ is a non-increasing function of r. Extend $\mathcal{G}$ to be zero on $M\backslash\Omega'$, and observe that $\Delta_p\mathcal{G} \geq -\delta_o$ on M. Let $0 \leq \phi \in \mathrm{Lip}_c(M\backslash\{o\})$. For a given $\bar{q} \geq p$, we integrate $\Delta_p\mathcal{G} \geq 0$ against the test function $\phi^p\mathcal{G}^{\bar{q}-p+1} \in \mathrm{Lip}_c(M)$ and apply Cauchy–Schwarz inequality to get

$$\int \phi^p\mathcal{G}^{\bar{q}-p}|\nabla\mathcal{G}|^p \leq \left|\frac{p}{\bar{q}-p+1}\right| \int \phi^{p-1}\mathcal{G}^{\bar{q}-p+1}|\nabla\mathcal{G}|^{p-1}|\nabla\phi|.$$

Therefore, by Hölder inequality we obtain

$$\int \phi^p\mathcal{G}^{\bar{q}-p}|\nabla\mathcal{G}|^p \leq \left|\frac{p}{\bar{q}-p+1}\right|^p \int \mathcal{G}^{\bar{q}}|\nabla\phi|^p. \tag{4.42}$$

Using (4.31) with $\psi = \phi \mathcal{G}^{\frac{\bar{q}}{p}} \in \mathrm{Lip}_0(\Omega')$ and (4.42), we compute

$$\mathscr{S}_{p,\nu}^{-\frac{1}{p}} \left\| \phi \mathcal{G}^{\frac{\bar{q}}{p}} \right\|_{\frac{\nu p}{\nu - p}} \leq \left\| \nabla \left(\phi \mathcal{G}^{\frac{\bar{q}}{p}} \right) \right\|_p \leq \left\| \mathcal{G}^{\frac{\bar{q}}{p}} |\nabla \phi| + |\bar{q}/p| \phi \mathcal{G}^{\frac{\bar{q}-p}{p}} |\nabla \mathcal{G}| \right\|_p$$

$$\leq \left\| \mathcal{G}^{\frac{\bar{q}}{p}} |\nabla \phi| \right\|_p + \left| \frac{\bar{q}}{p} \right| \left\| \phi \mathcal{G}^{\frac{\bar{q}-p}{p}} |\nabla \mathcal{G}| \right\|_p \qquad (4.43)$$

$$\leq \left[1 + \left| \frac{\bar{q}}{\bar{q} - p + 1} \right| \right] \left\| \mathcal{G}^{\frac{\bar{q}}{p}} |\nabla \phi| \right\|_p.$$

We use Moser iteration to estimate $\mathcal{G}$ from above on annuli, and for this reason we set for convenience

$$t = r(x), \qquad A_\infty = B_{t(2-\theta)} \setminus B_t, \qquad T = \theta t, \qquad A_0 = B_T(A_\infty) = B_{2t} \setminus B_{(1-\theta)t}$$

and

$$k = \frac{\nu}{\nu - p} > 1, \qquad q = kp = \frac{\nu p}{\nu - p}.$$

For $i \in \{0, 1, 2, \ldots\}$ define

$$r_i = T \left(2 - \sum_{j=0}^{i} 2^{-j} \right), \qquad A_i = B_{r_i}(A_\infty)$$

and

$$\eta_i(t) = \begin{cases} 1 & \text{if } t \in [0, r_{i+1}) \\ 1 - \frac{2^{i+1}}{T}(t - r_{i+1}) & \text{if } t \in [r_{i+1}, r_i) \\ 0 & \text{if } t \geq r_i. \end{cases}$$

Set $\phi_i = \eta_i(r)$, with $r(x) = \mathrm{dist}(A_\infty, x)$. Choosing $\phi = \phi_i$, $\bar{q} = q_i = qk^i$ and $|\nabla \phi_i| \leq T^{-1} 2^{i+1}$ in (4.43) we deduce

$$\left(\int_{A_{i+1}} \mathcal{G}^{qk^{i+1}} \right)^{\frac{1}{k}} \leq \mathscr{S}_{p,\nu} \left[1 + \left| \frac{qk^i}{qk^i - p + 1} \right| \right]^p 2^{p(i+1)} T^{-p} \int_{A_i} \mathcal{G}^{qk^i}$$

$$\leq \mathscr{S}_{p,\nu} [1 + p]^p \, 2^{p(i+1)} T^{-p} \int_{A_i} \mathcal{G}^{qk^i},$$

where, in the last step, we used that the function $q/(q-p+1)$ is decreasing for $q \geq p$. Taking the k^i-th root, iterating and explicitly computing the sums, we infer

$$\sup_{A_\infty} \mathcal{G}^q = \lim_{i \to \infty} \left(\int_{A_i} \mathcal{G}^{qk^{i+1}} \right)^{\frac{1}{k^{i+1}}}$$

$$\leq (\mathscr{S}_{p,v}[1+p]^p)^{\sum_{j=0}^\infty k^{-j}} [2k]^p \sum_{j=0}^\infty (j+1)k^{-j} T^{-p \sum_{j=0}^\infty k^{-j}} \int_{A_0} \mathcal{G}^q$$

$$= (\mathscr{S}_{p,v}[1+p]^p)^{\frac{k}{k-1}} [2k]^{\frac{pk^2}{(k-1)^2}} T^{-\frac{kp}{k-1}} \int_{A_0} \mathcal{G}^q.$$

Extracting the q-th root and using the definition of A_∞, A_0, k, we get

$$\|\mathcal{G}\|_{L^\infty(\partial B_t)} \leq \left(\mathscr{S}_{p,v}\bar{C}_{p,v}\right)^{\frac{v-p}{p^2}} (\theta t)^{-\frac{v-p}{p}} \left(\int_{B_{2t} \backslash B_{(1-\theta)t}} \mathcal{G}^{\frac{vp}{v-p}} \right)^{\frac{v-p}{vp}}, \qquad (4.44)$$

where we set

$$\bar{C}_{p,v} = 2^v[1+p]^p.$$

Plugging in the Sobolev inequality (4.31) the test function $\psi = \min\{\mathcal{G}, \ell\} \in \mathrm{Lip}_c(\Omega)$, and using the fact that $\psi = \mathcal{G}$ on $\Omega \backslash B_{(1-\theta)t} \subset \{\mathcal{G} \leq \ell\}$ together with (4.41), we get

$$\left(\int_{B_{2t} \backslash B_{(1-\theta)t}} \mathcal{G}^{\frac{vp}{v-p}} \right)^{\frac{v-p}{v}} \leq \left(\int \psi^{\frac{vp}{v-p}} \right)^{\frac{v-p}{v}} \leq \mathscr{S}_{p,v} \int |\nabla\psi|^p$$

$$= \mathscr{S}_{p,v} \int_{\{\mathcal{G} \leq \ell\}} |\nabla\mathcal{G}|^p = \mathscr{S}_{p,v}\ell.$$

Let $\|\cdot\|_s$ denote the L^∞ norm on ∂B_s. Inserting into (4.44) and letting $\ell \downarrow \|\mathcal{G}\|_{(1-\theta)t}$,

$$\|\mathcal{G}\|_t \leq \mathscr{S}_{p,v}^{\frac{v}{p^2}} \bar{C}_{p,v}^{\frac{v-p}{p^2}} t^{-\frac{v-p}{p}} \theta^{-\frac{v-p}{p}} \|\mathcal{G}\|_{(1-\theta)t}^{\frac{1}{p}}. \qquad (4.45)$$

Fix $\xi \in (0,1)$ and consider a sequence $\{\sigma_k\}_{k \geq 0} \subset [1, \infty)$ with the property that

$$\sigma_{k+1} > \sigma_k \qquad \text{for } k \geq 0,$$

to be specified later, and construct inductively sequences $\{t_k\}$, $\{\theta_k\}$ for $k \geq 0$ as follows:

$$t_0 = t, \qquad \theta_0 = 1 - \xi^{\sigma_1}, \qquad \theta_k = 1 - \xi^{\sigma_{k+1} - \sigma_k} \quad \text{for } k \geq 1,$$

$$t_{k+1} = (1 - \theta_k)t_k = t\xi^{\sigma_{k+1}}.$$

Set for convenience

$$\hat{C} = \mathscr{S}_{p,v}^{\frac{v}{p^2}} \bar{C}_{p,v}^{\frac{v-p}{p^2}}.$$

We iterate (4.45) i-times for the chosen θ_k, t_k to deduce

$$\|\mathcal{G}\|_{t_0} \leq \hat{C} t_0^{-\frac{v-p}{p}} \theta_0^{-\frac{v-p}{p}} \|\mathcal{G}\|_{t_1}^{\frac{1}{p}}$$

$$\leq \hat{C}^{1+p^{-1}} \left[t_0 t_1^{p^{-1}} \right]^{-\frac{v-p}{p}} \left[\theta_0 \theta_1^{p^{-1}} \right]^{-\frac{v-p}{p}} \|\mathcal{G}\|_{t_2}^{\frac{1}{p^2}} \tag{4.46}$$

$$\leq \ldots \leq \hat{C}^{\sum_{k=0}^{i} p^{-k}} \left[\prod_{k=0}^{i} (t_k \theta_k)^{p^{-k}} \right]^{-\frac{v-p}{p}} \|\mathcal{G}\|_{t_{i+1}}^{\frac{1}{p^{i+1}}}.$$

We shall find a suitable sequence $\{\sigma_k\}$ such that

$$P_1 \doteq \prod_{k=0}^{\infty} (t_k \theta_k)^{p^{-k}} = t^{\frac{p}{p-1}} (1 - \xi^{\sigma_1}) \prod_{k=1}^{\infty} \left[\xi^{\sigma_k} - \xi^{\sigma_{k+1}} \right]^{\frac{1}{p^k}}$$

converges with nice estimates as $p \to 1$. Taking the logarithm, this amounts to estimating from below the sum

$$\sum_{k=1}^{\infty} \frac{1}{p^k} \log \left[\xi^{\sigma_k} - \xi^{\sigma_{k+1}} \right]$$

by $\mathcal{O}(\frac{1}{p-1})$. For fixed $\tau > 1$, we choose σ_k inductively by taking

$$\sigma_1 = 1, \qquad \sigma_{k+1} = \frac{\log (\xi^{\sigma_k} - \xi^{\tau})}{\log \xi},$$

so in particular,

$$\xi^{\sigma_k} - \xi^{\sigma_{k+1}} = \xi^{\tau}, \qquad \text{hence} \qquad \sigma_{k+1} > \sigma_k > \ldots > \sigma_1 = 1.$$

Note also that $\sigma_k \in (1, \tau)$ for every k, since $\xi^\tau < \xi^{\sigma_k}$ and therefore $t_k = t\xi^{\sigma_k} \geq t\xi^\tau$. With such a choice,

$$\sum_{k=1}^{\infty} \frac{1}{p^k} \log\left[\xi^{\sigma_k} - \xi^{\sigma_{k+1}}\right] = \tau \log \xi \sum_{k=1}^{\infty} \frac{1}{p^k} = \frac{\tau \log \xi}{p-1}$$

and thus

$$P_1 = t^{\frac{p}{p-1}}(1 - \xi^{\sigma_1}) \exp\left\{\frac{\tau \log \xi}{p-1}\right\} = t^{\frac{p}{p-1}}(1 - \xi)\xi^{\frac{\tau}{p-1}}.$$

Recalling that $\|\mathcal{G}\|_r$ is a non-increasing function of r, $\|\mathcal{G}\|_{t_{i+1}} \leq \|\mathcal{G}\|_{t\xi^\tau}$ and therefore

$$\lim_{i \to \infty} \|\mathcal{G}\|_{t_{i+1}}^{\frac{1}{p^{i+1}}} \leq \lim_{i \to \infty} \|\mathcal{G}\|_{t\xi^\tau}^{\frac{1}{p^{i+1}}} = 1.$$

Thus, letting $i \to \infty$ and computing the sum at the exponent of $\hat{C}$, we deduce from (4.46) the upper bound

$$\|\mathcal{G}\|_t \leq \hat{C}^{\frac{p}{p-1}}\left[\prod_{k=0}^{\infty}(t_k\theta_k)^{p^{-k}}\right]^{-\frac{v-p}{p}} = \hat{C}^{\frac{p}{p-1}}t^{-\frac{v-p}{p-1}}\left[(1 - \xi)\xi^{\frac{\tau}{p-1}}\right]^{-\frac{v-p}{p}}.$$

Finally, letting $\tau \to 1$ and maximizing in $\xi \in (0, 1)$ we obtain

$$\max_{\xi \in (0,1)} (1 - \xi)\xi^{\frac{1}{p-1}} = \frac{p-1}{p}p^{-\frac{1}{p-1}},$$

and therefore, recalling the definition of $\hat{C}$,

$$\|\mathcal{G}\|_t \leq \left(\frac{p}{p-1}\right)^{\frac{v-p}{p}} \mathcal{S}_{p,v}^{\frac{v}{p(p-1)}} \bar{C}_{p,v}^{\frac{v-p}{p(p-1)}} t^{-\frac{v-p}{p-1}} p^{\frac{v-p}{p(p-1)}}.$$

Estimate (4.37) then follows from the definition of $\bar{C}_{p,v}$. □

Remark 4.27 In the above proof, we performed a non-standard iteration to obtain a constant $C_{p,v}$ which remains bounded as $p \to 1$. As can be easily checked, this is not possible with a standard dyadic iteration. Although not essential for the purposes of the present book, the fact that $C_{p,v}$ remains bounded as $p \to 1$ turns out to be crucial in applications to the Inverse Mean Curvature Flow, see [156].

Boundary Value Problems for Nonlinear ODEs

5

At the beginning of Chap. 4, we observed that to find radial solutions of $(P_\geq)$ and $(P_\leq)$ one is lead to solve the following ODE:

$$\left[v_g \varphi(w')\right]' = v_g \beta f(w) l(|w'|) \tag{5.1}$$

on an interval of $\mathbb{R}_0^+$, where we have extended φ to an odd function on all of $\mathbb{R}$. The functions v_g and β are bounds, respectively, for the volume of geodesic spheres of M and for b. We devote this section to the study of (5.1). Regarding φ, f, l we assume the following:

$$\begin{cases} \varphi \in C(\mathbb{R}_0^+), \ \varphi(0) = 0, \ \varphi(t) > 0 \text{ for all } t \in \mathbb{R}^+, \\ \varphi \text{ is strictly increasing on } \mathbb{R}^+, \\ f \in C(\mathbb{R}), \\ f \geq 0 \quad \text{in } (0, \eta_0), \text{ for some } \eta_0 \in (0, \infty), \\ l \in C(\mathbb{R}_0^+), \quad l \geq 0 \text{ in } \mathbb{R}_0^+. \end{cases} \tag{5.2}$$

We point out that no monotonicity is needed neither on f nor on l. In some results, we also require the validity of the next conditions:

$$\begin{cases} \varphi \in C^1(\mathbb{R}^+), \ \varphi' > 0 \text{ on } \mathbb{R}^+, \\ \dfrac{t\varphi'(t)}{l(t)} \in L^1(0^+). \end{cases} \tag{5.3}$$

© The Author(s), under exclusive license to Springer Nature Switzerland AG 2021
B. Bianchini et al., *Geometric Analysis of Quasilinear Inequalities on Complete Manifolds*, Frontiers in Mathematics, https://doi.org/10.1007/978-3-030-62704-1_5

Set

$$\varphi(\infty) = \lim_{t \to \infty} \varphi(t) \in (0, \infty],$$

and define K and F respectively as in (2.7) and (2.8).

In what follows, we find convenient to normalize the interval where we study (5.1), say on $[0, T_0]$, and for this reason we introduce two functions $a, \wp$ that will be related to, respectively, β and v_g, in a way that may depend from the geometrical problem at hand. We require

$$\begin{cases} \wp \in C^1([0, T_0]), & \wp > 0 \text{ on } [0, T_0], \\ a \in C([0, T_0]), & a > 0 \text{ on } [0, T_0]. \end{cases} \tag{5.4}$$

5.1   The Dirichlet Problem

We first investigate the existence and the qualitative properties of C^1 weak solutions of the singular two-points boundary value problem

$$\begin{cases} [\wp \, \varphi(w')]' = \wp a f(w) l(|w'|) & \text{on } (0, T), \\ w(0) = 0, \quad w(T) = \eta, \\ 0 \le w \le \eta, \quad w' \ge 0 \text{ on } (0, T), \end{cases} \tag{5.5}$$

where $\eta > 0$, $T \in (0, T_0)$ are given.

The results of this section are inspired by Chapters 4 and 8 of [194], and we also borrow some of the main ideas of the proof of Proposition 3.1 in [95] and the appendix of Chapter 4 in [194], but with several improvements in the spirit of [33, Thm. 4.1].

One of the main points in our investigation is to determine under which assumptions on f, l, φ and a, solutions of (5.5) satisfy $w'(0) = 0$ or $w'(0) > 0$, that is, whether or not w can be pasted to the zero function on $(-\infty, 0)$ in a C^1 way. As we shall see, such assumptions will be substantially related to the integrability condition (KO$_0$).

For $\eta, \xi > 0$ set

$$a_0 = \min_{[0, T_0]} a, \qquad\qquad a_1 = \max_{[0, T_0]} a;$$

$$\wp_0 = \min_{[0, T_0]} \wp, \qquad\qquad \wp_1 = \max_{[0, T_0]} \wp;$$

$$f_\eta = \max_{[0, \eta]} f, \qquad\qquad l_\xi = \max_{[0, \xi]} l; \tag{5.6}$$

$$\Theta(T) = \sup_{[0, T]} \frac{1}{\wp(t)} \int_0^t \wp(s) a(s) ds.$$

Note that $\Theta(T) \to 0$ as $T \to 0$.

We aim to prove the following existence result.

Theorem 5.1 *Assume (5.2), (5.4) and*

$$f(0)l(0) = 0.$$

Fix $\xi > 0$, let $T \in (0, T_0)$ and $\eta \in (0, \eta_0)$, with η_0 as in (5.2), satisfying

$$\frac{\wp_1}{\wp_0} \varphi\left(\frac{\eta}{T}\right) + 2\Theta(T) f_\eta l_\xi < \varphi(\xi). \tag{5.7}$$

Then, problem (5.5) admits a weak solution $w \in C^1([0, T])$ such that

$$0 \le w' \le \varphi^{-1}\left(\frac{\wp_1}{\wp_0} \varphi\left(\frac{\eta}{T}\right) + 2\Theta(T) f_\eta l_\xi\right). \tag{5.8}$$

In particular, $0 \le w' < \xi$.

We begin with the following auxiliary result, see Lemma 4.1.3 of [194].

Lemma 5.2 *Under assumptions (5.2) and (5.4), suppose that φ is extended on all of $\mathbb{R}$ in such a way that $t\varphi(t) > 0$ on $\mathbb{R}\backslash\{0\}$. Then, any weak solution $w \in C^1([0, T])$ of*

$$\begin{cases} \operatorname{sign}(w) \cdot \left[\wp\varphi(w')\right]' \ge 0 & \text{in } (0, T), \\ w(0) = 0, \quad w(T) = \eta > 0 \end{cases} \tag{5.9}$$

is such that

$$w \ge 0, \qquad w' \ge 0 \qquad \text{in } [0, T].$$

Moreover, there exists $t_0 \in [0, T)$ such that

$$w \equiv 0 \ \text{ in } [0, t_0]; \qquad w > 0, \ w' > 0 \ \text{ in } (t_0, T], \tag{5.10}$$

Proof We first claim that $w \ge 0$ on $[0, T]$. Otherwise, by contradiction there exist t_0, t_1, with $0 \le t_0 < t_1 < T$, such that $w(t_0) = w(t_1) = 0$ and $w < 0$ on (t_0, t_1). Using the non-negative, Lipschitz test function $\psi = -w$ on $[t_0, t_1]$, $\psi = 0$ otherwise, we get

$$\int_{t_0}^{t_1} \wp\varphi(w')w' \le 0.$$

Since $s\varphi(s) > 0$ in $\mathbb{R}\backslash\{0\}$ by assumption, the integrand is strictly positive. This gives the desired contradiction.

Let $\mathcal{J} = \{t \in (0, T) : w'(t) > 0\}$. Since $w \in C^1([0, T])$ and $w(T) > w(0)$, $\mathcal{J} \neq \emptyset$ and $\mathcal{J}$ is open in $(0, T)$. Let $t_0 = \inf \mathcal{J} \in [0, T)$, so that $w \equiv 0$ on $[0, t_0]$. For each fixed $t \in (t_0, T)$, there necessarily exists $\bar{t} \in (t_0, t)$ with $w'(\bar{t}) > 0$. Integrating (5.9) on $[\bar{t}, t]$ we deduce

$$\wp(t)\varphi(w'(t)) \geq \wp(\bar{t})\varphi(w'(\bar{t})) > 0,$$

which implies that $w'(t) > 0$. Hence, $w' > 0$ on $(t_0, T]$. Integrating again we obtain $w > 0$ on $(t_0, T]$, concluding the proof. $\qquad\qquad\qquad\qquad\qquad\qquad\qquad\qquad\qquad\qquad\qquad\qquad\qquad\square$

Remark 5.3 Note that the a priori knowledge of $w \geq 0$ in $[0, T]$ allows us to directly apply the second part of the proof of Lemma 5.2 and conclude that $w' \geq 0$ on $(0, T)$, and so $0 \leq w \leq \eta$.

We are now ready to solve the singular two-points boundary value problem (5.5).

Proof of Theorem 5.1 Redefine f and l on the complements of, respectively, $[0, \eta]$ and $[0, \xi]$, in such a way that

$$0 \leq f(s) \leq f_\eta \quad \text{for } s \geq \eta, \qquad f(s) = 0 \quad \text{for } s < 0,$$
$$0 < l(s) \leq l_\xi \quad \text{for } s \geq \xi. \tag{5.11}$$

Note that we can change l as above still keeping the validity of $l \in C(\mathbb{R}_0^+)$, while, with this procedure, we can only ensure that $f \in C(\mathbb{R}\backslash\{0\})$, since f has a jump discontinuity at $s = 0$ when $f(0) > 0$. The modifications will not affect the conclusions of the theorem since any ultimate solution with $w' \geq 0$ satisfies $0 \leq w \leq \eta$ and $|w'| \leq \xi$. However, the region $\{w = 0\}$ needs a special care. We extend φ to a continuous function defined on all of $\mathbb{R}$ in such a way that

$$\begin{cases} \varphi < 0 \quad \text{on } (-\infty, 0), \qquad \varphi(t) = -\varphi(-t) \quad \text{if } t \in \left[-\varphi^{-1}\left(\dfrac{\eta}{T}\right), 0\right], \\ \varphi \quad \text{is strictly increasing on } \mathbb{R}; \\ \lim_{t \to -\infty} \varphi(t) = -\infty. \end{cases} \tag{5.12}$$

Let

$$\mu_1 = \frac{\wp_1}{\wp_0}\varphi\left(\frac{\eta}{T}\right) + \Theta(T)f_\eta l_\xi < \varphi(\xi),$$

where the last inequality is due to (5.7), and set

$$I = [-\wp_1\mu_1, \wp_0\mu_1].$$

To show the existence of solutions of (5.5), following the approach in [194] we use Browder's version of the Leray–Schauder theorem (see Theorem 11.6 of [101]), an idea that is attributed by the authors in [194] to M. Montenegro. We consider the parametric family of boundary value problems

$$\begin{cases} [\wp\varphi(w')]' = \sigma\wp a f(w)l(|w'|) & \text{on } (0, T), \\ w(0) = 0, \qquad w(T) = \sigma\eta \geq 0, \end{cases}$$

for $\sigma \in [0, 1]$. In our case, however, the presence of a non-constant l makes things more subtle than in [194]. To tackle the problem we let X be the Banach space

$$X = (C^1([0, T]), \|\cdot\|),$$

where $\|\cdot\|$ is the C^1 norm:

$$\|w\| = \|w\|_\infty + \|w'\|_\infty \qquad \forall\, w \in X.$$

Define $\mathcal{H} : X \times [0, 1] \to X$ as follows:

$$\mathcal{H}(w, \sigma)(t) = \sigma\eta - \int_t^T \varphi^{-1}\left(\frac{1}{\wp(s)}\left[\delta \right.\right.$$
$$\left.\left. + \sigma \int_0^s \wp(\tau)a(\tau)f(w(\tau))l(|w'(\tau)|)d\tau\right]\right)ds, \tag{5.13}$$

where $\delta = \delta(w, \sigma) \in I$ and δ is chosen in such a way that

$$\mathcal{H}(w, \sigma)(0) = 0.$$

We claim that such a choice of δ is possible, and in fact it is unique. First, we check that $\mathcal{H}(w, \sigma)$ is well defined for each fixed $(w, \sigma) \in X \times [0, 1]$ and $\delta \in I$. This follows from the next chain of inequalities, where we use the definition of $\Theta(T)$ and (5.7)

$$\frac{1}{\wp(s)}\left[\delta + \sigma\int_0^s \wp(\tau)a(\tau)f(w(\tau))l(|w'(\tau)|)d\tau\right] \leq \frac{\wp_0\mu_1}{\wp_0} + \Theta(T)f_\eta l_\xi \tag{5.14}$$

$$= \mu_1 + \Theta(T)f_\eta l_\xi < \varphi(\xi).$$

We remark that, by construction, φ is a homeomorphism of $(-\infty, 0)$ onto itself. Moreover, if $\delta = -\wp_1\mu_1$, recalling that φ^{-1} is increasing on $\mathbb{R}$ we have

$$\mathcal{H}(w,\sigma)(0) = \sigma\eta - \int_0^T \varphi^{-1}\left(\frac{1}{\wp(s)}\left[-\wp_1\mu_1 + \sigma\int_0^s \wp(\tau)a(\tau)f(w(\tau))l(|w'(\tau)|)d\tau\right]\right)ds$$

$$\geq \sigma\eta - \int_0^T \varphi^{-1}\left(-\frac{\wp_1\mu_1}{\wp(s)} + \Theta(T)f_\eta l_\xi\right)ds$$

$$\geq \sigma\eta - \int_0^T \varphi^{-1}\left(-\mu_1 + \Theta(T)f_\eta l_\xi\right)ds \geq \sigma\eta,$$

where the last inequality follows from (5.12) and since $\mu_1 \geq \Theta(T)f_\eta l_\xi$. On the other hand, for $\delta = \wp_0\mu_1$, for all $(w,\sigma) \in X \times [0,1]$ we find

$$\mathcal{H}(w,\sigma)(0) = \sigma\eta - \int_0^T \varphi^{-1}\left(\frac{1}{\wp(s)}\left[\wp_0\mu_1 + \sigma\int_0^s \wp(\tau)a(\tau)f(w(\tau))l(|w'(\tau)|)d\tau\right]\right)ds$$

$$\leq \eta - \int_0^T \varphi^{-1}\left(\frac{\wp_0\mu_1}{\wp(s)}\right)ds \leq \eta - \int_0^T \varphi^{-1}\left(\frac{\wp_1}{\wp(s)}\varphi\left(\frac{\eta}{T}\right)\right)ds$$

$$\leq \eta - \int_0^T \varphi^{-1}\left(\varphi\left(\frac{\eta}{T}\right)\right)ds = 0.$$

Now, the integrand in the RHS of (5.13) is a strictly increasing function of δ for (w,σ) fixed. It is therefore clear that there exists a unique $\delta = \delta(w,\sigma) \in I$ such that $\mathcal{H}(w,\sigma)(0) = 0$.

It remains to show that a fixed point of $w = \mathcal{H}(w,1)$ exists. To apply Browder's version of the Leray–Schauder theorem, we shall check that:

(i) $\mathcal{H}(w,0) = 0$, the zero function of X,
(ii) $\mathcal{H} : X \times [0,1] \to X$ is continuous and compact,
(iii) There exists a constant $\Upsilon > 0$ such that $\|w\| \leq \Upsilon$ for all $(w,\sigma) \in X \times [0,1]$, with $w = \mathcal{H}(w,\sigma)$.

Property (i) is immediate by the definition of $\mathcal{H}$, since $\delta(w,0) = 0$ for all $w \in X$. Regarding (iii), by construction each solution of $w = \mathcal{H}(w,\sigma)$ is of class $C^1([0,T])$ and has the property that $\varphi(w') \in \mathrm{Lip}([0,T])$. We claim that $w \geq 0$ on $[0,T]$. Indeed, if $w < 0$ somewhere, fix an interval $(t_1, t_2) \subset (0,T)$ such that $w < 0$ on (t_1, t_2), $w(t_1) = w(t_2) = 0$. From $f = 0$ on $(-\infty, 0)$ we deduce $(\varphi(w'))' = 0$ on (t_1, t_2); thus, integrating against the test function $(w + \varepsilon)_-$ and letting $\varepsilon \to 0$ we get

$$0 = \int_{t_1}^{t_2} \wp\varphi(w')w',$$

and because of the positivity of $s\varphi(s)$ on $\mathbb{R}\backslash\{0\}$ we deduce that w', and consequently w, vanishes identically on (t_1, t_2), which is a contradiction. By Remark 5.3, from $w \geq 0$ in $[0, T]$ we infer $w' \geq 0$ and $0 \leq w \leq \eta$ in $[0, T]$, and the identity

$$w'(t) = [\mathcal{H}(w, \sigma)]'(t)$$

$$= \varphi^{-1}\left(\frac{1}{\wp(t)}\left[\delta + \sigma \int_0^t \wp(\tau)a(\tau)f(w(\tau))l(|w'(\tau)|)d\tau\right]\right)$$

(5.15)

implies

$$0 \leq w'(t) \leq \varphi^{-1}\left(\frac{\delta}{\wp(t)} + \Theta(T)f_\eta l_\xi\right)$$

(5.16)

$$\leq \varphi^{-1}\left(\mu_1 + \Theta(T)f_\eta l_\xi\right) < \xi.$$

Hence, each solution of $w = \mathcal{H}(w, \sigma)$ enjoys the a priori estimate $\|w\| \leq \eta + \xi$, as required.

We are left to prove (ii). Let $\{(w_k, \sigma_k)\}_k$ be a bounded sequence in $X \times [0, 1]$, say $\|w_k\| \leq L$ for all k. Using that $\delta_k = \delta(w_k, \sigma_k) \in I$ and $0 \leq f(t) \leq f_\eta$ for all $t \in \mathbb{R}$, together with (5.14), we deduce that

$$\|\mathcal{H}(w_k, \sigma_k)'\|_\infty \leq \max\left\{\left|\varphi^{-1}\left(-\frac{\wp_1\mu_1}{\wp_0}\right)\right|, \varphi^{-1}\left(\mu_1 + \Theta(T)f_\eta l_\xi\right)\right\},$$

thus $\{\mathcal{H}(w_k, \sigma_k)\}_k$ is equi-bounded in X and equi-continuous in $[0, T] \times [0, 1]$. To show the equi-continuity of $\{\mathcal{H}(w_k, \sigma_k)\}_k$ in C^1, we shall estimate the difference

$$\left|\mathcal{H}(w_k, \sigma_k)'(t) - \mathcal{H}(w_k, \sigma_k)'(s)\right|$$

for $0 \leq s < t \leq T$. Set for convenience

$$x_k = \frac{1}{\wp(t)}\left(\delta_k + \sigma_k \int_0^t \wp(\tau)a(\tau)f(w_k(\tau))l(|w_k'(\tau)|)d\tau\right)$$

$$y_k = \frac{1}{\wp(s)}\left(\delta_k + \sigma_k \int_0^s \wp(\tau)a(\tau)f(w_k(\tau))l(|w_k'(\tau)|)d\tau\right),$$

and note that

$$\left|\mathcal{H}(w_k, \sigma_k)'(t) - \mathcal{H}(w_k, \sigma_k)'(s)\right| = \left|\varphi^{-1}(x_k) - \varphi^{-1}(y_k)\right|.$$

Fix $\varepsilon > 0$ and let $\varrho = \varrho(\varphi^{-1}, \varepsilon) > 0$ be the corresponding number of the uniform continuity of φ^{-1} in $[-\frac{\wp_1}{\wp_0}\mu_1, \mu_1]$. Set $c = \wp_1 a_1 f_\eta l_\xi$, and suppose that $|t - s| < \varrho/C$, where

$$C \doteq \frac{\wp_1 \mu_1}{\wp_0^2} \max_{\tau \in [0,T]} |\wp'(\tau)| + \Lambda, \qquad \Lambda \doteq \frac{c}{\wp_0}\left(\frac{T}{\wp_0} \max_{\tau \in [0,T]} |\wp'(\tau)| + 1\right). \tag{5.17}$$

This is possible by (5.4), since $\wp \geq \wp_0 > 0$ on $[0, T]$ and $\wp \in C^1(\mathbb{R}_0^+)$. Define

$$I_k(t) = \int_0^t \wp(\tau)a(\tau)f(w_k(\tau))l(|w_k'(\tau)|)d\tau, \qquad \mathcal{I}_k(t) = \frac{I_k(t)}{\wp(t)},$$

and note that for each k

$$0 \leq I_k(t) - I_k(s) \leq c(t - s) \qquad \text{and} \qquad \lim_{t \to 0^+} \mathcal{I}_k(t) = 0.$$

Using $\wp > 0$ in $\mathbb{R}_0^+$ and $\wp \in C^1(\mathbb{R}_0^+)$ we get

$$\left|\sigma_k \mathcal{I}_k(s) - \sigma_k \mathcal{I}_k(t)\right| \leq \left|\mathcal{I}_k(s) - \mathcal{I}_k(t)\right| = \left|\frac{\wp(t)I_k(s) - \wp(s)I_k(t)}{\wp(s)\wp(t)}\right|$$

$$\leq \frac{|\wp(t) - \wp(s)|}{\wp(s)\wp(t)}I_k(s) + \frac{|I_k(s) - I_k(t)|}{\wp(t)}$$

$$\leq \frac{c}{\wp_0}\left(\frac{T}{\wp_0} \max_{\tau \in [0,T]} |\wp'(\tau)| + 1\right)|t - s| = \Lambda|t - s|.$$

Since $\delta_k \in I$, by (5.17) we estimate

$$|x_k - y_k| \leq |\delta_k|\frac{|\wp(t) - \wp(s)|}{\wp(s)\wp(t)} + \left|\sigma_k \mathcal{I}_k(s) - \sigma_k \mathcal{I}_k(t)\right|$$

$$\leq \left(\frac{\wp_1 \mu_1}{\wp_0^2} \max_{\tau \in [0,T]} |\wp'(\tau)| + \Lambda\right)|t - s| = C|t - s| < \varrho.$$

In conclusion,

$$\left|\mathcal{H}(w_k, \sigma_k)'(t) - \mathcal{H}(w_k, \sigma_k)'(s)\right| = \left|\varphi^{-1}(x_k) - \varphi^{-1}(y_k)\right| < \varepsilon$$

provided that $|t - s| < \varrho/C$, independently of k. As an immediate consequence of the Ascoli–Arzelà Theorem, $\mathcal{H}$ maps bounded sequences of $X \times [0, 1]$ into relatively compact sequences of X.

Finally, $\mathcal{H}$ is continuous in $X \times [0, 1]$. Indeed, if a sequence $\{(w_k, \sigma_k)\}_k$ converges to some (w, σ) in $X \times [0, 1]$, then $w_k \to w$ and $w'_k \to w'$ uniformly in $[0, T]$, and $\delta_k \to \delta = \delta(w, \sigma)$ as $k \to \infty$. Therefore, $f(w_k)l(|w'_k|) \to f(w)l(|w'|)$ uniformly in $[0, T]$, since the modified function f is continuous in $\mathbb{R}$, and so

$$\mathcal{H}(w_k, \sigma_k) \to \mathcal{H}(w, \sigma) \qquad \text{pointwise in } [0, T],$$

by (5.13) and the dominated convergence theorem. The arguments used above show that $\mathcal{H}(w_k, \sigma_k) \to \mathcal{H}(w, \sigma)$ and $\mathcal{H}(w_k, \sigma_k)' \to \mathcal{H}(w, \sigma)'$ uniformly in $[0, T]$. In other words, $\mathcal{H}(w_k, \sigma_k) \to \mathcal{H}(w, \sigma)$ in X, as required.

The Leray–Schauder fixed point theorem can therefore be applied and the mapping $\mathcal{H}(w, 1)$ has a fixed point w that, by (5.16), satisfies inequality (5.8).

To conclude, we prove that w solves the ODE in (5.5) with the original f, l. Differentiating (5.15) with $\sigma = 1$, we see that w is a weak solution of

$$\big(\wp\varphi(w')\big)' = \wp a \bar{f}(w)\bar{l}(|w'|) \qquad \text{on } (0, T),$$

where now we have denoted with $\bar{f}, \bar{l}$ the modifications of f, l in (5.11). Clearly, $\bar{l}(w') = l(w')$ on $(0, T)$, since $0 \le w' < \xi$. From $0 \le w \le \eta$ and Lemma 5.2, there exists $t_0 \in [0, T)$ such that $w = 0$ on $[0, t_0]$ and $w > 0$, $w' > 0$ on $(t_0, T]$, and since $\bar{f} = f$ on $(0, \eta)$ we deduce

$$\big(\wp\varphi(w')\big)' = \wp a f(w)l(|w'|) \tag{5.18}$$

on (t_0, T). On the other hand, because of our assumption $f(0)l(0) = 0$ the function $w = 0$ solves (5.18) on $(0, t_0)$. Since $w \in C^1([0, T])$, (5.18) holds weakly on all of $(0, T)$, as claimed. □

Remark 5.4 The requirement $f(0)l(0) = 0$ is crucial for the validity of the above theorem, because otherwise w might be negative somewhere.

Given f, φ, l satisfying (5.2), define K, F as in (2.7), (2.8). We next present two auxiliary calculus results. The first is similar to Lemma 4.4.1–(i) in [194]. We recall that the notion of a C-increasing function is given in Definition 2.12.

Lemma 5.5 *Assume that f is C-increasing in $(0, \eta_0)$, for some $\eta_0 > 0$. Then, for each $\sigma \in [0, 1]$ we have $F(\sigma t) \le C \sigma F(t)$ for all $t \in [0, \eta_0)$.*

Proof Fix $\sigma \in [0, 1]$. Since f is C-increasing in $(0, \eta_0)$, we have $\sigma f(\sigma t) \leq C \sigma f(t)$ for each $t \in [0, \eta_0)$, and thus

$$F(\sigma t) = \int_0^{\sigma t} f(s)\,ds = \int_0^t \sigma f(\sigma \tau)\,d\tau \leq C \sigma \int_0^t f(\tau)\,d\tau = C\sigma F(t),$$

as claimed.

$\square$

The second lemma concerns the preservation of the validity of (KO_0) and (KO_∞) when we replace f with σf, $\sigma \in \mathbb{R}^+$. Similar results have been proved in [194, Lemma 4.1.2], [95] (remark on page 523) and [157].

Lemma 5.6 *Let f, l satisfy (5.2), (5.3) and suppose $l > 0$ on $\mathbb{R}^+$.*

(i) Assume that f is positive and C-increasing in $(0, \eta_0)$, for $\eta_0 > 0$ in (5.2). Then

$$\frac{1}{K^{-1}(F(s))} \in L^1(0^+) \qquad \Longleftrightarrow \qquad \frac{1}{K^{-1}(\sigma F(s))} \in L^1(0^+)$$

for some (equivalently, any) $\sigma \in \mathbb{R}^+$.

(ii) Assume that f is positive and C-increasing on $(\bar{\eta}_0, \infty)$, for some $\bar{\eta}_0 > 0$. Having defined $F(t) = \int_{\bar{\eta}_0}^t f$, it holds

$$\frac{1}{K^{-1}(F(s))} \in L^1(\infty) \qquad \Longleftrightarrow \qquad \frac{1}{K^{-1}(\sigma F(s))} \in L^1(\infty)$$

for some (equivalently, any) $\sigma \in \mathbb{R}^+$.

Proof For the ease of notation we denote with $(KO_0)(\sigma)$ and $(KO_\infty)(\sigma)$, respectively, the integrability conditions

$$\int_{0^+} \frac{ds}{K^{-1}(\sigma F(s))} < \infty, \qquad \int^\infty \frac{ds}{K^{-1}(\sigma F(s))} < \infty.$$

We prove (i), beginning with the implication

$$(KO_0) \Rightarrow (KO_0)(\sigma). \tag{5.19}$$

If $\sigma \geq 1$, (5.19) is immediate from the monotonicity of F and K.
If $\sigma \in (0, 1)$, we apply Lemma 5.5 with σ/C replacing σ (note that $C \geq 1$) to deduce

$F(\sigma t/C) \le \sigma F(t)$ for each $t \in (0, \eta_0)$. Integrating and changing variables,

$$\int_{0+} \frac{ds}{K^{-1}(\sigma F(s))} \le \int_{0+} \frac{ds}{K^{-1}(F(\sigma s/C))} = \frac{C}{\sigma} \int_{0+} \frac{d\tau}{K^{-1}(F(\tau))},$$

which proves (5.19). To show the reverse implication in (5.19), it is enough to observe that $(KO_0)(\sigma)$ is condition (KO_0) for the function $\bar{f} = \sigma f$, and to apply the previous estimates with $\bar{f}$ replacing f and σ^{-1} replacing σ.
The proof of (ii) is analogous. □

We are now ready to obtain further information on the solution of problem (5.5) given in Theorem 5.1. First, we investigate sufficient conditions to ensure that $w'(0) > 0$.

Proposition 5.7 *Assume (5.2), (5.3) and (5.4). Suppose that*

$f(0)l(0) = 0;$
f *is C-increasing on* $(0, \eta_0)$, *for* $\eta_0 > 0$ *as in (5.2),*

and that one of the following two sets of conditions is met:

(i) $f \equiv 0$ *on* $(0, \eta_0)$;
(ii) $f > 0$ *on* $(0, \eta_0)$, *and also*

- l *is C-increasing on* $(0, \xi_0)$, *for some* $\xi_0 > 0$.
- $\wp$ *is monotone on* $[0, T_0]$, *either increasing or decreasing* $(T_0$ *as in (5.4)),*

and

$$\frac{1}{K^{-1} \circ F} \notin L^1(0^+). \qquad\qquad (\neg KO_0)$$

Then, the solution w of problem (5.5), with $\eta \in (0, \eta_0)$, constructed in Theorem 5.1, has the further properties

$$w > 0 \quad on \ (0, T], \qquad w' > 0 \quad on \ [0, T]. \qquad\qquad (5.20)$$

Proof First, observe that if we prove that $w'(0) > 0$, then (5.20) follows by a direct application of Lemma 5.2. We prove $w'(0) > 0$ for cases (i) and (ii) separately.

Case (i). Since $f \equiv 0$ on $(0, \eta_0)$, then $(\wp\varphi(w'))' = 0$ in $(0, T)$ by (5.5) and the choice $\eta < \eta_0$. Hence, integrating

$$\wp(t)\varphi(w'(t)) = \wp(0)\varphi(w'(0)) \quad \text{for all } t \in (0, T]. \tag{5.21}$$

Suppose by contradiction that $w'(0) = 0$. From (5.2) we have $\varphi > 0$ on $\mathbb{R}^+$, and also $\wp > 0$ on $[0, T]$. Thus, (5.21) would imply that w', hence w, is identically zero in $[0, T]$, contradicting $w(T) = \eta$.

Case (ii). Theorem 5.1 guarantees that $\varphi(w') \in C^1([0, T])$, $w'(0) \geq 0$ and $0 \leq w \leq \eta$. Let us reason by contradiction and suppose that $w'(0) = 0$. We shall then prove that

$$\frac{1}{K^{-1} \circ F} \in L^1(0^+),$$

which contradicts $(\neg KO_0)$, completing the proof.

First, by Lemma 5.2 there exists $t_0 \in [0, T)$ such that $w(t) \equiv 0$ on $[0, t_0]$ while $w > 0$ on $(t_0, T]$. If $t_0 = 0$, then $w'(t_0) = 0$ by our assumption, while if $t_0 > 0$, then $w(t_0) = w'(t_0) = 0$ since w is $C^1([0, T])$. From (5.15) and since $\varphi' > 0$ on $\mathbb{R}^+$ we infer the existence of w'' in (t_0, T). Thus, w satisfies

$$\wp\varphi'(w')w'' + \wp'\varphi(w') = \wp a f(w)l(w') \qquad \text{in } (t_0, T). \tag{5.22}$$

We first suppose that $\wp' \geq 0$ on $[0, T_0]$. By (5.10) and (5.4), w is a solution of the inequality

$$w'\varphi'(w')w'' \leq af(w)w'l(w') \qquad \text{in } (t_0, T).$$

Integrating on $[t_0, t)$, with $t \in (t_0, T]$ we have

$$\int_{t_0}^{t} \frac{w'\varphi'(w')w''}{l(w')} d\tau \leq \int_{t_0}^{t} af(w)w'd\tau \leq a_1 \int_{t_0}^{t} f(w)w'd\tau.$$

Changing variables and using $w'(t_0) = w(t_0) = 0$ we deduce

$$K(w'(t)) = \int_{0}^{w'(t)} \frac{s\varphi'(s)}{l(s)}ds \leq a_1 \int_{0}^{w(t)} f(s)ds = a_1 F(w(t)). \tag{5.23}$$

Assumption $f > 0$ on $(0, \eta_0)$ implies that $F > 0$ on $(0, \eta_0)$. Having chosen $T_1 \in (t_0, T]$ in such a way that $a_1 F(w(T_1)) < K_\infty$, we apply K^{-1}, rearrange and integrate to obtain

$$\int_{t_0}^t \frac{w'(s) \, ds}{K^{-1}(a_1 F(w(s)))} \leq (t - t_0) \qquad \forall t \in (t_0, T_1].$$

Changing variables,

$$\int_0^{w(t)} \frac{d\tau}{K^{-1}(a_1 F(\tau))} \leq (t - t_0) \qquad \forall t \in (t_0, T_1].$$

By Lemma 5.6 property (KO$_0$) holds, as claimed.

We are left to consider the case $\wp' \leq 0$. Then, by (5.22) we deduce that $w'' \geq 0$ on (t_0, T), hence w' is increasing there. Integrating (5.22) on (t_0, t) and using the C-monotonicity of f, l together with $w'(t_0) = 0$, we get

$$\varphi(w'(t)) = \frac{1}{\wp(t)} \int_{t_0}^t \wp(s) a(s) f(w(s)) l(w'(s)) ds$$

(5.24)

$$\leq C^2 a_1 f(w(t)) l(w'(t)) \left[\frac{1}{\wp(t)} \int_0^t \wp(s) ds \right].$$

Now, consider the energy $E(t) = K(w'(t)) - a_1 F(w(t))$. Differentiating and using (5.24) and the definition of a_1, we obtain

$$E'(t) = \frac{w' \varphi'(w') w''}{l(w')} - a_1 f(w) w' = \frac{w'}{l(w')} \left[-\frac{\wp'}{\wp} \varphi(w') + a f(w) l(w') - a_1 f(w) l(w') \right]$$

$$\leq -\frac{\wp'}{\wp} \frac{w' \varphi(w')}{l(w')} \leq \left| \frac{\wp'(t)}{\wp(t)^2} \int_0^t \wp(s) ds \right| C^2 a_1 f(w) w'$$

$$\leq \frac{\wp_1 T_0 \|\wp'\|_{L^\infty([0, T_0])}}{\wp_0^2} C^2 a_1 f(w) w' = \bar{c} f(w) w'.$$

Integrating and using $w(t_0) = w'(t_0) = 0$,

$$K(w'(t)) \leq a_1 F(w(t)) + \bar{c} F(w(t)) = (a_1 + \bar{c}) F(w(t)).$$

Having obtained again an inequality like (5.23), to achieve the desired contradiction it is sufficient to repeat verbatim the last steps of the proof for $\wp' \geq 0$. □

Remark 5.8 When $\wp' \geq 0$, to reach the desired conclusion in (ii) we do not use the assumption that l is C-increasing.

Proposition 5.7 has a converse, at least if the threshold η in (5.5) is sufficiently small, namely (KO$_0$) implies that $w'(0) = 0$. To reach the goal, the idea is to compare w with an explicit supersolution of (5.5), whose construction generalizes the End Point Lemma in [194, Lemma 4.4.1].

Proposition 5.9 *Assume* (5.2), (5.3) *and* (5.4). *Suppose that*

$f(0)l(0) = 0;$
f *is C-increasing on* $(0, \eta_0)$, *for* $\eta_0 > 0$ *as in* (5.2)*;*
l *is C-increasing on* $(0, \xi_0)$, *for some* $\xi_0 > 0$.

If $f > 0$ *on* $(0, \eta_0)$ *and*

$$\frac{1}{K^{-1} \circ F} \in L^1(0^+), \qquad\qquad (KO_0)$$

then there exists η_1 sufficiently small that, for each $\eta \in (0, \eta_1)$, the solution w of problem (5.5) *constructed in Theorem 5.1 satisfies*

$$w'(0) = 0.$$

Proof For $\sigma \in (0, 1)$ to be determined, using (KO$_0$) and Lemma 5.6 we implicitly define $z(t)$ by setting

$$t = \int_0^{z(t)} \frac{ds}{K^{-1}(\sigma F(s))} \qquad \text{for } t \in [0, T).$$

Note that z is positive on $(0, T)$. Differentiating,

$$z' = K^{-1}(\sigma F(z)), \qquad\qquad (5.25)$$

whence $z' > 0$ on $(0, T)$ and $z'(0) = 0$. Evaluating K on both sides of (5.25) and differentiating once more we get

$$\frac{z'\varphi'(z')z''}{l(z')} = \sigma f(z)z' \qquad \text{on } (0, T).$$

Since φ', l, z' are positive, we deduce $z'' \geq 0$, and simplifying by z' we infer

$$\left(\varphi(z')\right)' = \sigma f(z)l(z').$$

Integrating, using $z'(0) = 0$ together with the C-increasing property of f and l (note that z' is increasing) we obtain

$$\varphi\big(z'(t)\big) = \sigma \int_0^t f\big(z(s)\big)l\big(z'(s)\big)\mathrm{d}s \leq \sigma C^2 T f\big(z(t)\big)l\big(z'(t)\big).$$

Summarizing, so far we have obtained,

$$\big[\wp\varphi(z')\big]' = \wp\big(\varphi(z')\big)' + \wp'\varphi(z') \leq \frac{\sigma}{a_0}\left[1 + C^2 T \frac{\|\wp'\|_{L^\infty([0,T_0])}}{\wp_0}\right]\wp a f(z)l(z')$$

$$= \frac{1}{2C}\wp a f(z)l(z'),$$

where we have defined σ in order to satisfy the last equality. Next, we fix $\eta_1 \leq z(T)$ small enough in such a way that each $\eta \in (0, \eta_1)$ meets the requirements in Theorem 5.1, to guarantee the existence of w. We claim that, for $\eta \in (0, \eta_1)$, the solution w in Theorem 5.1 satisfies $w \leq z$ on $(0, T)$. This, together with the already established $z'(0) = 0$, forces $w'(0) = 0$ and concludes our proof. By contradiction, suppose that $c = \max_{[0,T]}(w-z) > 0$ and let $\Gamma = \{w - z = c\}$. By construction, $\Gamma \Subset (0, T)$, and since w, z are C^1 we deduce $w' = z'$ on Γ. By continuity and since $w' > 0$ on $(0, T)$, we can choose $\delta \in (0, c)$ close enough to c in such a way that $l(z') \leq 2l(w')$ on the set $I_\delta = \{w - z > \delta\}$. On I_δ, using the C-increasing property we therefore have

$$\big[\wp\varphi(w')\big]' \geq a\wp f(w)l(w') \geq \frac{1}{2C}a\wp f(z)l(z') \geq \big[\wp\varphi(z')\big]' = \big[\wp\varphi((z+c)')\big]'$$

and $w = z+c$ on ∂I_δ. By standard comparison (one can apply, for instance Proposition 6.1 below to an appropriate radial model), $w \leq z + c$ on U_δ, contradiction. $\qquad\square$

5.2 The Mixed Dirichlet–Neumann Problem

We next move to investigate the problem

$$\begin{cases} \big[\wp\,\varphi(w')\big]' = \wp a f(w)l(|w'|) & \text{on } (0, T), \\[2mm] w'(0) = 0, \quad w(T) = \eta, \\[2mm] 0 \leq w \leq \eta, \quad w' \geq 0 \text{ on } (0, T), \end{cases}$$

for given $\eta > 0$, $T \in (0, T_0)$. We here extend and generalize in several directions the core of Corollary 1.4 of [95], *without requiring any monotonicity on l*, as well as the results of Section 4 of [33]. We assume (5.2) and (5.4), and we define $a_0, a_1, \wp_0, \wp_1, f_\eta, l_\xi$ and $\Theta(T)$ as in (5.6).

Theorem 5.10 *Assume (5.2) and (5.4), and that*

$$f > 0 \quad on \ \mathbb{R}^+, \qquad f(0) = 0;$$
$$l > 0 \quad on \ \mathbb{R}_0^+. \tag{5.26}$$

Then, for each $\eta, \xi > 0$ and $T \in (0, T_0)$ satisfying

$$\Theta(T) f_\eta l_\xi < \varphi(\xi), \tag{5.27}$$

the problem

$$\begin{cases} [\wp \varphi(w')]' = \wp a f(w) l(|w'|) & on \ (0, T) \\ w'(0) = 0, \quad w(T) = \eta, \\ 0 \le w \le \eta, \quad 0 \le w' < \xi \ on \ (0, T), \end{cases} \tag{5.28}$$

admits a solution $w \in C^1([0, T])$, and there exists $t_0 \in [0, T)$ such that

$$w(t) \equiv w(t_0) \ge 0 \quad on \ [0, t_0], \qquad w' > 0 \ on \ (t_0, T]. \tag{5.29}$$

Moreover, if

$$\varphi \in C^1(\mathbb{R}^+), \qquad \varphi' > 0 \quad on \ \mathbb{R}^+,$$

then $w \in C^2((t_0, T])$ and satisfies

$$\frac{\varphi'(w')}{l(w')} w'' = a \, f(w) - \frac{\wp'}{\wp} \cdot \frac{\varphi(w')}{l(w')} \qquad on \ (t_0, T).$$

All of the above conclusions still hold if condition $\wp > 0$ on $[0, T]$, in (5.4), is replaced by

$$\wp > 0 \quad on \ (0, T], \qquad \wp(0) = 0$$
$$\wp' \ge 0 \quad on \ [0, \delta), \quad for \ some \ \delta > 0. \tag{5.30}$$

Remark 5.11 Differently from the Dirichlet problem, if we allow l to vanish at $t = 0$ in the Neumann case we cannot guarantee that the solution of (5.28) be non-constant. This motivates the necessity to require $l > 0$ on $\mathbb{R}_0^+$.

Proof The strategy goes along the same lines as that for the Dirichlet problem. First, we redefine f outside of $[0, \eta]$ and l outside of $[0, \xi]$ in such a way that

$$f \in C(\mathbb{R}), \qquad 0 \leq f(s) \leq f_\eta \quad \text{for } s \geq \eta, \qquad f(s) = 0 \quad \text{for } s < 0,$$

$$l \in C(\mathbb{R}_0^+), \qquad 0 < l(s) \leq l_\xi \quad \text{for } s \geq \xi.$$

This will not affect the conclusion of the proposition, since any ultimate solution w of (5.28), with $w \geq 0$, $w' \geq 0$ in $[0, T]$, satisfies $0 \leq w \leq \eta$ and $|w'| < \xi$.

Denote with X the Banach space $X = C^1([0, T])$, endowed with the usual norm $\|w\| = \|w\|_\infty + \|w'\|_\infty$. Define the homotopy $\mathcal{H} : X \times [0, 1] \to X$ by

$$\mathcal{H}[w, \sigma](t) = \sigma\eta - \int_t^T \varphi^{-1}\left(\frac{\sigma}{\wp(s)} \int_0^s \wp(\tau)a(\tau)f(w(\tau))l(|w'(\tau)|)\mathrm{d}\tau\right)\mathrm{d}s. \qquad (5.31)$$

We claim that $\mathcal{H}$ is well defined and valued in X. Indeed, in our assumptions

$$0 \leq \frac{\sigma}{\wp(t)} \int_0^t \wp(\tau)a(\tau)f(w(\tau))l(|w'(\tau)|)\mathrm{d}\tau \leq \Theta(T)f_\eta l_\xi < \varphi(\xi), \qquad (5.32)$$

hence the term in round brackets in (5.31) lies in the domain of φ^{-1} and

$$\mathcal{H}[w, \sigma]'(t) = \varphi^{-1}\left(\frac{\sigma}{\wp(t)} \int_0^t \wp(\tau)a(\tau)f(w(\tau))l(|w'(\tau)|)\mathrm{d}\tau\right) \in [0, \xi). \qquad (5.33)$$

Furthermore, $\mathcal{H}[w, \sigma]'$ is continuous on $[0, T]$, hence $\mathcal{H}$ is valued in X. By construction, $\mathcal{H}[w, \sigma](T) = \sigma\eta$ and $\mathcal{H}[w, 0] = 0$. From

$$0 \leq \frac{1}{\wp(t)} \int_0^t \wp(\tau)a(\tau)f(w(\tau))l(|w'(\tau)|)\mathrm{d}\tau \leq f_\eta l_\xi \frac{1}{\wp(t)} \int_0^t \wp(\tau)a(\tau)\mathrm{d}\tau,$$

we deduce that $\mathcal{H}[w, \sigma]'(0) = 0$. Fix $\sigma \in (0, 1]$, and let w be a solution of $w = \mathcal{H}[w, \sigma]$. We claim that $w(0) \geq 0$: otherwise, since $w(T) = \sigma\eta > 0$ there would exist a first point $t_1 \in (0, T)$ such that $w < 0$ on $[0, t_1)$ and $w(t_1) = 0$, and therefore $f(w(t)) = 0$ on $[0, t_1]$. Thus, $w' \equiv 0$ on $[0, t_1]$ by (5.33), that is, w would be constant on $[0, t_1)$, contradicting $w(0) < 0 = w(t_1)$. From $w \geq 0$ we also deduce $w' \geq 0$ on $[0, T]$ by (5.33). Furthermore, (5.33) implies that $\varphi(w')$ is of class $C^1([0, T])$, and then from (5.31) that w is a classical weak solution of the problem

$$
\begin{cases}
[\wp\varphi(w')]' = \sigma\wp a f(w)l(|w'|) & \text{on } (0, T), \\
w'(0) = 0, \quad w(T) = \sigma\eta \\
0 \le w \le \sigma\eta, \qquad 0 \le w' < \xi & \text{on } (0, T).
\end{cases}
\tag{5.34}
$$

In particular, for $\sigma = 1$, w is the desired solution of (5.28). To prove (5.29), let $\sigma = 1$. From $w(T) = \eta$ and $w' \ge 0$ we infer the existence of a minimal $t_0 \in (0, T)$ such that $w > 0$ on $(t_0, T]$. Since $f > 0$ on $\mathbb{R}^+$, $l > 0$ on $\mathbb{R}_0^+$ and $a > 0$ on $[0, T]$, a solution of $w = \mathcal{H}[w, 1]$ satisfies

$$
w'(t) = \mathcal{H}[w, 1]'(t)
$$

$$
= \varphi^{-1}\left(\frac{1}{\wp(t)}\int_0^t \wp(\tau)a(\tau)f(w(\tau))l(|w'(\tau)|)d\tau\right) > 0 \qquad \forall t \in (t_0, T].
$$

If $t_0 \ne 0$, by the monotonicity and non-negativity of w we get $w = 0$ on $[0, t_0]$. To show (5.10), using $\varphi(w') \in C^1([0, T])$, $\varphi' > 0$ on $\mathbb{R}^+$ and $w' > 0$ on $(t_0, T]$ in (5.33) we deduce $w' \in C^1((t_0, T])$. Identity (5.10) immediately follows by expanding the derivative in (5.28).

We assert that a solution of $w = \mathcal{H}[w, 1]$ exists, using again the Browder version of the Leray–Schauder theorem (see [101, Thm 11.6]).

To begin with, as already observed $\mathcal{H}[w, 0] \equiv 0$ for all $w \in X$. We next show that $\mathcal{H}$ is continuous on $X \times [0, 1]$. Indeed, consider a sequence $\{(w_k, \sigma_k)\}_k \subset X \times [0, 1]$, with $w_k \to w$ in X and $\sigma_k \to \sigma$ as $k \to \infty$. By continuity, $\sigma_k f(w_k)l(|w_k'|) \to \sigma f(w)l(|w'|)$, and so $\mathcal{H}[w_k, \sigma_k] \to \mathcal{H}[w, \sigma]$ by (5.31) and Lebesgue convergence theorem, as required. Next we show that $\mathcal{H}$ is compact. To this aim, let $\{(w_k, \sigma_k)\}_k$ be a bounded sequence in $X \times [0, 1]$. From (5.33),

$$
\|\mathcal{H}[w_k, \sigma_k]'\|_\infty < \xi,
$$

and thus, since $\mathcal{H}[w_k, \sigma_k](T) = \sigma\eta \in [0, \eta]$, $\{\mathcal{H}[w_k, \sigma_k]\}_k$ is equi-bounded in X. We shall prove that $\{\mathcal{H}[w_k, \sigma_k]'\}_k$ is equi-continuous. Set

$$
I_k(t) = \int_0^t \wp(\tau)a(\tau)f(w_k(\tau))l(|w'(\tau)|)d\tau,
$$

$$
x_k = \frac{\sigma_k I_k(t)}{\wp(t)}, \qquad y_k = \frac{\sigma_k I_k(s)}{\wp(s)},
\tag{5.35}
$$

and note that

$$\left|\mathcal{H}[w_k, \sigma_k]'(t) - \mathcal{H}[w_k, \sigma_k]'(s)\right| = \left|\varphi^{-1}(x_k) - \varphi^{-1}(y_k)\right|, \tag{5.36}$$

and that, by (5.32),

$$0 \le x_k, y_k < \varphi(\xi), \qquad |I_k(s) - I_k(t)| \le c|s - t| \quad \text{with } c = \wp_1 a_1 f_\eta l_\xi. \tag{5.37}$$

Since $\wp > 0$ on $[0, T]$ and there it is C^1, we deduce

$$|x_k - y_k| = \sigma_k \left|\frac{\wp(t)I_k(s) - \wp(s)I_k(t)}{\wp(t)\wp(s)}\right| \le \left|\frac{\wp(t)I_k(s) - \wp(s)I_k(t)}{\wp(t)\wp(s)}\right|$$

$$\le \frac{|\wp(t) - \wp(s)|}{\wp(s)\wp(t)}I_k(s) + \frac{|I_k(s) - I_k(t)|}{\wp(t)} \tag{5.38}$$

$$\le \left[\frac{cT}{\wp_0^2}\left(\max_{[0,T]}\wp'\right) + \frac{c}{\wp_0}\right]|s - t| \doteq \Lambda|s - t|.$$

Given $\varepsilon > 0$, let $\varrho = \varrho(\varphi^{-1}, \varepsilon) > 0$ be given by the uniform continuity of φ^{-1} on $[0, \varphi(\xi)]$. If $|s - t| < \varrho/\Lambda$, then $|x_k - y_k| < \varrho$ and thus, by (5.36),

$$\left|\mathcal{H}[w_k, \sigma_k]'(t) - \mathcal{H}[w_k, \sigma_k]'(s)\right| = \left|\varphi^{-1}(x_k) - \varphi^{-1}(y_k)\right| < \varepsilon, \tag{5.39}$$

proving the (uniform) equi-continuity of $\{\mathcal{H}[w_k, \sigma_k]'\}_k$. The compactness of $\mathcal{H}$ then follows from the Ascoli–Arzelà Theorem.

To apply the Leray–Schauder theorem it remains to check the existence of a constant $\Upsilon > 0$ such that $\|w\| \le \Upsilon$ for each solution of $\mathcal{H}[w, \sigma] = w$. But this immediately follows from properties (5.34), and indeed $\|w\| \le \eta + \xi$.

It remains to consider the case when $\wp > 0$ on $[0, T]$ is replaced by assumption (5.30). From (5.33) and the monotonicity of $\wp$ on $[0, \delta)$, we deduce that for $t \in (0, \delta)$

$$\mathcal{H}[w, \sigma]'(t) = \varphi^{-1}\left(\frac{\sigma}{\wp(t)}\int_0^t \wp(\tau)a(\tau)f(w(\tau))l(|w'(\tau)|)d\tau\right)$$

$$\le \varphi^{-1}\left(\frac{f_\eta l_\xi a_1}{\wp(t)}\int_0^t \wp(\tau)d\tau\right) \le \varphi^{-1}\left(f_\eta l_\xi a_1 t\right).$$

Hence, $\mathcal{H}[w, \sigma]'$ is continuous up to $t = 0$ (i.e. $\mathcal{H}$ is valued in X) and $w'(0) = 0$ for each solution of $w = \mathcal{H}[w, \sigma]$. The rest of the proof follows verbatim, except for the equi-continuity of $\{\mathcal{H}[w_k, \sigma_k]'\}$ that we now consider. By the monotonicity of $\wp$,

$$|x_k| \leq \frac{|I_k(t)|}{\wp(t)} \leq a_1 f_\eta l_\xi t,$$

independently of k. Thus, given $\varepsilon > 0$ and the corresponding $\varrho = \varrho(\varphi^{-1}, \varepsilon)$ of the uniform continuity of φ^{-1} on $[0, \varphi(\xi)]$, we can choose $\vartheta \in (0, T/2)$ independent of k such that $|x_k| < \varrho/2$ if $t < 2\vartheta$ and $|y_k| < \varrho/2$ if $s < 2\vartheta$. Set

$$\hat{\wp}_0 \doteq \inf_{[\vartheta, T]} \wp > 0, \qquad \hat{\Lambda} \doteq \frac{cT}{\hat{\wp}_0^2} \left(\max_{[0,T]} \wp' \right) + \frac{c}{\hat{\wp}_0},$$

with c as in (5.37). Define

$$\bar{\vartheta} = \min \left\{ \frac{\varrho}{\hat{\Lambda}}, \vartheta \right\}.$$

Let $s, t \in [0, T]$ with $|s - t| < \bar{\vartheta}$. If $s, t \geq \vartheta$, then the chain of inequalities (5.38) holds verbatim with $\hat{\Lambda}, \hat{\wp}_0$ replacing $\Lambda, \wp_0$, respectively, and we deduce

$$|x_k - y_k| \leq \hat{\Lambda} |s - t| < \varrho.$$

On the other hand, if one of the two values s, t, say t, is less than ϑ, then from $\bar{\vartheta} \leq \vartheta$ we deduce $s < t + |s - t| < 2\vartheta$. Hence, $|x_k| < \varrho/2$ and $|y_k| < \varrho/2$, and thus

$$|x_k - y_k| \leq |x_k| + |y_k| < \varrho.$$

In both the cases, $|x_k - y_k| < \varrho$ and therefore (5.39) holds, proving the (uniform) equi-continuity of $\{\mathcal{H}[w_k, \sigma_k]'\}$. $\square$

The next result relates condition $w(0) = 0$ to (KO$_0$). In order to do so, we shall further require (5.3) in order to define K, F as in (2.7), (2.8).

Proposition 5.12 *In the assumptions of Theorem 5.10, suppose further the validity of* (5.3) *and that*

$$\wp' \geq 0 \qquad on \ (0, T),$$

$$f \ is \ C\text{-increasing on } (0, \eta_0), \ for \ some \ constant \ \eta_0 > 0.$$

If $w(0) = 0$, then (KO$_0$) *holds.*

Proof Because of the monotonicity of $\wp'$ and w, from (5.28) we deduce that

$$\big(\varphi(w')\big)' \leq a f(w) l(w') \leq a_1 f(w) l(w') \qquad \text{on } (0, T).$$

We now follow the steps in Proposition 5.7: differentiating on (t_0, T), we deduce

$$K'(w') w'' \leq a_1 f(w) w',$$

and integrating on (t_0, t) with the aid of $w(t_0) = w'(t_0) = 0$ we infer $K(w') \leq a_1 F(w)$. Let $t_1 \in (t_0, t)$ be such that $a_1 F(w) < K_\infty$ for $t \in (t_0, t_1)$. This is possible, by continuity, since $F(w(t_0)) = 0$. Applying K^{-1}, integrating and changing variables we get

$$\int_0^{w(t)} \frac{ds}{K^{-1}(a_1 F(s))} \leq (t - t_0) \qquad \text{on } (t_0, t_1),$$

and (KO_0) follows from Lemma 5.6.   □

We next investigate the maximal interval of definition of w. Assume the validity of (5.2) and (5.26). To tie w with (KO_∞), we assume (5.3) and

$$\frac{t\varphi'(t)}{l(t)} \notin L^1(\infty), \tag{5.40}$$

in order for K to be a homeomorphism of $\mathbb{R}_0^+$ onto itself. We further replace (5.4) and (5.30) with

$$\begin{cases} \wp \in C^1(\mathbb{R}_0^+), & \wp > 0, \quad \wp' \geq 0 \quad \text{on } \mathbb{R}^+, \\ a \in C(\mathbb{R}_0^+), & a > 0 \quad \text{on } \mathbb{R}_0^+. \end{cases} \tag{5.41}$$

Fix $T > 0$. Applying Theorem 5.10 we infer the existence of w solving (5.28) for each $\eta > 0$ sufficiently small (inequality (5.27) is always satisfied for small η since, by (5.26), $f_\eta \to 0$ as $\eta \to 0$). From $w'(T) > 0$, we conclude that w can be extended on a maximal interval $[0, R)$. Our next task is to prove that, if the Keller–Osserman condition (KO_∞) is violated, then $R = \infty$.

Proposition 5.13 *Assume (5.2), (5.3), (5.26) and (5.40). Let $a, \wp$ satisfy (5.41). For a fixed $T > 0$, consider the solution w of (5.28) for small positive η and let $[0, R)$ be the maximal interval where w is defined. If t_0 is as in Theorem 5.10, then*

$$w = w(t_0) \quad \text{on } [0, t_0], \qquad w > 0, \quad w' > 0 \quad \text{on } (t_0, R). \tag{5.42}$$

Furthermore, suppose that f is C-increasing on $(\bar{\eta}_0, \infty)$ for some $\bar{\eta}_0 \geq 0$. If

$$\frac{1}{K^{-1} \circ F} \notin L^1(\infty), \qquad\qquad (\neg KO_\infty)$$

then $R = \infty$.

Proof Taking into account the sign of f, l, a, by (5.28) $\wp\varphi(w')$ is C^1 and strictly increasing where w is positive, and from (5.29) it readily follows that $\wp\varphi(w') > 0$ on (t_0, R). Properties (5.42) are then immediate from (5.29), $w'(T) > 0$ and our assumptions on φ. Next, suppose by contradiction that $R < \infty$. We first claim that necessarily

$$w^* = \lim_{t \to R^-} w(t) = \infty, \qquad\qquad (5.43)$$

where the existence of the limit is guaranteed by the monotonicity of w. To prove (5.43), assume by contradiction that $w^* < \infty$. Because of (5.10) and since $\wp' \geq 0$,

$$\frac{\varphi'(w')w''}{l(w')} \leq af(w) \leq a_1 f(w) \qquad \text{on } (t_0, R),$$

where we have set $a_1 = \|a\|_{L^\infty([0,R])}$. Multiplying by w', integrating on (t_0, t) and changing variables we deduce $K(w') \leq a_1 F(w)$ (we recall that $w'(t_0) = 0$ by (5.28)). Thus, w' is bounded in $(R/2, R)$, namely $\|w'\|_\infty \leq K^{-1}(a_1 F(w^*)) = L$. For $t, s \in (R/2, R)$, define I_k, x_k, y_k as in (5.35), and note that

$$|x_k| + |y_k| \leq 2a_1 f_{w^*} l_L R = \bar{C}.$$

Given $\varepsilon > 0$ let $\varrho = \varrho(\varphi^{-1}, \varepsilon)$ be given by the uniform continuity of φ^{-1} on $[0, \bar{C}]$. Proceeding as in (5.38) we deduce the existence of $\Lambda > 0$ such that

$$|x_k - y_k| \leq \Lambda |s - t| \qquad \text{for each } s, t \in \left(\frac{R}{2}, R\right).$$

If $|s - t| < \varrho/\Lambda$, then $|x_k - y_k| < \varrho$ and so

$$|w'(t) - w'(s)| = |\varphi^{-1}(x_k) - \varphi^{-1}(y_k)| < \varepsilon.$$

In conclusion, w' is uniformly continuous on $(R/2, R)$, and can be therefore extended by continuity at $t = R$. By the existence theory for ODEs, w would be further extendible beyond R, which is a contradiction. This proves that $w(R^-) = \infty$. Now, fix $T_1 \in (0, R)$ large enough that $w(T_1) > \bar{\eta}_0$. Applying K^{-1} to inequality $K(w') \leq a_1 F(w)$ on (T_1, R), rearranging, integrating on $[T_1, t)$ and changing variables we get

$$\int_{w(T_1)}^{w(t)} \frac{ds}{K^{-1}(a_1 F(s))} \leq t - T_1. \tag{5.44}$$

Since $(\neg KO_\infty)$ holds and, by assumption, f is C-increasing on $(\bar{\eta}_0, \infty)$, applying Lemma 5.6 we deduce that the left-hand side of (5.44) is unbounded as $t \to R^-$ while the right-hand side is not, contradiction. $\qquad\square$

Comparison Results and the Finite Maximum Principle

6

6.1 Basic Comparisons and a Pasting Lemma

In this section, we collect two comparison theorems and a "pasting lemma" for $\mathrm{Lip_{loc}}$ solutions that will be repeatedly used in the sequel. Throughout the section, we assume

$$\varphi \in C(\mathbb{R}_0^+), \qquad \varphi(0) = 0, \qquad \varphi > 0 \text{ on } \mathbb{R}^+. \tag{6.1}$$

The first comparison is Proposition 6.1 of [182], see also Theorem 2.4.1 of [194].

Proposition 6.1 *Assume* (6.1) *and that φ is strictly increasing on $\mathbb{R}^+$. Let $\Omega \subset M$ be open, and suppose that $u, v \in \mathrm{Lip_{loc}}(\Omega) \cap C(\overline{\Omega})$ solve*

$$\begin{cases} \Delta_\varphi u \geq \Delta_\varphi v & \text{weakly in } \Omega, \\ u \leq v & \text{on } \partial\Omega, \end{cases}$$

and $\displaystyle \limsup_{x \in \Omega,\, x \to \infty} \big(u(x) - v(x)\big) \leq 0$ *if Ω has non-compact closure. Then, $u \leq v$ in Ω.*

Our second comparison result is a special case of [11, Thm. 5.6], see also [194, Thm. 3.6.5].

© The Author(s), under exclusive license to Springer Nature Switzerland AG 2021
B. Bianchini et al., *Geometric Analysis of Quasilinear Inequalities on Complete Manifolds*, Frontiers in Mathematics, https://doi.org/10.1007/978-3-030-62704-1_6

Proposition 6.2 *Let φ, f, l satisfy*

$$\varphi \in C(\mathbb{R}_0^+) \cap C^1(\mathbb{R}^+), \qquad \varphi(0) = 0, \qquad \varphi' > 0 \quad \text{on } \mathbb{R}^+$$

$$f \in C(\mathbb{R}), \qquad f \text{ is non-decreasing on } \mathbb{R},$$

$$l \in C(\mathbb{R}_0^+) \cap \text{Lip}_{\text{loc}}(\mathbb{R}^+), \qquad l > 0 \text{ on } \mathbb{R}^+.$$

Let $\Omega \subset M$ be an open subset of a Riemannian manifold M, and fix $0 < b(x) \in C(\overline{\Omega})$. If $u, v \in \text{Lip}_{\text{loc}}(\Omega) \cap C(\overline{\Omega})$ solve

$$\begin{cases} \Delta_\varphi u \geq b(x) f(u) l(|\nabla u|) & \text{on } \Omega, \\[2mm] \Delta_\varphi v \leq b(x) f(v) l(|\nabla v|) & \text{on } \Omega, \\[2mm] u \leq v & \text{on } \partial\Omega \\[2mm] \text{ess inf}_K \left\{ |\nabla v| + |\nabla u| \right\} > 0 & \text{for each } K \Subset \Omega. \end{cases}$$

Then, $u \leq v$ on Ω.

Remark 6.3 Condition $\text{ess inf}_K \left\{ |\nabla v| + |\nabla u| \right\} > 0$ for each $K \Subset \Omega$ cannot be avoided, as the counterexample in Remark 1, p. 79 of [194] shows. However, the restriction can be removed if Δ_φ is strictly elliptic, see Section 3.5 of [194] for definitions and relevant results.

Remark 6.4 The underlying metric is not required to be smooth, and indeed a metric whose local matrix g_{ij} is continuous is sufficient.

Remark 6.5 It is worth to comment on [11, Thm. 5.6]. There, the authors consider solutions of more general quasilinear inequalities of the form

$$\text{div}\mathbf{A}(x, \nabla u) \geq \mathcal{B}(x, u, \nabla u) \qquad \text{and} \qquad \text{div}\mathbf{A}(x, \nabla v) \leq \mathcal{B}(x, v, \nabla v),$$

for suitable Caratheódory maps $\mathbf{A}$, $\mathcal{B}$. In our setting,

$$\mathbf{A}(x, \xi) = \frac{\varphi(|\xi|)}{|\xi|} \xi,$$

thus the positivity of $\varphi(s)/s$ and $\varphi'(s)$ on $\mathbb{R}^+$ implies that the tangent map $\mathbf{A}_*$ of $\mathbf{A}$ is uniformly positive definite on compacta of fibres of $T\Omega\backslash\{\mathbf{0}\} \to \Omega$, a condition needed to apply Lemma 5.7 in [11]. The regularity of $\mathcal{B}$, defined at the end of p. 592 therein, is equivalent to condition $l \in \text{Lip}_{\text{loc}}(\mathbb{R}^+)$.

To conclude, we discuss the pasting lemma. It is well known that the maximum of two subharmonic functions is still subharmonic. For subsolutions of more general operators, the situation is more delicate, and we refer to [141] for a very general result, and to [25, Appendix] for an alternative approach via obstacle problems, in the setting of homogeneous operators. For our purposes, it is sufficient to consider the case in which one of the solutions is constant. The technique goes back to T. Kato in [132] and has been generalized to a large class of quasilinear operators in [64]. The next result is special case of [64, Thm. 2.1].

Lemma 6.6 *Assume* (6.1), *and let* $f \in C(\mathbb{R})$, $l \in C(\mathbb{R}_0^+)$ *with* $f(0)l(0) = 0$. *Suppose furthermore than* $b \in C(M)$. *If* $u \in \mathrm{Lip}_{\mathrm{loc}}(M)$ *is a nontrivial solution of*

$$\Delta_\varphi u \geq b(x)f(u)l(|\nabla u|) \qquad on\ M, \tag{6.2}$$

then $u_+ = \max\{u, 0\}$ *is a* $\mathrm{Lip}_{\mathrm{loc}}(M)$, *non-negative solution of*

$$\Delta_\varphi u_+ \geq b(x)f(u_+)l(|\nabla u_+|) \qquad on\ M.$$

Remark 6.7 Note that $u = 0$ is a solution of (6.2) since $f(0)l(0) = 0$.

6.2   The Finite Maximum Principle

We now prove Theorem 2.13 in the Introduction. For the convenience of the reader, we rewrite the assumptions and restate the result. We require

$$\begin{cases} \varphi \in C(\mathbb{R}_0^+) \cap C^1(\mathbb{R}^+), & \varphi(0) = 0, \qquad \varphi' > 0 \text{ on } \mathbb{R}^+; \\[2mm] \dfrac{t\varphi'(t)}{l(t)} \in L^1(0^+) \\[2mm] f \in C(\mathbb{R}), \qquad f \geq 0 \quad \text{on } (0, \eta_0); \\[2mm] l \in C(\mathbb{R}_0^+), \qquad l > 0 \text{ on } \mathbb{R}^+. \end{cases} \tag{6.3}$$

Theorem 6.8 *Let* M *be a Riemannian manifold, and assume that* φ, f *and* l *satisfy* (6.3), *and moreover,*

- $f(0)l(0) = 0$;
- f *is C-increasing on* $(0, \eta_0)$, *with* η_0 *as in* (6.3);
- l *is C-increasing on* $(0, \xi_0)$, *for some* $\xi_0 > 0$.

Fix a domain $\Omega \subset M$, and let $0 < b \in C(\Omega)$. Then, (FMP) holds on Ω if and only if either

$$f \equiv 0 \qquad on \ (0, \eta_0), \tag{6.4}$$

for some $\eta_0 > 0$, or

$$f > 0 \quad on \ (0, \eta_0), \quad and \quad \frac{1}{K^{-1} \circ F} \notin L^1(0^+). \tag{6.5}$$

Proof We recall that the validity of (FMP) means that for any solution $u \in C^1(\Omega)$ of

$$\begin{cases} \Delta_\varphi u \leq b(x) f(u) l(|\nabla u|) & \text{on } \Omega \\ u \geq 0 & \text{on } \Omega, \end{cases} \tag{6.6}$$

if $u(x_0) = 0$ for some $x_0 \in \Omega$, then $u \equiv 0$. The argument follows the lines of the proof in [194], with the help of a trick from [157]. We prove separately the sufficiency and necessity of (6.4), (6.5). First, having fixed a point $o \in \Omega$ to be specified later, we choose $R, \kappa > 0$ such that $\overline{B_{2R}(o)}$ does not intersect cut(o) and

$$\text{Ric}(\nabla r, \nabla r) \geq -(m-1)\kappa^2 \qquad \text{on } \overline{B_{2R}(o)} \backslash \{o\},$$

where $r(x) = \text{dist}(x, o)$. By the Laplacian comparison theorem (see Theorem 3.8), denoting with $v_{g_\kappa}(r)$ the volume of a geodesic sphere of radius r in a model manifold of sectional curvature $-\kappa^2$

$$\Delta r \leq \frac{v'_{g_\kappa}(r)}{v_{g_\kappa}(r)} = (m-1)\kappa \coth(\kappa r) \qquad \text{on } \overline{B_{2R}(o)} \backslash \{o\}. \tag{6.7}$$

Sufficiency of conditions (6.4) and (6.5).

For $o \in \Omega$ and having set R, κ as above, fix $a_1 \in \mathbb{R}^+$ such that $b(x) \leq a_1$ on $B_{2R}(o)$. Let $C \geq 1$ be the constant defining the C-increasing property of f. Define $\wp(t) = v_{g_\kappa}(2R - t)$, $a(t) = a_1$, $T_0 = 3R/2$ and $T = R$. We claim that we can suitably reduce R and choose $\eta \in (0, \eta_0)$ small enough, in such a way that

$$\frac{\wp_1}{\wp_0} \varphi\left(\frac{\eta}{R}\right) + 4C\Theta(R) f_\eta l_\xi < \varphi(\xi), \tag{6.8}$$

where f_η, l_ξ, $\wp_0$, $\wp_1$ and Θ are defined as in (5.6). Indeed, since $v'_{g_\kappa} \geq 0$, by definition $\wp_1 = v_{g_\kappa}(2R)$, $\wp_0 = v_{g_\kappa}(R/2)$, and thus $\wp_1/\wp_0 \to 4^m$ and $\Theta(R) \to 0$ as $R \to 0$. Hence, we can first reduce R to guarantee

$$4C\Theta(R) f_{\eta_0} l_\xi < \frac{\varphi(\xi)}{2}$$

and then choose $\eta \in (0, \eta_0)$ small enough to satisfy (6.8). Applying Theorem 5.1, there exists a solution $z(t)$ of

$$\begin{cases} \left(\wp(t)\varphi(z_t)\right)_t = 2C\wp a_1 f(z)l(z_t) & \text{on } (0, R), \\ z(0) = 0, \qquad z(R) = \eta, \end{cases}$$

where the subscript t indicates differentiation with respect to t. Furthermore, combining Theorem 5.1 and Proposition 5.7 (note that here $\wp' < 0$), the solution z satisfies the following properties:

$$z > 0 \text{ on } (0, R], \qquad z_t > 0 \text{ on } [0, R], \qquad \|z_t\|_\infty < \xi.$$

Taking into account that we have extended φ on $\mathbb{R}$ in such a way that $\varphi(-s) = -\varphi(s)$, the function $w(r) = z(2R - r)$ satisfies

$$\begin{cases} [v_g\varphi(w')]' = 2Cv_g a_1 f(w)l(|w'|) & \text{on } (R, 2R), \\ w(2R) = 0, \quad w(R) = \eta, \quad w' < 0 \text{ on } [R, 2R]. \end{cases} \tag{6.9}$$

Define $v(x) = w(r(x))$. Using (6.7), together with $\varphi(w') < 0$ and (6.9), we deduce that v solves

$$\Delta_\varphi v = \left(\varphi(w')\right)' + \varphi(w')\Delta r \geq v_g^{-1}(r)\left[v_g(r)\varphi(w')\right]'$$

$$= v_g(r)^{-1}\left(2Cv_g(r)a_1 f(w)l(|w'|) \geq 2Cb(x)f(v)l(|\nabla v|),$$

on the annulus $E_R(o) = B_{2R}(o)\backslash \overline{B_R(o)}$. Moreover, denoting with ν the outward pointing unit normal from $\partial B_{2R}(o)$,

$$\langle \nabla v, \nu \rangle = w'(2R) < 0 \qquad \text{on } \partial B_{2R}(o). \tag{6.10}$$

Following E. Hopf's argument, we now prove that u solving (6.6) is identically zero provided that $u(x_0) = 0$ at some x_0. Suppose by contradiction that this is not the case, and let $\Omega_+ = \{x \in \Omega : u(x) > 0\}$. Choose $x_1 \in \Omega_+$ in such a way that $\text{dist}(x_1, \partial\Omega_+) > \text{dist}(x_1, \partial\Omega)$, and let $B(x_1) \subset \Omega_+$ be the largest ball contained in Ω_+. Then, $u > 0$ in $B(x_1)$, while $u(\bar{x}) = 0$ for some $\bar{x} \in \partial B(x_1) \cap \Omega$. Clearly, $\nabla u(\bar{x}) = 0$, since $\bar{x}$ is an absolute minimum for u. Take a unit speed minimizing geodesic $\gamma : [0, \text{dist}(x_1, \bar{x})] \to \Omega$ from x_1 to $\bar{x}$. Up to choosing the arclength parameter s sufficiently close to $\text{dist}(x_1, \bar{x})$ and setting $2R = \text{dist}(x_1, \bar{x}) - s$, the closure of the ball $B_{2R}(o)$ centred at $o = \gamma(s)$ does not intersect $\text{cut}(o)$, and $B_{2R}(o) \subset B(x_1)$, with $\bar{x} \in \partial B_{2R}(o)$. We consider the function v

constructed above on $E_R(o) \subset B_{2R}(o)$, with η small enough to satisfy (6.8) and also

$$\eta < \inf_{\partial B_R(o)} u.$$

We claim that $u \geq v$ on $E_R(o)$. Otherwise, suppose that

$$\max_{E_R(o)} (v - u) = \bar{\delta} > 0,$$

and let Γ be the set of maximum points of $v - u$. Note that $\Gamma \Subset E_R(o)$ and that $\nabla u = \nabla v$ for each $x \in \Gamma$. For $\delta \in (0, \bar{\delta})$, set $U_\delta = \{v - u > \delta\}$. By construction, there exists $\varepsilon > 0$ such that $\varepsilon \leq |\nabla v| \leq 1$ on $E_R(o)$, and since $l > 0$ on $\mathbb{R}^+$, we deduce that the quotient $l(|\nabla u|)/l(|\nabla v|)$ is continuous on $E_R(o)$ and equal to 1 on Γ. A compactness argument shows that, for δ sufficiently close to $\bar{\delta}$, $l(|\nabla u|)/l(|\nabla v|) \leq 2$ on U_δ. Taking into account that the C-increasing property of f implies $f(v) \geq C^{-1} f(u)$ on U_δ, we deduce

$$\Delta_\varphi u \leq b(x) f(u) l(|\nabla u|) \leq 2Cb(x) f(v) l(|\nabla v|) \leq \Delta_\varphi v \qquad \text{on } U_\delta.$$

From $v = u + \delta$ on ∂U_δ, by the comparison Proposition 6.1, we get $v \leq u + \delta$ on U_δ, contradicting the very definition of U_δ. Hence, $v \leq u$ on $E_R(o)$, and in particular, $\langle \nabla(u - v), v \rangle \leq 0$ at $\bar{x}$, which is impossible by (6.10) and by the fact that $\nabla u(\bar{x}) = 0$. This contradiction concludes the proof of the sufficiency part.

Necessity of conditions (6.4) *and* (6.5).

Suppose the failure of both (6.4) and (6.5), or equivalently (recall that f is C-increasing) the validity of

$$f > 0 \quad \text{on } (0, \eta_0), \qquad \frac{1}{K^{-1} \circ F} \in L^1(0^+). \tag{KO$_0$}$$

For each $o \in M$, we shall now construct on $B_{2R}(o)$ (with R as in the beginning of the proof) a C^1 non-negative, nonzero solution u of $(P_\leq)$ with $u = 0$ on $\overline{B_R(o)}$, contradicting (FMP). Set $T_0 = 2R$, $T = R$,

$$\wp(t) = v_{g_\kappa}(t + R), \qquad a(t) = \inf_{B_{2R}(o) \setminus B_R(o)} b,$$

where $v_{g_\kappa}(r)$ is the volume of a geodesic sphere of radius r in a model of curvature $-\kappa^2$, as defined at the beginning of the proof. Since $f(0)l(0) = 0$, we can choose η small enough to satisfy (5.7) in Theorem 5.1, whence there exists $w \in C^1([0, T])$ non-decreasing and solving

$$
\begin{cases}
[\wp\varphi(w')]' = \wp af(w)l(w') & \text{on } (0, T), \\
0 \le w \le \eta, \quad w' \ge 0 & \text{on } [0, T], \\
w(0) = 0, \qquad w(T) = \eta.
\end{cases}
\tag{6.11}
$$

Up to reducing η further, we can apply Proposition 5.9 to deduce that $w'(0) = 0$. Set $u(x) = w(r(x) - R)$. By the Laplacian comparison theorem and since $w' \ge 0$, on $B_{2R}(o) \backslash B_R(o)$, it holds

$$
\Delta_\varphi u = \big(\varphi(w')\big)' + \varphi(w')\Delta r \le \big(\varphi(w')\big)' + \varphi(w')\frac{v'_{g_\kappa}(r)}{v_{g_\kappa}(r)}
$$

$$
\le \Big(\wp^{-1}[\wp\varphi(w')]'\Big)_{r(x)-R} \le af(w)l(w')
$$

$$
\le b(x)f(u)l(|\nabla u|).
$$

Extending u to be zero on $B_R(o)$ defines a nonzero, C^1-solution of $(P_\le)$ on all of $B_{2R}(o)$, which clearly violates (FMP). $\qquad\square$

Remark 6.9 The function u in the proof of the necessity part, defined on $B_{2R}(o)$ and vanishing identically on a smaller ball $B_R(o)$, is an example of a *dead core* (super) solution. For a thorough investigation of dead core problems, we refer the reader to [194] and the references therein. We mention that u can even be constructed to be positive on $B_{2R}(o) \backslash \overline{B_R(o)}$. Indeed, it is enough to replace the solution w of (6.11) with the supersolution z defined in the proof of Proposition 5.9, which is known to be positive on $(0, T]$.

Remark 6.10 With a similar technique, one could consider the (FMP) for more general equations of the type

$$
\Delta_\varphi u \le b(x)f(u)l(|\nabla u|) + \bar{b}(x)\bar{f}(u)\bar{l}(|\nabla u|),
$$

for suitable $\bar{b}, \bar{f}, \bar{l}$. The prototype case

$$
\Delta_p u \le f(u) + |\nabla u|^q, \qquad \text{with } p > 1
$$

has been considered in [194, Thm. 5.4.1] and [196] when $q \ge p - 1$, and in [88] for $q < p - 1$.

Weak Maximum Principle and Liouville's Property 7

7.1 The Equivalence Between (WMP_∞) and (L)

In this section, we prove a more general version of Proposition 2.19 that describes the relationship between (WMP_∞) and the Liouville property (L). We begin by introducing another form of the weak maximum principle.

Definition 7.1 Assume (2.3) and fix b, l satisfying (2.5). We say that

- the *open weak maximum principle* (OWMP_∞) holds for $(bl)^{-1}\Delta_\varphi$ if, for each $f \in C(\mathbb{R})$, for each open set $\Omega \subset M$ with $\partial\Omega \neq \emptyset$ and for each $u \in C(\overline{\Omega}) \cap \mathrm{Lip}_{\mathrm{loc}}(\Omega)$ solving

$$
\begin{cases}
\Delta_\varphi u \geq b(x) f(u) l(|\nabla u|) & \text{on } \Omega, \\
\sup_\Omega u < \infty,
\end{cases}
\tag{7.1}
$$

we have that

$$
\text{either} \quad \sup_\Omega u = \sup_{\partial\Omega} u, \quad \text{or} \quad f\left(\sup_\Omega u\right) \leq 0.
\tag{7.2}
$$

The open weak maximum principle at infinity has been introduced in [7] in the study of immersed submanifolds of warped product ambient spaces and parallels Ahlfors' definition of parabolicity. For a detailed investigation and an extensive bibliography, we refer to Chapters 3 and 4 of [6].

Remark 7.2 The recent [154] contains a different approach to maximum principles at infinity in the spirit of (OWMP$_\infty$), called there *the Ahlfors' property*. The approach is based on viscosity theory and enables to consider classes of fully nonlinear operators of geometric interest. The use of viscosity solutions is very well suited to treat weak and strong maximum principles in a unified way, and especially to investigate principles like the classical Ekeland's quasi-maximum principle or (SMP$_\infty$), where a gradient condition on u appears.

Remark 7.3 Companion to the above form of (WMP$_\infty$) on sets with boundary, there is a notion of parabolicity for manifolds with boundary. It efficiently applies, for instance, to study minimal graphs defined on unbounded domains of a complete manifold. The interested reader is suggested to consult [128] for parabolicity and [3, 7] for (OWMP$_\infty$).

Proposition 7.4 *Let φ and b, f, l satisfy, respectively, the assumptions in (2.3) and (2.5). Then, the following properties are equivalent:*

(i) $(bl)^{-1}\Delta_\varphi$ satisfies (WMP$_\infty$);
(ii) (L) holds for $\mathrm{Lip}_{\mathrm{loc}}$ solutions, for some (equivalently, every) f satisfying

$$f(0) = 0, \qquad f > 0 \quad on \ \mathbb{R}^+;$$

(iii) each non-constant $u \in \mathrm{Lip}_{\mathrm{loc}}(M)$ solving $(P_\geq)$ on M and bounded above satisfies $f(u^*) \leq 0$, with $u^* = \sup_M u$;
(iv) $(bl)^{-1}\Delta_\varphi$ satisfies (OWMP$_\infty$).

Proof We prove the chain of implications $(i) \Rightarrow (iii) \Rightarrow (ii) \Rightarrow (i)$ and then $(iv) \Leftrightarrow (i)$. When the "some-every" alternative occurs, we always assume the weaker property and prove the stronger.
$(i) \Rightarrow (iii)$.
Let $u \in \mathrm{Lip}_{\mathrm{loc}}(M)$ be a non-constant solution of $(P_\geq)$ that is bounded from above, and assume by contradiction that $f(u^*) = 2K > 0$. By continuity, there exists $\eta < u^*$ sufficiently close to u^* such that $f(u) \geq K$ on $\Omega_\eta = \{u > \eta\}$, thus

$$\Delta_\varphi u \geq Kb(x)l(|\nabla u|) \qquad on \ \Omega_\eta.$$

The definition of (WMP$_\infty$) then implies $K \leq 0$, contradiction.
$(iii) \Rightarrow (ii)$.
Let $u \in \mathrm{Lip}_{\mathrm{loc}}(M)$ be a non-constant, bounded, non-negative solution of $(P_\geq)$. Then, $f(u^*) \leq 0$ by (iii). However, from $f > 0$ on $\mathbb{R}^+$, we get $u^* \equiv 0$, that is, $u \equiv 0$ is constant, contradiction.

$(ii) \Rightarrow (i)$.

Let us consider a problem $(P_\geq)$ with $f > 0$ on $\mathbb{R}^+$, $f(0) = 0$ for which (L) holds. Suppose by contradiction that (i) is not satisfied, that is, there exists a non-constant $u \in \mathrm{Lip}_{\mathrm{loc}}(M)$ with

$$\Delta_\varphi u \geq K b(x) l(|\nabla u|) \qquad \text{on some } \Omega_{\bar\eta},$$

for some $K > 0$ and $\bar\eta < u^*$. Since $f(0) = 0$, we can choose $\eta \in (\bar\eta, u^*)$ in such a way that $f < K$ on $(0, u^* - \eta)$. Hence, $u_\eta = u - \eta$ solves

$$\Delta_\varphi u_\eta \geq K b(x) l(|\nabla u_\eta|) \geq b(x) f(u_\eta) l(|\nabla u_\eta|) \qquad \text{on } \Omega_\eta.$$

Thanks to Lemma 6.6, $w = \max\{u_\eta, 0\}$ is a non-constant, non-negative, bounded solution of $\Delta_\varphi w \geq b(x) f(w) l(|\nabla w|)$ on M, contradicting property (L) for such an f.

$(i) \Rightarrow (iv)$.

If u solves (7.1), but none of the properties in (7.2) holds, then by continuity we can choose $\eta \in (\sup_{\partial\Omega} u, \sup_\Omega u)$ such that $f(u) \geq K > 0$ on $\{u > \eta\}$. Thus, u solves $\Delta_\varphi u \geq K b(x) l(|\nabla u|)$ on $\{u > \eta\} \subset \Omega$, contradicting (WMP$_\infty$).

$(iv) \Rightarrow (i)$.

Assuming that (WMP$_\infty$) does not hold, take a non-constant u that is bounded above and solves $\Delta_\varphi u \geq K b(x) l(|\nabla u|)$ on some set $\Omega_\eta = \{u > \eta\}$, for some $K > 0$. If η is close enough to u^*, $\partial\Omega_\eta \neq \emptyset$; thus, clearly u contradicts (OWMP$_\infty$) on Ω_η with the choice $f = K$.   □

7.2   Volume Growth and (WMP$_\infty$)

In this section, we explore geometric conditions that ensure the validity of (WMP$_\infty$), in the form given by (iii) of Proposition 2.19, that is, each non-constant $u \in \mathrm{Lip}_{\mathrm{loc}}(M)$ solving $(P_\geq)$ and bounded above satisfies $f(u^*) \leq 0$. Here, as usual, $u^* = \sup_M u$, and similarly, we will use the notation $u_* = \inf_M u$.

When $l \equiv 1$, the problem has been tackled in a series of papers [131, 181, 182, 207] by means of integral methods, and in particular, we refer to [182] for a thorough discussion. Since then, in a manifolds setting, the first results that we are aware of allowing a nontrivial l appeared in [157], where l is assumed to be C-increasing (in fact, polynomial in Thm. 5.1 therein). In particular, the relevant case when l can vanish both as $t \to 0^+$ and as $t \to \infty$ seems to be still open even in Euclidean space, although it has been recently considered for Carnot groups in [5]. It is a remarkable feature that the results quoted above ensure $f(u^*) \leq 0$ not only when u is a priori bounded above, but also when u does not grow too fast at infinity, in the sense that in this case u^* is also shown to be finite. This is a natural condition, and its origin is related to the growth of an explicit Khas'minskii-type potential for the operator considered, see Section 4 of [182] for more information. Related interesting results on $\mathbb{R}^m$ can be found in [63, 86, 87, 195] and will be described in more detail later.

The next theorem improves on [157, Thm 5.1] and [5, Thm. 2.1]. Throughout this section, we assume the following growth conditions:

there exist constants $p, \bar{p} > 1$, $C, \bar{C} > 0$ such that

$$\varphi(t) \le Ct^{p-1} \text{ on } [0, 1], \qquad \varphi(t) \le \bar{C}t^{\bar{p}-1} \text{ on } [1, \infty). \tag{7.3}$$

If $p = \bar{p}$, (7.3) is called in [63] the *weak p-coercivity* of Δ_φ. We will explain in depth the different roles played by p and $\bar{p}$ for the validity of (WMP$_\infty$).

Theorem 7.5 *Let M be a complete Riemannian manifold, and consider φ, b, f and l meeting the requirements in (2.5), (2.3) and the bounds (7.3) for some $p, \bar{p} > 1$. Assume that, for some $\mu, \chi \in \mathbb{R}$ satisfying*

$$\chi \ge 0, \qquad \mu \le \chi + 1, \tag{7.4}$$

the following inequalities hold:

$$
\begin{aligned}
b(x) &\ge C(1 + r(x))^{-\mu} &&\text{on } M, \\
f(t) &\ge C &&\text{for } t \gg 1, \\
l(t) &\ge C\frac{\varphi(t)}{t^\chi} &&\text{on } \mathbb{R}^+,
\end{aligned}
\tag{7.5}
$$

for some constant $C > 0$. Let $u \in \mathrm{Lip}_{\mathrm{loc}}(M)$ be a non-constant, weak solution of $(P_\ge)$ such that either

(i) u is bounded above and one of the following properties hold:

$$
\begin{aligned}
\mu < \chi + 1 \text{ and } &\quad \liminf_{r \to \infty} \frac{\log|B_r|}{r^{\chi+1-\mu}} < \infty \quad (= 0 \text{ if } \chi = 0); \\
\mu = \chi + 1 \text{ and } &\quad \liminf_{r \to \infty} \frac{\log|B_r|}{\log r} < \infty \quad (\le p \text{ if } \chi = 0).
\end{aligned}
\tag{7.6}
$$

(ii) u satisfies

$$u_+(x) = o(r(x)^\sigma) \qquad \text{as } r(x) \to \infty, \tag{7.7}$$

for some $\sigma > 0$ such that

$$\chi\sigma \le \chi + 1 - \mu, \tag{7.8}$$

and one of the following properties hold:

$$\chi\sigma < \chi + 1 - \mu, \quad \liminf_{r\to\infty} \frac{\log|B_r|}{r^{\chi+1-\mu-\chi\sigma}} < \infty \ (= 0 \text{ if } \chi = 0);$$

$$\chi\sigma = \chi + 1 - \mu, \ \chi > 0, \quad \liminf_{r\to\infty} \frac{\log|B_r|}{\log r} < \infty; \tag{7.9}$$

$$\chi\sigma = \chi + 1 - \mu, \quad \liminf_{r\to\infty} \frac{\log|B_r|}{\log r} \le \begin{cases} p - \sigma(p-1) \text{ if } \sigma \le 1, \\ \bar{p} - \sigma(\bar{p}-1) \text{ if } \sigma > 1. \end{cases}$$

Then, u is bounded above on M and $f(u^*) \le 0$. If moreover u satisfies $(P_=)$,

$$t f(t) \ge C|t| \qquad \text{for } |t| >> 1, \tag{7.10}$$

and either (i) or (ii) holds both for u_+ and for u_-, then

$$u \in L^\infty(M) \qquad \text{and} \qquad f(u^*) \le 0 \le f(u_*). \tag{7.11}$$

Remark 7.6 In the third case of (7.9), that is, when $\chi\sigma = \chi + 1 - \mu$ and the volume growth of B_r is suitably small with respect to $p, \bar{p}, \sigma$, the result still holds under the weaker assumption

$$u_+(x) = O\big(r(x)^\sigma\big) \qquad \text{as } r(x) \to \infty. \tag{7.12}$$

As a consequence of Theorem 7.5, we deduce Theorem 2.22 of the Introduction.

Proof of Theorem 2.22 Consider a non-constant solution $u \in \mathrm{Lip}_{\mathrm{loc}}(M)$ with $u^* < \infty$ of $\Delta_\varphi u \ge K b(x) l(|\nabla u|)$ on some upper level set $\Omega_\eta = \{x \in M : u(x) > \eta\}$. We shall prove that $K \le 0$. Suppose that this is not the case. By adding a constant to u, we can suppose that $\eta < 0$ but sufficiently near to 0 so that $\Omega_0 = \{x \in M : u(x) > 0\} \ne \emptyset$. Choose $f \in C(\mathbb{R})$ such that $0 \le f \le K$, $f(0) = 0$ and $f(t) = K$ for $t > u^*/2$. Then, u solves

$$\Delta_\varphi u \ge b(x) f(u) l(|\nabla u|) \tag{7.13}$$

on $\Omega_\eta \supset \Omega_0$, and with the aid of Lemma 6.6, we can assume that $u \ge 0$ solves (7.13) on the whole of M. Moreover, $f(u^*) = K$. To reach a contradiction, we just need to check the requirements to apply Theorem 7.5, case (i) and conclude $f(u^*) \le 0$. This, by Proposition 2.19, implies the (WMP$_\infty$).

First, observe that $\mu \leq \chi - \alpha/2$ in (2.29) can be rewritten as

$$1 + \frac{\alpha}{2} \leq \chi + 1 - \mu, \tag{7.14}$$

and from $\alpha \geq -2$, it implies $\mu \leq \chi + 1$. We examine the validity of (7.6). If $\alpha > -2$ and $\chi > 0$, then $\mu < \chi + 1$ by (7.14), and the volume assumption (2.28) implies the first in (7.6). If $\alpha = -2$ and $\chi > 0$, then again by (2.28) both of (7.6) are met, respectively, when $\mu < \chi + 1$ and $\mu = \chi + 1$. Suppose now that $\chi = 0$ and $\mu < \chi - \alpha/2$. Then, $\mu < \chi + 1$, and for each $\alpha \geq -2$, the strict inequality in (7.14) coupled with (2.28) guarantees the first in (7.6). If $\chi = 0$ and $\mu = \chi - \alpha/2$, according to whether $\alpha > -2$ or $\alpha = -2$, the requirement $V_\infty = 0$, respectively, $V_\infty \leq p$, in (2.29) is precisely what is needed to deduce the validity of (7.6), concluding the proof.   □

It is worth to postpone the proof of Theorem 7.5 and comment on various aspects of its statement.

- $p, \bar{p}$ play no explicit role in (7.4). However, a bound on χ in terms of p alone is hidden, in some cases, in the requirement $l(t) \geq \varphi(t)/t^\chi$ on $\mathbb{R}^+$: indeed, if $\varphi(t) \asymp t^{p-1}$ near $t = 0$, the continuity of l at zero forces

$$\chi \leq p - 1.$$

- For $l \equiv 1$, that is, when no gradient appears, the third in (7.5) forces $\varphi(t) \leq C^{-1} t^\chi$ on $\mathbb{R}^+$. If we suppose that $\varphi(t) \asymp t^{p-1}$ on $[0, 1]$ and $\varphi(t) \asymp t^{\bar{p}-1}$ on $[1, \infty)$, the above theorem can be applied provided

$$\bar{p} - 1 \leq \chi \leq p - 1.$$

Therefore, when the operator is the p-Laplacian, the gradientless case $l \equiv 1$ is recovered with the choice $\chi = p - 1$. On the other hand, for the mean curvature operator, $p \leq 2$ and $\bar{p} > 1$ can be chosen arbitrarily close to 1, and the gradientless case $l \equiv 1$ is recovered for any choice of $\chi \in [0, 1]$. For a fixed μ, in case (i) or in (ii) with $\sigma \leq 1$, it is evident that the choice $\chi = 1$, $p = 2$ gives the best result, while in case (ii) for $\sigma > 1$ the best choice is $\chi = 0$ and $\bar{p}$ approaching 1.
- The second in (7.9) is the only place where p and $\bar{p}$ appear. Its validity forces an upper bound for σ, since the right-hand side in the second of (7.9) is non-negative if and only if

$$\sigma \leq \frac{\bar{p}}{\bar{p} - 1}.$$

- The third in (7.9) supports and makes rigorous the next idea: in the sublinear range $\sigma < 1$, the region where $|\nabla u|$ is close to zero should be, somehow, larger than the one where $|\nabla u|$ is big, and consequently, the growth of φ on $[1, \infty)$ (if still polynomial) might be neglectable with respect to the behaviour of φ on $[0, 1]$. If $\sigma > 1$, the situation reverses, and now the main contribution should be given by φ on $[1, \infty)$.

  We feel interesting and a bit surprising that the method to prove Theorem 7.5 is able to detect, in some sense, the size of the regions where $|\nabla u|$ is small or large. In particular, if u is bounded above, this might suggest that the above proof could be refined to show that, under the same assumptions, the *strong* maximum principle (SMP$_\infty$) is true, see Problem 2 in the Introduction.

As we will show at the end of the section, Theorem 7.5 is sharp in the following sense: under the validity of the range of the parameters χ, μ in (7.4), for almost each condition on σ and $|B_r|$, we are able to find a non-constant solution of $(P_\geq)$ with $f \equiv 1$ satisfying all the remaining assumptions but the chosen one.

If f has a unique zero, from (7.11), we deduce a Liouville type theorem for slowly growing solutions u of $(P_=)$ that fits very well with some results obtained, in the Euclidean setting, by Farina and Serrin in [86,87]. For the sake of comparison, we state their theorems renaming their parameters to agree with our notation:

Theorem 7.7 ([87, Thms. 11 and 12]) *On Euclidean space, consider φ, b, f and l meeting the requirements in (2.5), (2.3) and the bounds (7.3) for some $p = \bar{p} > 1$. Assume that, for some $\mu, \chi, \omega \in \mathbb{R}$ satisfying*

$$0 < \omega \leq \chi \leq p - 1,$$

the following inequalities hold:

$$b(x) \geq C\left(1 + r(x)\right)^{-\mu} \quad \text{on } \mathbb{R}^m,$$
$$tf(t) \geq C|t|^{\omega+1} \qquad \text{for } t \in \mathbb{R}, \tag{7.15}$$
$$l(t) \geq Ct^{p-1-\chi} \qquad \text{on } \mathbb{R}^+,$$

for some constant $C > 0$. Then, a solution $u \in \mathrm{Lip}_{\mathrm{loc}}(\mathbb{R}^m)$ of $(P_=)$ satisfying

$$|u(x)| = O\left(r(x)^\sigma\right) \qquad \text{as } r(x) \to \infty,$$

for some σ > 0, is constant provided one of the following cases occur:

(i) $\omega < \chi$, $\mu < \chi + 1$,

$$(p-1)(1+\chi-\mu) \geq (p-m)(\chi-\omega) \quad and \quad \sigma \in \left(0, \frac{\chi+1-\mu}{\chi-\omega}\right); \qquad (7.16)$$

(ii) $\omega < \chi$, $m < p$,

$$(p-1)(1+\chi-\mu) \leq (p-m)(\chi-\omega) \quad and \quad \sigma \in \left(0, \frac{p-m}{p-1}\right); \qquad (7.17)$$

(iii) $\omega = \chi$, $\mu < \chi + 1$, *independently of* $\sigma > 0$,
(iv) $\omega = \chi$, $m < p$,

$$\mu \geq \chi + 1 \quad and \quad \sigma \in \left(0, \frac{p-m}{p-1}\right).$$

Remark 7.8 Note that the bounds in (i) and (ii) well match when equality holds in the first of (7.16) and (7.17). If $l \equiv 1$, that is, $\chi = p - 1$, (i) and (iii) have been proved, respectively, in Theorems B and A in [86], and their sharpness is discussed in Examples 5 and 4, Section 11 therein. It is interesting to observe that the first in (7.16) is not required in [86, Thm. B] but appears in discussing the sharpness of (i). More precisely, Example 5 in [86] stresses that the conclusion of Theorem 7.7 in case (i) does not hold when $\sigma = \frac{\chi+1-\mu}{\chi-\omega}$ in (7.16), provided that the first in (7.16) is strengthened to

$$(p-1)(1+\chi-\mu) > (p-m)(\chi-\omega) \qquad with \ \chi = p-1. \qquad (7.18)$$

On the contrary, perhaps surprisingly, the conclusion of (ii) still holds for $\sigma = \frac{p-m}{p-1}$, see the next Corollary 7.9. This is, clearly, not in contradiction with [86] in view of the incompatibility of (7.18) with (7.17), see also the discussion [195, p. 677]. Further results for solutions of ($P_{\geq}$) that are a priori bounded or vanishing at infinity are given in Theorems D, E and F in [86] and Theorems 1 and 2 in [219]. Inspection shows that they fit very well with the case when u is bounded above in Theorem 7.5.

First, we compare Theorem 7.5 with case (i) in Theorem 7.7, and we therefore assume $M = \mathbb{R}^m$, $p = \bar{p}$ in (7.3), $0 < \chi \leq p - 1$ and $\mu < \chi + 1$. It is apparent that, for each $\omega > 0$, condition $tf(t) \geq C|t|^{\omega+1}$ implies both (7.10) and $tf(t) > 0$ on $\mathbb{R}^+$. Theorem 7.5 then gives the constancy of solutions of ($P_=$) on $\mathbb{R}^m$ under the assumption

$$u(x) = o\left(r(x)^\sigma\right) \quad as \ r(x) \to \infty \quad and \quad \sigma \in \left[0, \frac{\chi+1-\mu}{\chi}\right), \qquad (7.19)$$

for any dimension m and any $p > 1$. The upper bound for σ is smaller than the one in (7.16), as a counterpart of the stronger requirement $tf(t) \geq C|t|^{\omega+1}$, but (7.16) converges to (7.19) as $\omega \to 0^+$. Moreover, the first in (7.16) is not needed, in accordance with [86, Thm. B] and [219, Thm 2 (i)]. Next, we investigate the relationship with (iii), (iv) in Theorem 7.7, where $\omega = \chi$ is assumed. Since in our case $\omega = 0$, (iii) and (iv) should be compared with case $\chi = 0$ of Theorem 7.5. Observe that $\chi = 0$ includes an interesting class of borderline inequalities such as

$$\Delta_p u \geq b(x) f(u) |\nabla u|^{p-1}, \tag{7.20}$$

for which we have the next corollary; since, for $\chi = 0$, any $\sigma > 0$ satisfies (7.8), when $\mu < 1$, we obtain a Liouville theorem for solutions u with polynomial growth (i.e. satisfying $|u| = O(r^\sigma)$ as $r \to \infty$ for some $\sigma > 0$).

Corollary 7.9 *Let M be a complete Riemannian manifold, and consider φ, b, f meeting the requirements in (2.5), (2.3) and the bounds (7.3) for some $p, \bar{p} > 1$. Assume that, for some $\mu \leq 1$, the following inequalities hold:*

$$\begin{aligned} b(x) &\geq C \big(1 + r(x)\big)^{-\mu} && \text{on } M, \\ f(t) &\geq C && \text{for } t \gg 1 \end{aligned} \tag{7.21}$$

and for some constant $C > 0$. Let $u \in \mathrm{Lip}_{\mathrm{loc}}(M)$ be a non-constant, weak solution of

$$\Delta_\varphi u \geq b(x) f(u) \varphi(|\nabla u|) \qquad \text{on } M. \tag{7.22}$$

Suppose that $u_+(x) = O\big(r(x)^\sigma\big)$, for some $\sigma > 0$, and that

(1) $\mu < 1$, $\displaystyle\liminf_{r \to \infty} \frac{\log |B_r|}{r^{1-\mu}} = 0$ *or*

(2) $\mu = 1$, $\displaystyle\liminf_{r \to \infty} \frac{\log |B_r|}{\log r} = d_0$, *and* $0 < \sigma \leq \begin{cases} \dfrac{p - d_0}{p - 1}, & \text{if } d_0 \geq 1, \\[2mm] \dfrac{p - d_0}{\bar{p} - 1}, & \text{if } d_0 < 1. \end{cases}$

$$\tag{7.23}$$

Then, u is bounded above on M and $f(u^*) \leq 0$. If moreover u satisfies $(P_=)$,

$$tf(t) \geq C|t| \qquad \text{for } |t| >> 1, \tag{7.24}$$

and either (i) or (ii) of Theorem 7.5 holds for u_+ and for u_-, then

$$u \in L^\infty(M) \qquad \text{and} \qquad f(u^*) \leq 0 \leq f(u_*). \tag{7.25}$$

Proof of Corollary 7.9, Assuming Theorem 7.5 It is enough to choose $\chi = 0$ and $\sigma > 0$ in Theorem 7.5. An algebraic manipulation shows that the condition on σ appearing in (2) of (7.23) is equivalent to

$$
d_0 \leq
\begin{cases}
p - \sigma(p-1) & \text{if } \sigma \leq 1; \\
\bar{p} - \sigma(\bar{p}-1) & \text{if } \sigma > 1.
\end{cases}
$$

While Theorem 7.5 requires $u_+ = o(r^\sigma)$, when $\mu > 1$ no problem arises as we can enlarge σ a bit to match this last requirement. On the other hand, if $\mu = 1$, thanks to Remark 7.6, we can still reach the conclusion in Theorem 7.5 when $u_+ = O(r^\sigma)$. In particular, the upper bound for σ in the second of (7.23) can be achieved. $\qquad\qquad\square$

Remark 7.10 Comparing with (iii), (iv) in Theorem 7.7, we readily see that (7.23) fits very well with the assumptions in [87], and we can also capture the case $\sigma = \frac{p-m}{p-1}$. On the other hand, it should be remarked that our result is restricted to $\mu \leq 1$.

Remark 7.11 In the Euclidean setting $M = \mathbb{R}^m$ and for $p = \bar{p} \geq m$, other interesting Liouville theorems for slowly growing solutions of $(P_=)$ can be found in [195, Thm. 1.1] and [87, Thm. 10], where the case $tf(t) \geq 0$ ($\equiv 0$ in the second) is considered. There, the authors obtain the constancy of solutions of $(P_=)$ on $\mathbb{R}^m$ whenever $p > m$ and[1]

$$
u(x) = o\left(r(x)^{\frac{p-m}{p-1}}\right) \qquad \text{as } r(x) \to \infty. \tag{7.26}
$$

Condition (7.26) is sharp and related to the growth of the fundamental solution for the p-Laplacian (see [195] and Remark 10.3 in [64]). Further interesting results covering $p \geq m$ can be found in [64] (Theorems 10.1 and 10.4 therein).

Next, we show how Corollary 7.9 applies in the setting of the mean curvature operator to obtain the next

Theorem 7.12 *Let M be a complete manifold satisfying*

$$
\liminf_{r \to \infty} \frac{\log |B_r|}{r^{1-\mu}} = 0, \tag{7.27}
$$

---

[1]It should be observed that assumption (1.3) in [195], when rephrased for $(P_\geq)$, gives necessarily $\varphi(t) = Ct^{p-1}$. However, the above restriction does not appear in Theorem 10 of [87], which considers the case $f(t) \equiv 0$.

for some $\mu < 1$. Let b, f, l satisfy (2.5) and

$$b(x) \geq C\big(1 + r(x)\big)^{-\mu} \qquad on\ M,$$

$$l(t) \geq C\frac{t}{\sqrt{1+t^2}} \qquad on\ \mathbb{R}^+, \tag{7.28}$$

$$f\ is\ non\text{-}decreasing\ on\ \mathbb{R}\ and\ f \not\equiv 0$$

for some $C > 0$. If $u \in \mathrm{Lip}_{\mathrm{loc}}(M)$ is a non-constant solution of

$$\mathrm{div}\left(\frac{\nabla u}{\sqrt{1+|\nabla u|^2}}\right) = b(x)f(u)l(|\nabla u|) \qquad on\ M \tag{7.29}$$

with polynomial growth, then u solves the minimal surface equation

$$\mathrm{div}\left(\frac{\nabla u}{\sqrt{1+|\nabla u|^2}}\right) = 0 \qquad on\ M$$

and u is bounded on one side. In particular, if (7.27) is strengthened to

$$\mathrm{Ric} \geq 0 \qquad on\ M, \tag{7.30}$$

then a solution of (7.29) with polynomial growth, if any, is constant for each $\mu < 1$.

Remark 7.13 The assumption $f \not\equiv 0$ is necessary, as the example of affine functions solving the minimal graph equation on $\mathbb{R}^m$ shows.

Remark 7.14 When $l \equiv 1$, $b \equiv 1$ and $M = \mathbb{R}^m$, V. Tkachev [225] proved Theorem 7.12 for C^2 solutions of (7.29) without any growth restriction, see also Thm. 10.4 of A. Farina's survey [83]. The result has first been extended for mean curvature type operators by Y. Naito and Y. Usami [168, Thm. 1], with a different argument using radialization in a way related to the one in Sect. 10. An improvement of [168, 225] for (7.29) with a nontrivial gradient dependence is shown in Theorem 10.30 below, while generalizations to a larger class of quasilinear operators can be found in [219, Thm. 3], with the same method as in [225], and in [194, Thm. 8.1.3] with an approach, by means of Khas'minskii potentials, close to the one in [168].

Proof of Theorem 7.12 Since one of the sets $\{t \in \mathbb{R} : f(t) > 0\}$ and $\{t \in \mathbb{R} : f(t) < 0\}$ is non-empty, the monotonicity of f implies that either $f(t) \geq C_1$ for $t >> 1$, or $f(t) \leq -C_1$ for $t << -1$, for some constant $C_1 > 0$. Without loss of generality, we can suppose $f(t) \geq C_1$ for $t >> 1$. Using Corollary 7.9, since u is non-constant we deduce

$u^* < \infty$ and $f(u^*) \leq 0$. Again by the monotonicity of f, either $f = 0$ on $(-\infty, u^*]$ or $f < 0$ somewhere. In the second case, $f(t) \leq -C$ for $t \ll -1$, and thus, applying again Corollary 7.9 with $-u$ replacing u and $-f(-t)$ replacing $f(t)$, we obtain $f(u_*) \geq 0$. Therefore, $f = 0$ on $[u_*, u^*]$. Summarizing, in both cases, u solves the minimal surface equation

$$\text{div}\left(\frac{\nabla u}{\sqrt{1 + |\nabla u|^2}}\right) = 0 \qquad \text{on } M \tag{7.31}$$

(hence, u is smooth), and u is bounded on one side. If (7.27) is replaced by (7.30), the volume comparison implies $|B_r| \leq Cr^m$, and therefore, (7.27) holds for each $\mu < 1$. The constancy of u is then a consequence of the Bernstein theorem in [59]. □

We now come to the proof of Theorem 7.5. The result follows in a more or less direct way from the following refined maximum principle for slowly growing solutions of

$$\Delta_\varphi u \geq K(1 + r)^{-\mu} \frac{\varphi(|\nabla u|)}{|\nabla u|^\chi} \qquad \text{on } \Omega_\gamma, \tag{7.32}$$

where $\Omega_\gamma = \{x \in M : u(x) > \gamma\}$ is the superlevel set of u at height $\gamma \in \mathbb{R}$.

Theorem 7.15 *Let M be a complete Riemannian manifold and let the growth (7.3) be met for some $p, \bar{p} > 1$. Let $\mu, \chi \in \mathbb{R}$ verify (7.4), and let $u \in \text{Lip}_{\text{loc}}(M)$ be a function for which*

$$\hat{u} = \limsup_{r(x) \to \infty} \frac{u_+(x)}{r(x)^\sigma} < \infty, \tag{7.33}$$

for some $\sigma \in \mathbb{R}_0^+$ satisfying

$$0 \leq \chi\sigma \leq \chi + 1 - \mu. \tag{7.34}$$

Suppose that either one of the following assumptions is met:

$$\chi\sigma < \chi + 1 - \mu \quad \text{and} \quad \liminf_{r \to \infty} \frac{\log|B_r|}{r^{\chi + 1 - \mu - \chi\sigma}} = d_0 < \infty;$$

or $\tag{7.35}$

$$\chi\sigma = \chi + 1 - \mu \quad \text{and} \quad \liminf_{r \to \infty} \frac{\log|B_r|}{\log r} = d_0 < \infty.$$

If, for some $\gamma \in \mathbb{R}$, the open set Ω_γ is non-empty and u is a non-constant, weak solution of (7.32) in Ω_γ, then

$$K \le H \cdot \hat{u}^\chi,$$

where setting

$$\zeta = \chi + 1 - \mu, \qquad \text{and} \quad d^* = \min\left\{p - \sigma(p-1), \bar{p} - \sigma(\bar{p}-1)\right\}, \qquad (7.36)$$

the constant $H = H(\sigma, \chi, p, \bar{p}, \mu, d_0)$ is given by

$$H = \begin{cases} (i) & 0 \quad \text{if} \quad \chi > 0, \quad \sigma = 0; \\[2mm] (ii) & d_0\big[\zeta - \chi\sigma\big]^{\chi+1} \quad \text{if} \quad \chi > 0, \quad 0 < \chi\sigma < \dfrac{\chi}{\chi+1}\zeta; \\[2mm] (iii) & d_0\sigma^\chi(\zeta - \chi\sigma) \quad \text{if} \quad \chi > 0, \quad \dfrac{\chi}{\chi+1}\zeta \le \chi\sigma < \zeta; \\[2mm] (iv) & d_0(1-\mu) \quad \text{if} \quad \chi = 0, \quad \mu < 1, \quad \sigma \ge 0 \\[2mm] (v) & \sigma^\chi(d_0 - d^*)_+ \quad \text{if} \quad \chi\sigma = \zeta > 0 \quad \text{or} \quad \chi\sigma = \zeta = 0, \quad \chi = 0. \end{cases} \qquad (7.37)$$

Proof We can suppose $K > 0$, otherwise the estimate is trivial. Note that (7.32) is invariant with respect to translations $u \mapsto u_s = u + s$. Fix $\beta > \hat{u}$. We claim that a suitable translated u_s satisfies

$$u_s \le \beta(1+r)^\sigma \qquad \text{on } M, \qquad u_s > 0 \text{ somewhere.}$$

Indeed, if $\sigma > 0$ and $u^* = \infty$, (7.33) implies that $u < \beta(1+r)^\sigma$ outside a large compact set Ω, and translating u downwards, we can achieve the same inequality also in Ω, still keeping $u_s > 0$ somewhere. On the other hand, the claim is obvious if $\sigma = 0$. In this last case, note that here we do *not* claim that $\hat{u}$ is not attained: this would follow from a strong maximum principle that to the best of our knowledge is unknown under the sole assumption (7.3). Using that the resulting u_s is positive somewhere, we can also assume $\gamma > 0$. Hereafter, computations will be performed with $u = u_s$. Choose $\alpha > \beta$ and define

$$v(x) = \alpha\big(1 + r(x)\big)^\sigma - u(x),$$

so that

$$(\alpha - \beta)(1+r)^\sigma \le v \le \alpha(1+r)^\sigma \qquad \text{on } \Omega_\gamma. \qquad (7.38)$$

Hereafter, with $C_1, C_2, C_3, \ldots$, we will denote positive absolute constants, that is, independent of σ, μ, χ. Fix a function $\lambda \in C^1(\mathbb{R})$ such that

$$0 \leq \lambda \leq 1, \quad \lambda \equiv 0 \ \text{ on } (-\infty, \gamma], \qquad \lambda > 0 \ \text{ on } (\gamma, \infty), \quad \lambda' \geq 0,$$

and a cut-off function $\psi \in \mathrm{Lip}_c(M)$, $0 \leq \psi \leq 1$, whose support has nontrivial intersection with Ω_γ. Next, consider $F \in C^1(\mathbb{R}^2)$, $F = F(r, v)$, satisfying

$$F(r, v) > 0, \qquad F_v = \frac{\partial F}{\partial v}(r, v) < 0. \tag{7.39}$$

Suppose first that $\bar{p} \geq p$. We insert the test function

$$\psi^{\bar{p}} \lambda(u) F(v, r) \in \mathrm{Lip}_c(\overline{\Omega}_\gamma) \tag{7.40}$$

in the weak definition of (7.32). Using $\lambda' \geq 0$ together with the Cauchy–Schwarz inequality, we obtain

$$K \int \psi^{\bar{p}} \lambda F (1 + r)^{-\mu} \frac{\varphi(|\nabla u|)}{|\nabla u|^\chi} \leq \bar{p} \int \psi^{\bar{p}-1} \lambda F \varphi(|\nabla u|) |\nabla \psi| + \int \psi^{\bar{p}} \lambda F_v \varphi(|\nabla u|) |\nabla u|$$
$$+ \int \psi^{\bar{p}} \lambda \varphi(|\nabla u|) \left| \alpha \sigma (1 + r)^{\sigma-1} F_v + F_r \right|.$$

Rearranging,

$$\int \psi^{\bar{p}} \lambda |F_v| \frac{\varphi(|\nabla u|)}{|\nabla u|^\chi} B(x, u) \leq \bar{p} \int \psi^{\bar{p}-1} \lambda F \varphi(|\nabla u|) |\nabla \psi|, \tag{7.41}$$

with

$$B(x, u) = K(1 + r)^{-\mu} \frac{F}{|F_v|} + |\nabla u|^{\chi+1} - |\nabla u|^\chi \left| -\alpha \sigma (1 + r)^{\sigma-1} + \frac{F_r}{|F_v|} \right|. \tag{7.42}$$

Let us assume the validity of the following

$$\text{claim:} \quad B(x, u) \geq \Lambda |\nabla u|^{\chi+1}$$

$$\text{for some } \Lambda \in (0, 1] \text{ independent of } r. \tag{7.43}$$

Plugging into (7.41) gives

$$\frac{\Lambda}{\bar{p}} \int \psi^{\bar{p}} \lambda |F_v| \varphi(|\nabla u|) |\nabla u| \leq \int \psi^{\bar{p}-1} \lambda F \varphi(|\nabla u|) |\nabla \psi|. \tag{7.44}$$

We now split the integrals on the subsets $\{|\nabla u| < 1\}$ and $\{|\nabla u| \geq 1\}$, where we apply different Young inequalities. Letting p', $\bar{p}'$ be, respectively, the conjugate exponents to p and $\bar{p}$, we can rewrite (7.44) as follows:

$$\frac{\Lambda}{\bar{p}} \int_{\{|\nabla u| < 1\}} \psi^{\bar{p}} \lambda |F_v| \left[\frac{\varphi(|\nabla u|)}{|\nabla u|^{p-1}} \right] |\nabla u|^p + \frac{\Lambda}{\bar{p}} \int_{\{|\nabla u| \geq 1\}} \psi^{\bar{p}} \lambda |F_v| \left[\frac{\varphi(|\nabla u|)}{|\nabla u|^{\bar{p}-1}} \right] |\nabla u|^{\bar{p}}$$

$$\leq \int_{\{|\nabla u| < 1\}} \left[\psi^{\bar{p}} \lambda |F_v| \left(\frac{\varphi(|\nabla u|)}{|\nabla u|^{p-1}} \right) |\nabla u|^p \right]^{\frac{1}{p'}} \left[\psi^{\bar{p}-1-\frac{\bar{p}}{p'}} \lambda^{\frac{1}{p}} F |F_v|^{-\frac{1}{p'}} \left(\frac{\varphi(|\nabla u|)}{|\nabla u|^{p-1}} \right)^{\frac{1}{p}} |\nabla \psi| \right]$$

$$+ \int_{\{|\nabla u| \geq 1\}} \left[\psi^{\bar{p}} \lambda |F_v| \left(\frac{\varphi(|\nabla u|)}{|\nabla u|^{\bar{p}-1}} \right) |\nabla u|^{\bar{p}} \right]^{\frac{1}{\bar{p}'}} \left[\lambda^{\frac{1}{\bar{p}}} F |F_v|^{-\frac{1}{\bar{p}'}} \left(\frac{\varphi(|\nabla u|)}{|\nabla u|^{\bar{p}-1}} \right)^{\frac{1}{\bar{p}}} |\nabla \psi| \right].$$

Observe that $\bar{p} \geq p$ implies the non-negativity of the exponent $\bar{p} - 1 - \frac{\bar{p}}{p'} \geq 0$ for ψ above. We apply Young's inequality $ab \leq (a\varepsilon)^{p'}/p' + (b/\varepsilon)^p/p$ to the first term in the right-hand side of the above inequality and an analogous one with $\bar{\varepsilon}, \bar{p}, \bar{p}'$ to the second one, of course with $\varepsilon, \bar{\varepsilon} > 0$. Rearranging, we obtain

$$\left(\frac{\Lambda}{\bar{p}} - \frac{\varepsilon^{p'}}{p'} \right) \int_{\{|\nabla u| < 1\}} \psi^{\bar{p}} \lambda |F_v| \left[\frac{\varphi(|\nabla u|)}{|\nabla u|^{p-1}} \right] |\nabla u|^p$$

$$+ \left(\frac{\Lambda}{\bar{p}} - \frac{\bar{\varepsilon}^{\bar{p}'}}{\bar{p}'} \right) \int_{\{|\nabla u| \geq 1\}} \psi^{\bar{p}} \lambda |F_v| \left[\frac{\varphi(|\nabla u|)}{|\nabla u|^{\bar{p}-1}} \right] |\nabla u|^{\bar{p}}$$

$$\leq \frac{\varepsilon^{-p}}{p} \int_{\{|\nabla u| < 1\}} \psi^{\bar{p}-p} \lambda F \left[\frac{F}{|F_v|} \right]^{p-1} \left(\frac{\varphi(|\nabla u|)}{|\nabla u|^{p-1}} \right) |\nabla \psi|^p$$

$$+ \frac{\bar{\varepsilon}^{-\bar{p}}}{\bar{p}} \int_{\{|\nabla u| \geq 1\}} \lambda F \left[\frac{F}{|F_v|} \right]^{\bar{p}-1} \left(\frac{\varphi(|\nabla u|)}{|\nabla u|^{\bar{p}-1}} \right) |\nabla \psi|^{\bar{p}}.$$

Choose $\varepsilon, \bar{\varepsilon}$ in such a way that both the coefficients in round brackets in the left-hand side are $\Lambda/(2\bar{p})$. From $\bar{p} \geq p$, $\psi \leq 1$ and $\Lambda \leq 1$, using (7.3), we infer the inequality

$$\int \psi^{\bar{p}} \lambda |F_v| \varphi(|\nabla u|) |\nabla u|$$
$$\leq C_1 \left[\int \lambda F \left(\frac{F}{|F_v|} \right)^{p-1} |\nabla \psi|^p + \int \lambda F \left(\frac{F}{|F_v|} \right)^{\bar{p}-1} |\nabla \psi|^{\bar{p}} \right], \tag{7.45}$$

where C_1 is some constant depending on Λ and on $p, \bar{p}, C, \bar{C}$ in (7.3). Fix $R_0 \geq 1$ large enough that u is not constant on $\Omega_\gamma \cap B_{R_0} \neq \emptyset$. Then, clearly ∇u is not identically zero on $\Omega_\gamma \cap B_{R_0}$, because otherwise u would be constant on connected components of Ω_γ, which would imply that $\Omega_\gamma \equiv M$ and u be constant, contradiction. Fix $\delta \in (1/2, 1)$. For

$R > 2R_0 \geq 2$, we choose ψ in such a way that

$$0 \leq \psi \leq 1, \quad \psi \equiv 1 \text{ on } B_{\delta R}, \quad \text{supp}(\psi) \subset B_R, \quad |\nabla \psi| \leq \frac{C_2}{(1-\delta)R}, \tag{7.46}$$

for some absolute constant C_2. Inserting into (7.45) and recalling that $\lambda \leq 1$, we obtain

$$\int_{\Omega_\gamma \cap B_{R_0}} \lambda |F_v| \varphi(|\nabla u|)|\nabla u|$$

$$\leq C_3 \left[\frac{1}{R^p} \int_{(B_R \setminus B_{\delta R}) \cap \Omega_\gamma} F\left(\frac{F}{|F_v|}\right)^{p-1} + \frac{1}{R^{\bar{p}}} \int_{(B_R \setminus B_{\delta R}) \cap \Omega_\gamma} F\left(\frac{F}{|F_v|}\right)^{\bar{p}-1} \right], \tag{7.47}$$

for some $C_3 = C_3(p, \bar{p}, C, \bar{C}, \Lambda, \delta)$. In the complementary case $\bar{p} \leq p$, we achieve (7.47) by simply exchanging the role of p, $\bar{p}$ and the related inequalities on $\{|\nabla u| < 1\}$, $\{|\nabla u| \geq 1\}$.

We now need to check the validity of the claim in (7.43), for a suitable choice of F. Observe that the expression of B in (7.42) is a function of the type

$$g(s) = P + s^{\chi+1} - Qs^\chi,$$

for $s = |\nabla u|$ and positive parameters

$$P = K(1+r)^{-\mu} \frac{F}{|F_v|}, \qquad Q = \left| -\alpha\sigma(1+r)^{\sigma-1} + \frac{F_r}{|F_v|} \right|$$

depending on r. It is a calculus exercise to check that $g(s) \geq \Lambda s^{\chi+1}$ on $\mathbb{R}_0^+$ when either

$$\begin{cases} \chi = 0, \quad Q \leq P \quad \text{and} \quad \Lambda \leq 1, \quad \text{or} \\ \chi > 0, \quad \dfrac{Q^{\chi+1}}{P} \leq \dfrac{(\chi+1)^{\chi+1}}{\chi^\chi} \quad \text{and} \quad \Lambda \leq 1 - \dfrac{\chi}{(\chi+1)^{\frac{\chi+1}{\chi}}} \left(\dfrac{Q^{\chi+1}}{P}\right)^{1/\chi}. \end{cases} \tag{7.48}$$

Having fixed a parameter $\theta \in (0, 1)$, we will choose F in order to satisfy the next relations between Q and P:

$$\begin{cases} \text{if } \chi = 0 \text{ we want } Q = P, & \text{and in this case set } \Lambda = 1; \\ \text{if } \chi > 0 \text{ we want } \dfrac{Q^{\chi+1}}{P} = \dfrac{(\chi+1)^{\chi+1}}{\chi^\chi} \theta^\chi & \text{and in this case set } \Lambda = 1 - \theta. \end{cases} \tag{7.49}$$

In this way, (7.48), and thus, (7.43) is met. Observe that the first case in (7.49) can be obtained by letting $\chi \to 0$ and then $\theta \to 0$ in the second one. To meet the identities in (7.49), we necessarily need an upper bound for Q/P or $Q^{\chi+1}/P$ that does not depend on r, and this suggests our choice of F, which will be different from case to case. Set for convenience

$$\eta = \mu + (\sigma - 1)(\chi + 1) = (\chi + 1)\sigma - \zeta, \qquad (7.50)$$

where ζ is as in (7.36), and note that

$$\sigma - \eta = \zeta - \chi\sigma \geq 0. \qquad (7.51)$$

Analysis of case (ii)**, case** (i) **for** $\zeta > 0$**, and case** (iv) **for** $\sigma < 1 - \mu$**:**

$$\chi > 0, \quad 0 \leq \chi\sigma < \frac{\chi}{\chi + 1}\zeta, \qquad \text{or} \qquad \chi = 0, \quad \mu < 1, \quad 0 \leq \sigma < 1 - \mu.$$

Using the definition of η, these cases correspond to

$$\chi > 0, \qquad \sigma \geq 0, \quad \eta < 0.$$

Note that $\sigma > \eta$. We choose

$$F(v, r) = \exp\left\{-\tau v(1 + r)^{-\eta}\right\}, \qquad (7.52)$$

for a real number $\tau > 0$ that will be specified later in order to satisfy the identity for P, Q in (7.49). Then, on Ω_γ,

$$\frac{F}{|F_v|} = \frac{(1 + r)^\eta}{\tau}, \qquad \frac{F_r}{|F_v|} = \frac{v\eta}{(1 + r)},$$

and hence, by (7.38) and using $\sigma > \eta$, $\eta < 0$,

$$-\alpha(\sigma - \eta)(1 + r)^{\sigma - 1} \leq -\alpha\sigma(1 + r)^{\sigma - 1} + \frac{F_r}{|F_v|} \leq 0.$$

Plugging into (7.42), we get

$$B(x, u) \geq \frac{K}{\tau}(1 + r)^{\eta - \mu} + |\nabla u|^{\chi+1} - |\nabla u|^\chi \alpha(\sigma - \eta)(1 + r)^{\sigma - 1}. \qquad (7.53)$$

In view of (7.50), to satisfy (7.49) with

$$P = \frac{K}{\tau}(1 + r)^{\eta - \mu}, \qquad Q = \alpha(\sigma - \eta)(1 + r)^{\sigma - 1}, \qquad (7.54)$$

we need the identities

$$\frac{[\alpha(\sigma - \eta)]^{\chi+1}\tau}{K} = \theta^{\chi}\frac{(\chi + 1)^{\chi+1}}{\chi^{\chi}} \qquad \text{if } \chi > 0,$$

$$\frac{\alpha(\sigma - \eta)\tau}{K} = 1 \qquad\qquad\qquad \text{if } \chi = 0.$$

(7.55)

According to whether $\chi = 0$ or > 0, we then define τ as the value such that (7.55) holds. With this choice, (7.43) is satisfied with $\Lambda = 1 - \theta$ if $\chi > 0$, or $\Lambda = 1$ if $\chi = 0$. In view of our choice of F, (7.47) becomes

$$\int_{B_{R_0}\cap\Omega_\gamma} \lambda F(1+r)^{-\eta}\varphi(|\nabla u|)|\nabla u| \leq C_4 \left[\frac{1}{R^p}\int_{(B_R\setminus B_{\delta R})\cap\Omega_\gamma} F(1+r)^{\eta(p-1)}\right.$$

$$\left. +\frac{1}{R^{\bar{p}}}\int_{(B_R\setminus B_{\delta R})\cap\Omega_\gamma} F(1+r)^{\eta(\bar{p}-1)}\right],$$

(7.56)

where C_4 also depends on τ. Up to increasing $\bar{p}$, a change that does not alter the validity of (7.3), we can suppose that $\bar{p} \geq p$. On $(B_R\setminus B_{\theta R}) \cap \Omega_\gamma$, (7.38) and $\sigma - \eta > 0$, $\eta < 0$ give

$$F(v, r) \leq \exp\left(-\tau(\alpha - \beta)(\delta R)^{\sigma-\eta}\right), \qquad (1+r)^{\eta(\bar{p}-1)} \leq (1+r)^{\eta(p-1)} \leq 1.$$

Inserting into (7.56) and using $R^{\bar{p}} \geq R^p$, we eventually get

$$0 < \int_{\Omega_\gamma\cap B_{R_0}} \lambda F(1+r)^{-\eta}\varphi(|\nabla u|)|\nabla u| \leq \frac{C_5}{R^p}\exp\left(-\tau(\alpha - \beta)(\delta R)^{\sigma-\eta}\right)|B_R|. \quad (7.57)$$

Because of (7.35) and (7.51), for each $d > d_0$ there exists a sequence $\{R_k\} \uparrow \infty$ such that

$$|B_{R_k}| \leq \exp\left\{dR_k^{\sigma-\eta}\right\}.$$

Substituting into (7.57) and letting $k \to \infty$,

$$0 < \int_{\Omega_\gamma\cap B_{R_0}} \lambda F(1+r)^{-\eta}\varphi(|\nabla u|)|\nabla u|$$

$$\leq C_5 \limsup_{k\to\infty} \left(\frac{\exp\left\{-\tau(\alpha - \beta)\delta^{\sigma-\eta}R_k^{\sigma-\eta} + dR_k^{\sigma-\eta}\right\}}{R_k^p}\right).$$

(7.58)

Being the left-hand side of the above inequality strictly bigger than zero, we deduce that necessarily $d \geq \tau(\alpha - \beta)\delta^{\sigma - \eta}$, and letting $\delta \to 1$ and $d \downarrow d_0$, we get

$$d_0 \geq \tau(\alpha - \beta).$$

Substituting the expression of τ in (7.55), setting $\alpha = t\beta$ with $t > 1$ and letting $\theta \to 1$, we deduce

$$K \leq d_0 \frac{\chi^\chi}{(\chi+1)^{\chi+1}} (\sigma - \eta)^{\chi+1} \frac{t^{\chi+1}}{t-1} \beta^\chi \quad \text{if } \chi > 0;$$

$$K \leq d_0(\sigma - \eta)\frac{t}{t-1} \quad \text{if } \chi = 0.$$

Minimizing with respect to $t \in (1, \infty)$ and letting $\beta \downarrow \hat{u}$, we eventually get

$$
\begin{aligned}
K &\leq d_0(\sigma - \eta)^{\chi+1}\hat{u}^\chi = d_0[\zeta - \chi\sigma]^{\chi+1}\hat{u}^\chi \quad \text{if } \chi > 0; \\
K &\leq d_0(\sigma - \eta) = d_0(1 - \mu) \quad \text{if } \chi = 0,
\end{aligned}
\tag{7.59}
$$

as claimed. We conclude by investigating part of case (i), that is, when

$$\chi > 0, \quad \sigma = 0 < \zeta.$$

Observe that a downward translation u_s of u still satisfies (7.32) with the same constant K (without loss of generality, we can suppose $\gamma = 0$). Hence, by (7.59),

$$K \leq H\hat{u}_s^\chi \quad \text{with} \quad H \leq d_0[\zeta - \chi\sigma]^{\chi+1}. \tag{7.60}$$

If $K > 0$, since u is bounded above and $\chi > 0$, we can choose u_s^χ positive and small enough to contradict (7.60). Hence, necessarily $K = 0$, and a posteriori we can choose $H = 0$ in (7.37) as required. At the end of the present proof, with the same trick we investigate the remaining case of (i), that is, when $\sigma = \zeta = 0$. Note that the trick is not possible if $\chi = 0$, being $u_s^\chi \equiv 1$.

Analysis of case (iii), and case (iv) for $\sigma \geq 1 - \mu$:

$$\chi > 0, \quad \frac{\chi}{\chi+1}\zeta \leq \chi\sigma < \zeta, \quad \text{or} \quad \chi = 0, \quad \sigma \geq 1 - \mu > 0.$$

Again from the definition of η, these cases correspond to the range

$$0 \leq \eta < \sigma,$$

for which we choose

$$F(v, r) = \exp\left\{-\tau v^{\frac{\sigma-\eta}{\sigma}}\right\}. \tag{7.61}$$

Also in these cases, we increase $\bar{p}$ in order for $\bar{p} \geq p$ to hold, since ultimately the size of $\bar{p}$ will not affect the conclusion of the theorem. Performing computations analogous to those giving (ii), we obtain the desired estimate

$$K \leq d_0 \sigma^\chi (\sigma - \eta) \hat{u}^\chi = d_0 \sigma^\chi [\zeta - \chi\sigma] \hat{u}^\chi \quad \text{if } \chi > 0;$$

$$K \leq d_0 (\sigma - \eta) = d_0 (1 - \mu) \quad\quad\quad\quad \text{if } \chi = 0.$$

Analysis of case (v), and case (i) for $\zeta = 0$:

$$\chi\sigma = \zeta > 0, \quad \text{or} \quad \chi\sigma = \zeta = 0.$$

In this case, by (7.50) it holds $\sigma - \mu = (\sigma - 1)(\chi + 1)$. We choose

$$F(v, r) = v^{-\tau},$$

$\tau > 0$ to be determined. Then, using (7.38),

$$B(x, u) \geq \frac{K}{\tau}(1 + r)^{-\mu} v + |\nabla u|^{\chi+1} - |\nabla u|^\chi \alpha\sigma(1 + r)^{\sigma-1}$$

$$\geq \frac{K(\alpha - \beta)}{\tau}(1 + r)^{\sigma-\mu} + |\nabla u|^{\chi+1} - |\nabla u|^\chi \alpha\sigma(1 + r)^{\sigma-1}, \tag{7.62}$$

and (7.49) applied with

$$P = \frac{K(\alpha - \beta)}{\tau}(1 + r)^{\sigma-\mu}, \quad\quad Q = \alpha\sigma(1 + r)^{\sigma-1}$$

implies the identities

$$\frac{(\alpha\sigma)^{\chi+1}\tau}{K(\alpha - \beta)} = \theta^\chi \frac{(\chi + 1)^{\chi+1}}{\chi^\chi} \quad \text{if } \chi > 0,$$

$$\frac{\alpha\sigma\tau}{K(\alpha - \beta)} = 1 \quad\quad\quad\quad \text{if } \chi = 0. \tag{7.63}$$

Having specified τ to satisfy (7.63), (7.43) is met and (7.47) with $\delta = 3/4$ now reads

$$0 < \int_{\Omega_\gamma \cap B_{R_0}} \lambda \, |F_v| \, \varphi(|\nabla u|)|\nabla u| \le C_4 \int_{(B_R \setminus B_{3R/4}) \cap \Omega_\gamma} \frac{v^{p-1-\tau}}{R^p} + \frac{v^{\bar{p}-1-\tau}}{R^{\bar{p}}}. \qquad (7.64)$$

Estimating v with the aid of (7.38), choosing the upper or lower bound according to the sign of $p - 1 - \tau$ and $\bar{p} - 1 - \tau$, the right-hand side is bounded from above by

$$C_5 |B_R| \Big(R^{-p+\sigma(p-1-\tau)} + R^{-\bar{p}+\sigma(\bar{p}-1-\tau)} \Big),$$

for a suitable constant $C_5 > 0$. Because of (7.35), for $d > d_0$ we consider a sequence $\{R_k\}$ for which $|B_{R_k}| \le R_k^d$. Evaluating (7.47) on R_k, and letting $k \to \infty$ in (7.64) and then $d \downarrow d_0$, we deduce that necessarily

$$d_0 \ge \min \big\{ p - \sigma(p - 1 - \tau), \bar{p} - \sigma(\bar{p} - 1 - \tau) \big\},$$

that is, by (7.36),

$$\tau\sigma \le d_0 - \min \Big\{ p - \sigma(p - 1), \bar{p} - \sigma(\bar{p} - 1) \Big\} = d_0 - d^*. \qquad (7.65)$$

(i) If $d_0 < d^*$ or $d_0 = d^*$ and $\sigma > 0$, then there exists no $\tau > 0$ satisfying (7.65). Thus, $K > 0$ leads to a contradiction, and we can therefore choose $H = 0$. This proves (v) for $d_0 < d_*$ and for $d_0 = d^*$, $\sigma > 0$, as well as (i) for $\chi > 0$, $\sigma = \zeta = 0$, $d_0 < d^*$.

(ii) If $d_0 \ge d^*$, then inserting the expression of τ obtained from (7.63) in (7.65), setting $\alpha = t\beta$ for $t > 1$, solving (7.65) with respect to K, letting $\theta \uparrow 1$ and $\beta \downarrow \hat{u}$, we deduce

$$K \le [d_0 - d^*] \frac{\chi^\chi}{(\chi + 1)^{\chi+1}} \hat{u}^\chi \sigma^\chi \frac{t^{\chi+1}}{t - 1} \qquad \text{if } \chi > 0;$$

$$K \le [d_0 - d^*] \frac{t}{t - 1} \qquad\qquad\qquad \text{if } \chi = 0,$$

and minimizing over $t \in (1, \infty)$, we get for both $\chi > 0$ and $\chi = 0$

$$K \le [d_0 - d^*] \sigma^\chi \hat{u}^\chi. \qquad (7.66)$$

This concludes the cases $d_0 > d^*$ and $\sigma > 0$, and $d_0 \ge d^*$ and $\chi = 0$. To deal with the remaining part of (i), that is, $\chi > 0$, $\sigma = \zeta = 0$ and $d_0 \ge d^*$, we can consider a downward translation u_s of u in place of u, and $\gamma = 0$. Then, u_s satisfies (7.32) with the same constant K; hence, (7.66) holds for each $\hat{u}_s$. However, from $\chi > 0$, $\hat{u}_s^\chi$ can be made as small as we wish, and since we have assumed $K > 0$, this would contradict (7.66). Concluding, necessarily $K \le 0$, and H can be chosen to be zero, as required. $\qquad \square$

We now prove Theorem 7.5.

Proof of Theorem 7.5 Suppose, by contradiction, that either $u^* = \infty$ or $f(u^*) > 0$. Because of the second in (7.21) and the continuity of f, in both of the cases there exists $\gamma < u^*$ sufficiently close to u^* such that $f(t) \geq C > 0$ on (γ, ∞), for some constant $C > 0$. By (7.21), u would solve

$$\Delta_\varphi u \geq K(1+r)^{-\mu} \frac{\varphi(|\nabla u|)}{|\nabla u|^\chi} \qquad \text{on } \Omega_\gamma.$$

For some $K > 0$. To apply Theorem 7.15, we shall consider

$$d^* = \min\left\{p - \sigma(p-1), \bar{p} - \sigma(\bar{p}-1)\right\}.$$

As said before, p (respectively, $\bar{p}$) in (7.3) can be reduced (resp. increased) as much as we wish, still keeping the validity of (7.3). Therefore,

- if $\sigma \leq 1$, we can increase $\bar{p}$ up to satisfy $\bar{p} \geq p$, which gives $d^* = p - \sigma(p-1)$. In particular, $d^* = p$ when $\sigma = 0$;
- if $\sigma > 1$, we can reduce p to satisfy $\bar{p} \geq p$, which now implies $d^* = \bar{p} - \sigma(\bar{p}-1)$.

The volume growth conditions in the last lines of (7.6) and (7.9) can therefore be rewritten as

$$\liminf_{r \to \infty} \frac{\log |B_r|}{\log r} \leq d^*.$$

We are in the position to apply Theorem 7.15 and deduce $0 < K \leq H\hat{u}^\chi$. We reach a contradiction by proving that either $H = 0$ or $\hat{u}^\chi = 0$. We split the argument into several cases:

- If $\sigma = 0$ and $\chi > 0$, then we are in case (i) of Theorem 7.15, and (7.35) is satisfied because of (7.6), thus $H = 0$, contradiction.
- If $\sigma \geq 0$ and $\chi = 0$, then we are either in case (iv) or in case (v) of Theorem 7.15, according to whether $\mu < 1$ or $\mu = 1$. In case (iv), our growth requirements (7.6) and (7.9) for $\chi = 0$, $\mu < 1$ imply that $d_0 = 0$ in (7.35). Applying Theorem 7.15, we get $H \leq d_0(1-\mu) = 0$, contradiction. In case (v), as said conditions (7.6) and (7.9) for $\chi = 0$, $\mu = 1$ are equivalent to $d_0 \leq d^*$. By Theorem 7.15, $H \leq \sigma^\chi(d_0 - d^*)_+ = 0$, contradiction.
- If $\sigma > 0$, $\chi > 0$, then by (7.7) we get $\hat{u}^\chi = 0$, contradiction.

We have thus proved $u^* < \infty$ and $f(u^*) \leq 0$. If now u satisfies $(P_=)$ and (7.10) is in force, we can apply the result both to u and to $\bar{u} = -u$, noting that

$$\Delta_\varphi \bar{u} = b(x)\bar{f}(\bar{u})l(|\nabla\bar{u}|), \qquad \bar{f}(t) = -f(-t),$$

and that $\bar{f}(t) \geq C$ for t large enough. From $\bar{f}(\bar{u}^*) \leq 0$, we get $f(u_*) \geq 0$, and (7.11) follows.   □

Remark 7.16 We now check Remark 7.6. Since the third in (7.9) corresponds to case (v) of Theorem 7.15, if $d_0 \leq d^*$, we achieve the contradiction $0 < K \leq H\hat{u}^\chi$ irrespectively to the vanishing of $\hat{u}$, as claimed.

7.3   Bernstein Theorems for Minimal and MCF Soliton Graphs

We now apply Theorem 7.5 to deduce some Bernstein type results for prescribed mean curvature graphs. We consider an ambient space $(\bar{M}^{m+1}, (\,,\,))$ with the warped product structure

$$\bar{M} = \mathbb{R} \times_h M, \qquad (\,,\,) = ds^2 + h(s)^2\langle\,,\,\rangle \qquad (7.67)$$

for some complete manifold $(M^m, \langle\,,\,\rangle)$ and some $0 < h \in C^\infty(\mathbb{R})$. Given $v : M \to \mathbb{R}$, we consider the graph

$$\Sigma = \Big\{(v(x), x) \in \bar{M} \ : \ x \in M\Big\}.$$

In the next theorems, we always consider *entire* graphs, that is, graphs defined on all of M. As explained in Chap. 1, we let Φ_t be the flow of the conformal field $X = h(s)\partial_s$, and we note that its flow parameter t starting from the slice $\{s = 0\}$ satisfies (1.10), that is,

$$t = \int_0^s \frac{d\sigma}{h(\sigma)}, \qquad t : \mathbb{R} \to t(\mathbb{R}) = I. \qquad (7.68)$$

Setting $s(t)$ for the inverse function, we define $\lambda(t) = h(s(t))$, $u(x) = t(v(x))$. Observe, in particular, that $u : M \to I$. We have

Theorem 7.17 *Let $(M^m, \langle\,,\,\rangle)$ be a complete manifold, and consider the warped product $\bar{M} = \mathbb{R} \times_h M$. Suppose that the warping function h satisfies:*

$$\begin{cases} h'(s)s \geq Cs & \text{for } |s| \geq s_0; \\[2mm] h^{-1} \in L^1(\infty) \cap L^1(-\infty), \end{cases} \qquad (7.69)$$

for some constants C, $s_0 > 0$. Assume the volume growth of geodesic balls in M satisfies

$$\liminf_{r \to \infty} \frac{\log |B_r|}{r^2} < \infty.$$

Then every entire, geodesic minimal graph Σ over M is bounded, and letting $v : M \to \mathbb{R}$ be the graph function, it holds

$$h'(v^*) \leq 0 \leq h'(v_*). \tag{7.70}$$

In particular, if h is strictly convex, Σ is the totally geodesic slice $\{s = s_1\}$, where s_1 is the unique minimum of h.

Proof By (1.12) and the minimality of Σ, u solves

$$\operatorname{div}\left(\frac{\nabla u}{\sqrt{1 + |\nabla u|^2}} \right) = m \frac{\lambda_t(u)}{\lambda(u)} \frac{1}{\sqrt{1 + |\nabla u|^2}}.$$

The second in (7.69) implies that I is a bounded interval, and thus, $u : M \to I$ is bounded. Taking into account that

$$f(t) = \frac{\lambda_t(t)}{\lambda(t)} = h'(s(t)) \geq C \qquad \text{for } t \geq t(s_0), \tag{7.71}$$

we get

$$\operatorname{div}\left(\frac{\nabla u}{\sqrt{1 + |\nabla u|^2}} \right) \geq \frac{mC}{\sqrt{1 + |\nabla u|^2}} \qquad \text{on } \{u > t(s_0)\}.$$

If the set $\{u > t(s_0)\}$ were non-empty, we can apply Theorem 7.15 with $\mu = 0$, $\chi = 1$, $p = \bar{p} = 2$ and $K = mC$ to obtain a contradiction. Therefore, u is bounded from above, and from Theorem 7.5, we deduce $f(u^*) \leq 0$. Analogously, considering $-u$, we infer $f(u_*) \geq 0$, and thus, (7.70) follows by changing variables. Note that, because of the first in (7.69), (7.70) implies the boundedness of v. If h is strictly convex, then by (7.69) it has a unique stationary point (a minimum) s_1, and (7.70) gives $v^* \leq s_1 \leq v_*$, that is, $v \equiv s_1$. We remark that each slice $\{s = s_2\}$ is totally umbilical in $\bar{M}$, with second fundamental form $h'(s_2)/h(s_2)\langle , \rangle$ in the direction $-\partial_s$, so $\{s = s_1\}$ is totally geodesic.

$\square$

Theorem 7.18 *Let* $\bar{M} = \mathbb{R} \times_h M$ *be as above, and suppose that h satisfies*

$$\begin{cases} h' > 0 \quad \text{on } \mathbb{R}, \\ h'(s) \geq C \qquad \text{for } s \geq s_0; \\ h^{-1} \in L^1(\infty), \end{cases} \tag{7.72}$$

for some constants $C, s_0 > 0$. *If M is complete and the volume growth of geodesic balls in M satisfies*

$$\liminf_{r \to \infty} \frac{\log |B_r|}{r^2} < \infty,$$

then there are no entire, geodesic minimal graphs over M.

Proof The proof is analogous. By the third in (7.72), $u : M \to I$ is bounded from above; hence, applying Theorems 7.15 and 7.5, we get $h'(v^*) \leq 0$, contradicting the first in (7.72). □

Proof of Theorem 2.25 It is immediate from Theorems 7.17 and 7.18, respectively, for cases (i) and (ii). Note that, in (i), the strict convexity of h and $h^{-1} \in L^1(-\infty) \cap L^1(\infty)$ implies that $sh'(s) \geq Cs$ for $|s|$ large enough. □

Remark 7.19 As a direct corollary, we deduce Do Carmo–Lawson's result [75] in the case $H = 0$, Theorem 1.3 in the Introduction: there are no geodesic, entire minimal graphs over horospheres, and the only geodesic, entire minimal graph over a totally geodesic hypersphere M is M itself. To see the first claim, apply Theorem 7.18 to the warped product $\mathbb{H}^{m+1} = \mathbb{R} \times_{e^s} \mathbb{R}^m$, and note that $h(s) = e^s$ satisfies all the requirements in (7.72). For the second claim, apply Theorem 7.17 to $\mathbb{H}^{m+1} = \mathbb{R} \times_{\cosh s} \mathbb{H}^m$.

The case of non-constant mean curvature will be investigated in Sect. 8. We now focus our attention on MCF solitons, starting with product ambient manifolds. Our first result is for self-translators, Theorem 2.26 in the Introduction:

Proof of Theorem 2.26 In the product case, $h(s) = 1$ and thus $t = s$, $\lambda(t) = 1$, $u(x) = t(v(x)) = v(x)$. By (1.15), a soliton for the field ∂_s, that is, a self-translator, satisfies the equation

$$\text{div}\left(\frac{\nabla u}{\sqrt{1 + |\nabla u|^2}}\right) = \frac{1}{\sqrt{1 + |\nabla u|^2}} \qquad \text{on } M.$$

We apply Theorem 7.5 with the choices $f(t) \equiv 1$, $b \equiv 1$, $\chi = 1$, $\mu = 0$, $p = \bar{p} = 2$: since the first two in (7.9) correspond to (2.30), we infer from Theorem 7.5 that u is bounded from above and $f(u^*) \leq 0$, contradiction. $\square$

We next examine more closely the case of self-translators in Euclidean space, in particular the case when the translation vector field Y differs from the vertical field ∂_s.

Theorem 7.20 *Let* $(\mathbb{R}^{m+1}, (,)) = \mathbb{R} \times \mathbb{R}^m$ *with coordinates* (s, x), *and let* $\Sigma = \{(v(x), x) : x \in \mathbb{R}^m\}$ *be an entire graph. Assume that*

$$\limsup_{r(x) \to \infty} \frac{|v(x)|}{r(x)} = \hat{v} < \infty. \tag{7.73}$$

Then, Σ *cannot be a self-translator with respect to any vector* Y *whose angle* $\vartheta \in (0, \pi/2)$ *with the horizontal hyperplane* $\mathbb{R}^m$ *satisfies*

$$\tan \vartheta > \hat{v}. \tag{7.74}$$

In particular, if $\hat{v} = 0$, Σ *cannot be a self-translator with respect to a vector* Y, *which is not tangent to the horizontal* $\mathbb{R}^m$.

Proof If we reflect Σ with respect to the horizontal hyperplane, the reflected graph is a self-translator with respect to the reflection of Y. Therefore, without loss of generality, we can assume that $(Y, \partial_s) > 0$, and that Y has unit norm, by time rescaling. Moreover, $Y \neq \partial_s$ since $\vartheta < \pi/2$. Therefore, up to a rotation of coordinates on $\mathbb{R}^m$, $Y = \cos \vartheta e_1 + \sin \vartheta \partial_s$, where e_1 is the gradient of the first coordinate function x_1 and $\sin \vartheta > 0$. In view of (1.11) and (1.12), and since $h = 1$, $t = s$, the soliton equation $mH = (Y, \nu)$ satisfied by $u(x) = t(v(x)) = v(x)$ reads

$$\mathrm{div}\left(\frac{\nabla u}{\sqrt{1 + |\nabla u|^2}}\right) = mH = \frac{1}{\sqrt{1 + |\nabla u|^2}} [\sin \vartheta - \cos \vartheta \langle \nabla u, e_1 \rangle].$$

Rearranging

$$\mathrm{div}\left(\frac{\nabla u}{\sqrt{1 + |\nabla u|^2}}\right) + \left\langle \frac{\nabla u}{\sqrt{1 + |\nabla u|^2}}, \nabla(x_1 \cos \vartheta) \right\rangle = \frac{\sin \vartheta}{\sqrt{1 + |\nabla u|^2}},$$

which we rewrite as

$$e^{-x_1 \cos \vartheta} \mathrm{div}\left(e^{x_1 \cos \vartheta} \frac{\nabla u}{\sqrt{1 + |\nabla u|^2}}\right) = \frac{\sin \vartheta}{\sqrt{1 + |\nabla u|^2}}. \tag{7.75}$$

The operator in the left-hand side is in divergence form if we consider the weighted volume measure $e^{x_1 \cos \vartheta} dx$, with dx the Euclidean volume. For these weighted operators, the proofs of Theorems 7.15 and 7.5 follow verbatim by replacing the Euclidean volume with the weighted volume

$$\mathrm{vol}_{x_1 \cos \vartheta}(B_r) = \int_{B_r} e^{x_1 \cos \vartheta} dx.$$

Explicit computation gives

$$\liminf_{r \to \infty} \frac{\log \mathrm{vol}_{x_1 \cos \theta}(B_r)}{r} = \cos \vartheta < \infty. \tag{7.76}$$

Suppose by contradiction that (7.74) holds. In particular, u is non-constant. We apply Theorem 7.15 to (7.75) with the choices

$$K = \sin \vartheta, \quad \sigma = 1, \quad \chi = 1, \quad \mu = 0, \quad d_0 = \cos \vartheta, \quad p = \bar{p} = 2,$$

to conclude from case (iii) in (7.37) that $\sin \vartheta \leq (\cos \vartheta)\hat{u}$. Since $u = v$, we eventually contradict (7.74). □

Remark 7.21 The requirement $\vartheta > 0$, that is, $(Y, \partial_s) \neq 0$, is essential in the above theorem because otherwise the slices $\{s = \mathrm{const}\}$ are trivial self-translators. Furthermore, condition (7.74) is sharp: the totally geodesic hyperplane $\{s = x_1 \tan \vartheta\}$ is a self-translator with respect to $\partial_1 + (\tan \vartheta)\partial_s$ that satisfies $\tan \vartheta = \hat{v}$.

Remark 7.22 A result related to Theorem 7.20 appears in [15], where the authors proved that there exists no nontrivial complete self-translator (i.e. not a hyperplane) whose Gauss image lies in a geodesic ball $\mathbb{B} \subset \mathbb{S}^m$ of radius $< \pi/2$. Their assumption implies that Σ is a graph with respect to the plane orthogonal to the centre of $\mathbb{B}$ (not necessary the direction of translation), and the graph function v has bounded gradient.[2] Our requirements (7.73) and (7.74) seem to be skew with theirs. However, it might be possible that a suitable gradient estimate guarantees that a self-translator satisfying (7.73) has automatically bounded gradient. If this is the case, the main result in [15] would imply that the graph Σ of v be a hyperplane, which would prove Theorem 7.20 in view of Remark 7.21.

---

[2]In fact, they also require that H be bounded, but this automatically follows from the self-translator equation $\vec{H} = Y^{\perp}$, being Y constant.

Our last application is for entire graphs $\Sigma^m \to \mathbb{R}^{m+1}$ that are self-expanders for the mean curvature flow, that is, they move under MCF along the integral curves of the position vector field

$$Y(\bar{x}) = (\bar{x}^j - q^j)\partial_j = \frac{1}{2}\bar\nabla|\bar{x} - q|^2 \qquad \forall \bar{x} \in \mathbb{R}^{m+1},$$

for some fixed origin $q \in \mathbb{R}^{m+1}$.

Theorem 7.23 *Let $\Sigma = \{(v(x), x) : x \in \mathbb{R}^m\}$ be an entire graph in $(\mathbb{R}^{m+1}, (\,,\,))$. If Σ is a self-expander for the MCF and v is bounded, then Σ is a hyperplane.*

Proof Without loss of generality, we can assume that the centre q of the homothetic field Y is placed at the origin. We let $(s, x) \in \mathbb{R} \times \mathbb{R}^m = \mathbb{R}^{m+1}$ be coordinates on $\mathbb{R}^{m+1}$. If $\rho(x) : \mathbb{R}^m \to \mathbb{R}$ denotes the distance to the origin in $\mathbb{R}^m$, then

$$Y\big(v(x), x\big) = x^j \partial_j + v\partial_s = \frac{1}{2}\nabla(\rho^2) + v\partial_s.$$

In view of (1.11) and (1.12), and since $h = 1$, $t = s$, the soliton equation $mH = (Y, \nu)$ satisfied by $u(x) = t(v(x)) = v(x)$ reads

$$\operatorname{div}\left(\frac{\nabla u}{\sqrt{1 + |\nabla u|^2}}\right) = mH = \frac{1}{\sqrt{1 + |\nabla u|^2}}\left[u - \langle \nabla u, \nabla\frac{\rho^2}{2}\rangle\right].$$

Rearranging

$$\operatorname{div}\left(\frac{\nabla u}{\sqrt{1 + |\nabla u|^2}}\right) + \left\langle \frac{\nabla u}{\sqrt{1 + |\nabla u|^2}}, \nabla\frac{\rho^2}{2}\right\rangle = \frac{u}{\sqrt{1 + |\nabla u|^2}},$$

that is,

$$e^{-\rho^2/2}\operatorname{div}\left(e^{\rho^2/2}\frac{\nabla u}{\sqrt{1 + |\nabla u|^2}}\right) = \frac{u}{\sqrt{1 + |\nabla u|^2}}.$$

Suppose that v is non-constant. An explicit computation shows that

$$\liminf_{r \to \infty} \frac{\log \operatorname{vol}_{e^{\rho^2/2}}(B_r)}{r^2} < \infty.$$

Since, by assumption, v is bounded, we can apply Theorem 7.5 (adapted to weighted volumes) with the choices $\chi = 1$, $\mu = 0$, $p = \bar{p} = 2$, $f(t) = t$ to deduce $f(u^*) \le 0$, that is, $u \le 0$. Applying the same theorem to $-u$, we infer $u \equiv 0$, that is, Σ is a hyperplane containing the origin.

$\square$

7.4   Counterexamples

To show the sharpness of Theorem 7.5, we consider a model manifold M_g^m with radial sectional curvature

$$\mathrm{Sec}_{\mathrm{rad}} = -\kappa^2 (1 + r^2)^{\alpha/2} \qquad \text{for } r(x) \geq 1,$$

for some $\kappa > 0$ and $\alpha \geq -2$. By Propositions 2.1 and 2.11 in [185], as $r(x) \to \infty$

$$\Delta r \geq \begin{cases} (m-1)\kappa r^{\alpha/2}(1 + o(1)) & \text{if } \alpha > -2; \\[2mm] \dfrac{(m-1)\bar{\kappa}}{r}(1 + o(1)) & \text{if } \alpha = -2, \end{cases} \qquad (7.77)$$

where $\bar{\kappa} = (1 + \sqrt{1 + 4\kappa^2})/2$, and thus,

$$\log |B_r| \sim \begin{cases} \dfrac{2\kappa(m-1)}{2+\alpha} r^{1+\frac{\alpha}{2}} & \text{if } \alpha > -2; \\[2mm] \big[(m-1)\bar{\kappa} + 1\big] \log r & \text{if } \alpha = -2, \end{cases} \qquad (7.78)$$

as $r \to \infty$. If $\alpha = -2$, letting $\kappa \to 0$, we deduce the classical expressions for Δr and $|B_r|$ for the Euclidean space. We shortly write the inequalities in (7.77) as $\Delta r \geq \zeta r^{\alpha/2}$, where ζ tends to $(m-1)\kappa$ (if $\alpha > -2$) or to $(m-1)\bar{\kappa}$ (if $\alpha = -2$) as $r(x) \to \infty$. We are going to find an operator Δ_φ meeting (2.3) and (7.3) with the following property: given $\mu \in \mathbb{R}$, $0 \leq \chi \leq p - 1$ and $\sigma > 0$, we will construct an unbounded, radial solution $u \in \mathrm{Lip}_{\mathrm{loc}}(M_g)$, increasing as a function of r, solving

$$\Delta_\varphi u \geq K(1 + r)^{-\mu} \frac{\varphi(|\nabla u|)}{|\nabla u|^\chi} \qquad (7.79)$$

if r is large enough (equivalently, for a high enough upper level set). The solution is $u(x) = r(x)^\sigma$ whenever one of the following conditions hold:

1) $\chi\sigma > \chi + 1 - \mu$;

2) $\chi\sigma = \chi + 1 - \mu$, and $\alpha > -2$;

3) $\chi\sigma = \chi + 1 - \mu$, $\alpha = -2$, $\sigma \in (0, 1]$ and

$$\lim_{r \to \infty} \frac{\log |B_r|}{\log r} > p - \sigma(p - 1);$$

$\qquad\qquad\qquad\qquad\qquad\qquad\qquad\qquad\qquad\qquad\qquad\qquad (7.80)$

4) $\chi\sigma < \chi + 1 - \mu$, $\alpha > -2$, and

$$\lim_{r \to \infty} \frac{\log |B_r|}{r^{\chi+1-\mu-\chi\sigma}} > 0.$$

Moreover, the solution is $u(x) = r(x)^\sigma / \log r(x)$ when either

$$5) \quad \chi\sigma < \chi + 1 - \mu, \qquad \chi > 0, \qquad \text{and}$$

$$\lim_{r \to \infty} \frac{\log |B_r|}{r^{\chi + 1 - \mu - \chi\sigma}} = \infty, \qquad \text{or}$$

$$6) \quad \chi\sigma < \chi + 1 - \mu, \qquad \chi = 0, \qquad \text{and}$$

$$\lim_{r \to \infty} \frac{\log |B_r|}{r^{\chi + 1 - \mu - \chi\sigma}} \in (0, \infty).$$

(7.81)

In any of (1), ..., (6), observe that the bound $\mu \leq \chi + 1$ s not needed. Once we establish (7.79) on Ω_γ for large enough $\gamma > 0$, we can choose $f \in C(\mathbb{R})$ increasing and satisfying $f \equiv 0$ on $(0, \gamma)$, $f = K$ on $(2\gamma, \infty)$, $0 \leq f \leq K$ on $\mathbb{R}$. By the pasting Lemma 6.6, $\bar{u} = \max\{u, \gamma\}$ is a $\mathrm{Lip}_{\mathrm{loc}}$ solution of

$$\Delta_\varphi \bar{u} \geq (1 + r)^{-\mu} f(\bar{u}) \frac{\varphi(|\nabla \bar{u}|)}{|\nabla \bar{u}|^\chi} \qquad \text{on } M \tag{7.82}$$

(here, we used $\chi \leq p - 1$, to guarantee that $\varphi(t)/t^\chi$ does not diverge as $t \to 0^+$ and hence that the constant function γ solves (7.82)). The existence of $\bar{u}$ under any of (1), ..., (6) above shows the sharpness of the parameter ranges (7.4) and (7.8), and of the growth conditions (7.7) for u and (7.9) for $|B_r|$. In particular,

- in (2), all the assumptions are satisfied but the second or third in (7.9), where the liminf is ∞, while in (3), the liminf in the third in (7.9) is finite but bigger than the threshold $p - \sigma(p - 1)$ for $\sigma \leq 1$;
- in (4) the requirements in the first of (7.9) are all met, but $u_+ = O(r^\sigma)$ instead of $u_+ = o(r^\sigma)$;
- in (5) and (6), $u_+ = o(r^\sigma)$, but the requirements in the first of (7.9) barely fail.

To show (7.79) first note that, because of (7.78), the volume growth conditions in (3) and (4) are equivalent, respectively, to

$$(m - 1)\bar{\kappa} + 1 > p - \sigma(p - 1), \qquad \text{and} \qquad \frac{\alpha}{2} + 1 \geq \chi + 1 - \mu - \chi\sigma. \tag{7.83}$$

Consider

$$\varphi(t) = \frac{t^{p-1}}{(1 + t)^{q-1}}, \qquad p > 1, \ 1 \leq q \leq p, \tag{7.84}$$

and we search for solutions of the form $u(x) = h(r(x))$, for some increasing $0 < h \in C^2(\mathbb{R}^+)$. Then,

$$\Delta_\varphi u = \varphi'(h')h'' + \varphi(h')\Delta r \geq \varphi'(h')h'' + \zeta\varphi(h')r^{\alpha/2}$$

$$= \frac{(h')^{p-2}}{(1+h')^q}\left\{\left[(p-1) + (p-q)h'\right]h'' + \zeta r^{\alpha/2}h'(1+h')\right\}.$$

(7.85)

Set $h(t) = t^\sigma$, $\sigma > 0$. Then,

$$\Delta_\varphi u \geq C_1\frac{r^{(\sigma-1)(p-1)-1}}{(1+\sigma r^{\sigma-1})^q}.$$

(7.86)

$$\times\left\{\left[(p-1) + \sigma(p-q)r^{\sigma-1}\right](\sigma-1) + \zeta r^{\alpha/2+1}(1+\sigma r^{\sigma-1})\right\}$$

for some $C_1(p,\sigma) > 0$. If $\sigma \geq 1$, getting rid of the first term in brackets and using the definition of ζ, we obtain

$$\Delta_\varphi u \geq C_2\frac{r^{(\sigma-1)(p-1)-1+(\alpha/2+1)}}{(1+\sigma r^{\sigma-1})^{q-1}}.$$

(7.87)

On the other hand, if $\sigma \in (0, 1]$ and $\alpha > -2$, the term in between brackets in (7.86) is

$$(p-1)(\sigma-1)(1+o(1)) + \zeta r^{\alpha/2+1}(1+o(1)) \geq C_3\zeta r^{\alpha/2+1},$$

(7.88)

for large enough r, and the last inequality of (7.87) still holds. If $\sigma \in (0, 1)$ and $\alpha = -2$, the term is

$$\left[(p-1)(\sigma-1) + \zeta\right](1+o(1)),$$

while if $\sigma = 1$, $\alpha = -2$, the term is simply 2ζ. If r is large enough, the last expression is bounded from below by a positive constant whenever $\lim_{r\to\infty}\zeta > -(\sigma-1)(p-1)$, that is, when the first in (7.83) is met (i.e. the volume growth in 3) of (7.80)). Summarizing, for each $\sigma > 0$ (if $\sigma \leq 1$ and $\alpha = -2$, under the growth condition in 3) of (7.80)), the function u turns out to solve (7.87) for a suitable constant $C_2 > 0$. Since

$$(1+r)^{-\mu}\frac{\varphi(|\nabla u|)}{|\nabla u|^\chi} \leq C_4\frac{r^{-\mu+(\sigma-1)(p-1-\chi)}}{(1+\sigma r^{\sigma-1})^{q-1}},$$

(7.79) holds for r large enough provided that

$$-\mu + (\sigma-1)(p-1-\chi) \leq (\sigma-1)(p-1) - 1 + (\alpha/2 + 1),$$

that is, simplifying, if $\frac{\alpha}{2}+1 \geq \chi +1-\mu - \chi \sigma$. The relation is automatically satisfied both in cases (1), (2) and (3) of (7.80), and in case (4), it is equivalent to the growth condition. We have thus shown (7.79), as required.

To prove (5) and (6), we use $h(t) = t^{\sigma}/\log t$ in (7.85) and we consider for convenience the p-Laplacian ($q = 0$) to obtain

$$\Delta_{\varphi}u \geq C_1 \frac{r^{(\sigma-1)(p-1)-1}}{\log^{p-1}t}\left\{(p-1)(\sigma-1)\big(1+o(1)\big) + \zeta r^{1+\frac{\alpha}{2}}\big(1+o(1)\big)\right\},$$

as $r \to \infty$, for some constant $C_1(\sigma, p) > 0$.

In order to meet the volume conditions in (5), (6) and $\chi \sigma < \chi +1 - \mu$, necessarily $\alpha > -2$, and thus,

$$\Delta_{\varphi}u \geq C_2 \frac{r^{(\sigma-1)(p-1)+\frac{\alpha}{2}}}{\log^{p-1}t}\big(1+o(1)\big)$$

as $r \to \infty$. On the other hand, by the definition of Δ_{φ} on Ω_{γ}, we have

$$(1+r)^{-\mu}\frac{\varphi(|\nabla u|)}{|\nabla u|^{\chi}} \leq C_3 \frac{r^{-\mu+(\sigma-1)(p-1)-\chi(\sigma-1)}}{\log^{p-1}r}\log^{\chi}r.$$

Combining the last two inequalities, we examine the two cases:

- In case (5), $\chi > 0$ and u solves (7.79) whenever α is big enough to satisfy

$$(\sigma-1)(p-1) + \frac{\alpha}{2} > -\mu + (\sigma-1)(p-1) - \chi(\sigma-1),$$

that is,

$$1 + \frac{\alpha}{2} > \chi + 1 - \mu - \chi \sigma > 0, \qquad (7.89)$$

that in view of (7.78) implies

$$\lim_{r \to \infty}\frac{\log|B_r|}{r^{\chi+1-\mu-\chi\sigma}} = \infty,$$

as required.

- In case (6), because of $\chi = 0$ we infer that u solves (7.79) provided that the weak inequality

$$1 + \frac{\alpha}{2} \geq \chi + 1 - \mu - \chi \sigma > 0,$$

is satisfied. Choosing α in order to meet the equality sign, we obtain

$$\lim_{r\to\infty} \frac{\log|B_r|}{r^{\chi+1-\mu-\chi\sigma}} \in (0,\infty),$$

as required.

Remark 7.24 Example 7.84 satisfies (7.3) with $\bar{p} = p - q$. The case

3') $\qquad \chi\sigma = \chi + 1 - \mu, \qquad \alpha = -2, \qquad \sigma > 1 \qquad$ and

$$\lim_{r\to\infty} \frac{\log|B_r|}{\log r} > \bar{p} - \sigma(\bar{p} - 1)$$

is not covered by our counterexamples.

Strong Maximum Principle and Khas'minskii Potentials

The aim of this section is to prove Theorem 2.17 in the Introduction. We observe that the argument is based on the existence of what we call a "Khas'minskii potential", according to the following

Definition 8.1 A Khas'minskii potential at $o \in M$ is a function $\bar{w}$ depending on the parameters $r_0, r_1, \eta, K, \varepsilon > 0$ satisfying the next requirements:

$$\begin{cases} \Delta_\varphi \bar{w} \leq K b(x) l(|\nabla \bar{w}|) & \text{on } M \backslash B_{r_0} \\ \bar{w} > 0 & \text{on } M \backslash B_{r_0}, \\ \bar{w} \leq \eta & \text{on } B_{r_1} \backslash B_{r_0}, \\ \bar{w}(x) \to \infty & \text{as } r(x) \to \infty, \\ |\nabla \bar{w}| \leq \varepsilon & \text{on } M \backslash B_{r_0}. \end{cases} \tag{8.1}$$

The strategy to prove Theorem 2.17 is as follows: assuming, by contradiction, that (SMP_∞) does not hold, the Khas'minskii potential w will be compared to a non-constant, bounded solution $u \in C^1(M)$ of

$$\Delta_\varphi u \geq K b(x) l(|\nabla u|)$$

on an appropriate subset of $\Omega_{\eta,\varepsilon}$ to reach the conclusion. This approach is very old and we can trace it back, for instance, to Phrágmen-Lindelöff in the realm of classical complex analysis. In more recent times, it has been used by Redheffer [202, 203], and later refined in [182, Sect. 6], [183, Thm. 18] and [6, Ch. 3]. A similar approach, although with a somehow different point of view, was systematically used by Serrin [218] and recently

© The Author(s), under exclusive license to Springer Nature Switzerland AG 2021
B. Bianchini et al., *Geometric Analysis of Quasilinear Inequalities on Complete Manifolds*, Frontiers in Mathematics, https://doi.org/10.1007/978-3-030-62704-1_8

by Pucci-Serrin [194, Thm. 8.1.1]. In all of the quoted results, the Khas'minskii potential (or variants thereof) is either a data of the problem or it is constructed "ad-hoc". These methods seem difficult to be applied when a nontrivial gradient term l is present. Thus, we provide the existence of an appropriate potential via the solution of an associated ODE problem, see Lemma 8.2 below, that will be coupled with a condition on the Ricci tensor to transplant the function on the original manifold.

8.1   Ricci Curvature and (SMP_∞)

To investigate the ODE problem mentioned above, we assume the following:

$$
\begin{cases}
\varphi \in C(\mathbb{R}_0^+) \cap C^1(\mathbb{R}^+), & \varphi(0) = 0, & \varphi' > 0 \text{ on } \mathbb{R}^+, \\
f \in C(\mathbb{R}), & f(0) = 0, & f > 0 \text{ on } \mathbb{R}^+, \\
l \in C(\mathbb{R}_0^+), & l > 0 \text{ on } \mathbb{R}^+, \\
\beta \in C(\mathbb{R}_0^+), & \beta > 0 \text{ on } \mathbb{R}_0^+,
\end{cases}
\tag{8.2}
$$

and the next growth conditions:

$$
\begin{cases}
\dfrac{t\varphi'(t)}{l(t)} \in L^1(0^+), & \\[2mm]
l(t) \geq C_1 \dfrac{\varphi(t)}{t^\chi} & \text{on } (0,1] \text{ for some } C_1 > 0, \; \chi \geq 0 \\[2mm]
\varphi(t) \leq C t^{p-1} & \text{on } [0,1], \text{ for some } C > 0, \; p > 1.
\end{cases}
\tag{8.3}
$$

Fix a "volume" function $v \in C^2(\mathbb{R}^+)$ such that

$$
v > 0 \text{ on } \mathbb{R}^+, \qquad v' \geq 0 \qquad \text{on } \mathbb{R}^+.
\tag{8.4}
$$

We are now ready to prove

Lemma 8.2 *Assume the validity of* (8.2), (8.3), (8.4) *and furthermore suppose that*

$$
\beta(r) \geq C(1+r)^{-\mu} \qquad \text{on } \mathbb{R}_0^+, \text{ for some } \mu \leq \chi + 1,
$$

$$
\limsup_{r \to \infty} \frac{1}{v(r)} \int_{r_0}^r \frac{v(s)}{(1+s)^\mu}\, ds < \infty,
\tag{8.5}
$$

for some (hence any) $r_0 > 0$, and that either

$$\mu < \chi + 1 \quad and \quad \liminf_{t \to \infty} \frac{\log \int_{r_0}^t v}{t^{\chi+1-\mu}} < \infty \quad (= 0 \ if \ \chi = 0), \quad or$$

$$\mu = \chi + 1 \quad and \quad \liminf_{t \to \infty} \frac{\log \int_{r_0}^t v}{\log t} < \infty \quad (\leq p \ if \ \chi = 0).$$

(8.6)

Then, for each $0 < r_0 < r_1$, $\eta > 0$ and $\varepsilon > 0$ there exists a function $w \in C^1([r_0, \infty))$ satisfying:

$$\begin{cases} \left[v\varphi(w') \right]' \leq v\beta f(w)l(|w'|) & on \ [r_0, \infty) \\ w > 0, \ w' > 0 & on \ [r_0, \infty), \\ w \leq \eta & on \ [r_0, r_1], \\ w(t) \to \infty & as \ t \to \infty, \\ |w'| \leq \varepsilon & on \ [r_0, \infty). \end{cases}$$

Proof Clearly, it is enough to prove the result for $\beta(r) = C(1 + r)^{-\mu}$ and for η and ε small enough. First of all, we modify f, l, φ and choose a suitable η. These adjustments will be essential to prove the L^∞-gradient bound for w. Fix $\xi \in (0, 1]$ and choose $\bar{l}(t)$ satisfying

$$\bar{l} \in C(\mathbb{R}_0^+), \qquad \bar{l} > 0 \ on \ \mathbb{R}^+, \qquad \bar{l} \leq 2\|l\|_{L^\infty([0,\xi])} \ on \ \mathbb{R}^+,$$

$$\bar{l}(t) = l(t) \ if \ t \in [0, \xi), \qquad \bar{l}(t)t^\chi \geq \bar{C}_2 \ on \ [\xi, \infty),$$

(8.7)

for some constant $\bar{C}_2$ (here, we use $\chi \geq 0$). Regarding φ, we choose $\bar{\varphi} \in C^1(\mathbb{R}_0^+)$ such that

$$\bar{\varphi}' > 0 \ on \ \mathbb{R}^+, \qquad \bar{\varphi} = \varphi \ on \ (0, \xi), \qquad \bar{\varphi} \leq \bar{C}_1 t^\chi \bar{l}(t) \ on \ [\xi, \infty)$$

note that this is possible if $\bar{C}_1$ is sufficiently large, by the last of (8.7). By construction, since $\varphi(0) = 0$,

$$\bar{\varphi}(t) \leq Ct^{p-1} \ on \ [0, 1], \quad \bar{\varphi}(t) \leq C_3 t^\chi \ on \ [\xi, \infty), \quad \bar{l}(t) \geq C_4 \frac{\bar{\varphi}(t)}{t^\chi} \ on \ \mathbb{R}^+,$$

(8.8)

for some constants $C_3, C_4 > 0$. For $\eta > 0$ and $\xi > 0$, we introduce the notation

$$\beta_0 = \min_{[r_0, r_1]} \beta, \qquad \beta_1 = \max_{[r_0, r_1]} \beta;$$

$$v_0 = \min_{[r_0, r_1]} v, \qquad v_1 = \max_{[r_0, r_1]} v;$$

$$f_{2\eta} = \max_{[0, 2\eta]} f, \qquad l_\xi = \max_{[0, \xi]} \bar{l}.$$

Given $\sigma \in (0, \xi)$ to be specified later, we choose $\eta_\sigma \in [0, \sigma)$ small enough in order to satisfy

$$\frac{v_1}{v_0} \left[(r_1 - r_0)\beta_1 f_{2\eta_\sigma} l_\xi + \bar{\varphi} \left(\frac{\eta_\sigma}{r_1 - r_0} \right) \right] < \bar{\varphi}(\sigma). \qquad (8.9)$$

This is possible because $\bar{\varphi}(0) = 0$ and $f_{2\eta_\sigma} \to 0$ as $\eta_\sigma \to 0$ (since $f(0) = 0$). We next choose $f_\sigma \in C(\mathbb{R}_0^+)$ satisfying

$$0 \le f_\sigma(t) \le \min\{f(t), 1\}, \qquad f_\sigma(t) = 0 \;\; \text{if } t \le \eta_\sigma,$$
$$f_\sigma > 0 \;\; \text{if } t > \eta_\sigma, \qquad f_\sigma(\eta_\sigma + t) \le K'(t) \;\; \text{for } t \in [0, \xi], \qquad (8.10)$$

where $K(t)$ is the function defined in (2.7). The last condition can be satisfied because of the positivity of φ' and l on $\mathbb{R}^+$, hence of K'. We consider the Dirichlet problem

$$\begin{cases} ([v\bar{\varphi}(w_\sigma')])' = \sigma v\beta f_\sigma(w_\sigma)\bar{l}(|w_\sigma'|) & \text{on } [r_0, r_1], \\ w_\sigma(r_0) = \eta_\sigma, \qquad w_\sigma(r_1) = 2\eta_\sigma, \\ \eta_\sigma \le w_\sigma \le 2\eta_\sigma, \qquad w_\sigma' > 0 \;\; \text{on } [r_0, r_1]. \end{cases} \qquad (8.11)$$

We claim that a solution $w_\sigma \in C^1([r_0, r_1])$ exists if $\sigma > 0$ is small enough. Indeed, one can apply Theorem 5.1 with the following choices:

$$t = r - r_0, \qquad T = r_1 - r_0, \qquad \wp(t) = v(t + r_0),$$
$$w(t) = w_\sigma(r_0 + t) - \eta_\sigma, \qquad a(t) = \beta(r_0 + t)$$

and $f(t), \varphi(t), l(t)$ in Theorem 5.1 replaced, respectively, by $\sigma f_\sigma(\eta_\sigma + t), \bar{\varphi}(t)$ and $\bar{l}(t)$ (note that $f(0) = 0$). It is easy to see that (8.9) implies (5.7) for each $\sigma \in (0, \xi]$, hence Theorem 5.1 can be applied to guarantee the existence of a solution w_σ of (8.11) together with the bound $|w_\sigma'| \le \sigma$. We are left to prove that $w_\sigma' > 0$ on $[r_0, r_1]$, provided that σ is small enough. Because of Lemma 5.2, it is enough to show $w_\sigma'(r_0) > 0$ and to this aim we follow the argument in Proposition 5.7, see also Remark 5.8. Indeed, suppose by contradiction that $w_\sigma'(r_0) = 0$. By Lemma 5.2, there exists $\bar{r}_\sigma \in [r_0, r_1)$ such that $w_\sigma = \eta_\sigma$

for $r \leq \bar{r}_\sigma$, $w'_\sigma(\bar{r}_\sigma) = 0$ and $w'_\sigma > 0$ on $(\bar{r}_\sigma, r_1]$. Expanding (8.11) and using $v' \geq 0$ we deduce

$$\frac{w'_\sigma \bar{\varphi}'(w'_\sigma)}{\bar{l}(w'_\sigma)} w''_\sigma \leq \sigma \beta_1 f_\sigma(w_\sigma) w'_\sigma,$$

and integrating on $(\bar{r}_\sigma, r)$ we get

$$\bar{K}(w'_\sigma) \leq \sigma \beta_1 F_\sigma(w_\sigma), \tag{8.12}$$

where

$$\bar{K}(t) = \int_0^t \frac{s \bar{\varphi}'(s)}{\bar{l}(s)} \, ds, \qquad F_\sigma(t) = \int_{\eta_\sigma}^t f_\sigma(s) \, ds.$$

Because of (8.10), for $t \in [0, \xi + \eta_\sigma]$

$$F_\sigma(t) \leq \int_{\eta_\sigma}^t K'(s - \eta_\sigma) \, ds = K(t - \eta_\sigma) \equiv \bar{K}(t - \eta_\sigma),$$

where in the last inequality we have used $\bar{\varphi} = \varphi$ and $\bar{l} = l$ on $[0, \xi]$. Choosing $\sigma \leq \beta_1^{-1}$ from (8.12) we deduce the inequality $\bar{K}(w'_\sigma) \leq \bar{K}(w_\sigma - \eta_\sigma)$, and therefore $w'_\sigma \leq w_\sigma - \eta_\sigma$. By Gronwall's inequality and $w'_\sigma(\bar{r}_\sigma) = 0$ we obtain $w_\sigma \equiv \eta_\sigma$ on $[\bar{r}_\sigma, r_1]$, contradicting $w_\sigma(r_1) = 2\eta_\sigma$.

We now let $[r_0, R)$, $R > r_1$ be the maximal interval where w_σ is defined. Integrating (8.11), and using $|w'_\sigma| \leq \sigma$ on $[r_0, r_1]$ and $\|f_\sigma\|_\infty \leq 1$, for $r \in [r_0, R)$ we get

$$\bar{\varphi}(w'_\sigma(r)) = \frac{1}{v(r)} \left[\bar{\varphi}(w'_\sigma(r_0)) v(r_0) + \sigma \int_{r_0}^r v(s) \beta(s) f_\sigma(w_\sigma) \bar{l}(w'_\sigma) \, ds \right]$$

$$\leq \frac{1}{v(r)} \left[\bar{\varphi}(\sigma) v(r_0) + \sigma \|f_\sigma\|_\infty \|\bar{l}\|_\infty \int_{r_0}^r v(s) \beta(s) \, ds \right] \tag{8.13}$$

$$\leq C(\bar{\varphi}(\sigma) + \sigma),$$

where $C > 0$ is a constant depending on r_0 but independent of R, σ, and where the last inequality follows from $v' \geq 0$, the second in (8.5) and $\beta(r) = (1 + r)^{-\mu}$. In particular, if σ is small enough then w_σ has bounded gradient. Since w_σ is also increasing, necessarily $R = \infty$, otherwise one could extend the solution beyond R. Up to a further reduction of σ, by (8.13) we can guarantee $|w'_\sigma| \leq \varepsilon$ on $[r_0, \infty)$, ε being the parameter in the statement of the Lemma that we can assume to belong to $(0, \xi)$. On the other hand, from the first line in (8.13) and the positivity of w'_σ on $[r_0, r_1]$ we deduce $w'_\sigma > 0$ on all of $[r_0, \infty)$. It is clear that, for each $\sigma < \xi$, by construction w_σ solves

$$\left[v\varphi(w_\sigma')\right]' = \left[v\bar\varphi(w_\sigma')\right]' = \sigma v \beta f_\sigma(w_\sigma)\bar{l}(w_\sigma') \le v \beta f(w_\sigma)l(w_\sigma'),$$

as required. We are left to prove that $w_\sigma(r) \to \infty$ as $r \to \infty$. Suppose, by contradiction, that $w_\sigma^* = \sup_{[r_0,\infty)} w_\sigma < \infty$, and consider the model manifold M_g with metric

$$ds_g^2 = dr^2 + g(r)^2 d\theta^2$$

and warping function $g \in C^2(\mathbb{R}_0^+)$ satisfying

$$g > 0 \quad \text{on } \mathbb{R}^+, \qquad g(r) = \begin{cases} r & \text{for } r \in (0, r_0/2) \\ v(r)^{\frac{1}{m-1}} & \text{for } r \ge r_0. \end{cases}$$

By construction, the radial function $w_\sigma(r)$ on M_g is a solution of

$$\Delta_{\bar\varphi} w_\sigma = \sigma \beta(r) f_\sigma(w_\sigma)\bar{l}(|\nabla w_\sigma|) \qquad \text{on } M_g \backslash \overline{B}_{r_0},$$

and from $w_\sigma' > 0$ we get $w_\sigma^* > \eta_\sigma$. Consider the Lipschitz extension of w_σ obtained by setting $w_\sigma = \eta_\sigma$ on B_{r_0}. An analogous reasoning as in Lemma 6.6 shows that $f_\sigma(\eta_\sigma) = 0$ and $w_\sigma' > 0$ guarantee that the extended function solves $\Delta_{\bar\varphi} w_\sigma \ge \sigma \beta(r) f_\sigma(w_\sigma)\bar{l}(|\nabla w_\sigma|)$ on M_g. Combining properties (8.5) (for b) and (8.8) (for $\bar\varphi, \bar{l}$) with the volume growth conditions (8.6), and fixing any $\bar{p} > \chi + 1$, we are in the position to apply case (i) of Theorem 7.5 to deduce that necessarily $f_\sigma(w_\sigma^*) \le 0$, hence $w_\sigma \le \eta_\sigma$ because of (8.10). This contradicts the previously established inequality $w_\sigma^* > \eta_\sigma$, and concludes the proof.

$\square$

Remark 8.3 We stress that no growth condition on φ at infinity is required. Indeed, we applied the weak maximum principle (Theorem 7.5) to the modification $\bar\varphi$, but a posteriori just the value of φ for sufficiently small t is needed to produce the solution.

Remark 8.4 The simultaneous validity of the second in (8.5) and of (8.6) requires a delicate balancing between μ and $v(r)$, since (8.5) is easier to satisfy for μ large while (8.6) forces an upper bound on μ. For various examples of $v(r)$ with geometric interest, in particular for those appearing in the next theorem, there is a non-empty interval of μ for which both the conditions are met.

We are now ready to prove Theorem 2.17. We report here the statement to facilitate the reading.

Theorem 8.5 *Let M^m be a complete Riemannian manifold of dimension $m \geq 2$ such that, for some origin $o \in M$, the distance function $r(x)$ from o satisfies*

$$\mathrm{Ric}(\nabla r, \nabla r) \geq -(m-1)\kappa^2(1+r^2)^{\alpha/2} \qquad \text{on } D_o, \tag{8.14}$$

for some $\kappa \geq 0$ and $\alpha \geq -2$. Let l, φ satisfy (8.2) and (8.3), for some $\chi \geq 0$ and $p > 1$. Consider $b \in C(M)$ such that

$$b(x) \geq C(1+r(x))^{-\mu} \qquad \text{on } M,$$

for some constants $C > 0$, $\mu \in \mathbb{R}$. Assume

$$\mu \leq \chi - \frac{\alpha}{2} \quad \text{and either} \quad \begin{cases} \alpha \geq -2 \quad \text{and} \quad \chi > 0, \quad \text{or} \\ \alpha = -2, \quad \chi = 0 \quad \text{and} \quad \bar{\kappa} \leq \frac{p-1}{m-1}, \end{cases} \tag{8.15}$$

with $\bar{\kappa} = \frac{1}{2}(1 + \sqrt{1 + 4\kappa^2})$. Then, the operator $(bl)^{-1}\Delta_\varphi$ satisfies (SMP$_\infty$).

Proof Suppose, by contradiction, that there exists a non-constant $u \in C^1(M)$ with $u^* = \sup_M u < \infty$, and $\eta, \varepsilon > 0$ such that

$$\Delta_\varphi u \geq 2Kb(x)l(|\nabla u|) \qquad \text{on } \Omega_{\eta,\varepsilon} = \{x : u(x) > \eta, \ |\nabla u(x)| < 2\varepsilon\},$$

for some constant $K > 0$. In particular, $\Delta_\varphi u \geq 0$ in $\Omega_{\eta,\varepsilon}$, and thus, since points realizing u^* belong to $\Omega_{\eta,\varepsilon}$, applying the finite maximum principle, Theorem 6.8, to $u^* - u$ with $f \equiv 0$ we deduce that u^* is never attained. Consequently, since $u \in C^1(M)$ we infer that $\Omega_{\eta,\varepsilon}$ is unbounded. In what follows, balls are always considered to be centred at o. Fix $r_0 > 0$ and choose $\gamma \in (0, \eta/2)$ in such a way that

$$u(x) < u^* - 4\gamma \qquad \text{for each } x \in B_{r_0}.$$

Next, we choose $\bar{x} \in \Omega_{\eta,\varepsilon}$ such that $u(\bar{x}) > u^* - \gamma$, and a large ball $B_{r_1} \Subset M$ containing $\overline{B_{r_0}} \cup \{\bar{x}\}$. By (8.14) and [185, Prop. 2.1],

$$\Delta r \leq (m-1)\frac{g'(r)}{g(r)} \qquad \text{weakly on } M, \tag{8.16}$$

for some $g(r) \in C^2(\mathbb{R}_0^+)$ increasing and satisfying (3.14), that is,

$$
g(r) \asymp
\begin{cases}
\exp\left\{\dfrac{2\kappa}{2+\alpha}(1+r)^{1+\frac{\alpha}{2}}\right\} & \text{if } \alpha \geq 0 \\[3ex]
r^{-\frac{\alpha}{4}}\exp\left\{\dfrac{2\kappa}{2+\alpha}r^{1+\frac{\alpha}{2}}\right\} & \text{if } \alpha \in (-2,0) \\[3ex]
r^{\bar{\kappa}}, \quad \bar{\kappa} = \dfrac{1+\sqrt{1+4\kappa^2}}{2} & \text{if } \alpha = -2
\end{cases}
\tag{8.17}
$$

for $r \in [1,\infty)$. Define $v_g(r) = \omega_{m-1}g(r)^{m-1}$, and note that

$$
\log \int_{r_0}^r v_g \sim
\begin{cases}
\dfrac{2\kappa(m-1)}{2+\alpha}r^{1+\frac{\alpha}{2}} & \text{if } \alpha > -2, \\[3ex]
\left[(m-1)\bar{\kappa}+1\right]\log r & \text{if } \alpha = -2
\end{cases}
\tag{8.18}
$$

as $r \to \infty$. For each $\alpha \geq -2$, set

$$
\beta(r) = C(1+r)^{-\bar{\mu}}, \qquad \text{with} \qquad \bar{\mu} = \chi - \frac{\alpha}{2}.
$$

Because of (8.15), $\mu \leq \bar{\mu}$ and therefore

$$
b(x) \geq \beta\big(r(x)\big).
\tag{8.19}
$$

Furthermore, by (8.17) and the fact that $\bar{\mu} \geq -\alpha/2$, we get

$$
\limsup_{r\to\infty} \frac{1}{v_g(r)} \int_{r_0}^r \frac{v_g(s)}{(1+s)^{\bar{\mu}}}\,ds < \infty
\tag{8.20}
$$

while, in view of (8.18) we obtain

$$
\text{if } \alpha > -2, \quad \text{then } \bar{\mu} < \chi + 1 \text{ and } \quad \lim_{r\to\infty} \frac{\log \int_{r_0}^r v}{r^{\chi+1-\bar{\mu}}} < \infty;
$$

$$
\text{if } \alpha = -2, \quad \text{then } \bar{\mu} = \chi + 1 \text{ and } \quad \lim_{r\to\infty} \frac{\log \int_{r_0}^r v}{\log r} < \infty \quad (\leq p \text{ if } \chi = 0).
$$

We are in the position to apply Lemma 8.2: for $\eta = \gamma_\varepsilon \leq \gamma$ small enough, there exists $w \in C^1([r_0, \infty))$ satisfying:

$$
\begin{cases}
\left[v_g \varphi(w')\right]' \leq v_g K \beta l(|w'|) & \text{on } [r_0, \infty) \\
w > 0, \; w' > 0 & \text{on } [r_0, \infty), \\
w \leq \gamma_\varepsilon & \text{on } [r_0, r_1], \\
w(r) \to \infty & \text{as } r \to \infty, \\
|w'| \leq \varepsilon & \text{on } [r_0, \infty).
\end{cases}
$$

We define the radial function $\bar{w}(x) = w(r(x))$, and we note that, because of (8.16), $w' > 0$ and (8.19), $\bar{w}$ solves

$$
\begin{cases}
\Delta_\varphi \bar{w} \leq K \beta(r) l(w'(r)) \leq K b(x) l(w'(r)) & \text{on } M \backslash B_{r_0} \\
\bar{w} > 0 & \text{on } M \backslash B_{r_0}, \\
\bar{w} \leq \gamma_\varepsilon & \text{on } B_{r_1} \backslash B_{r_0}, \\
\bar{w}(x) \to \infty & \text{as } r(x) \to \infty, \\
|\nabla \bar{w}| \leq \varepsilon & \text{on } M \backslash B_{r_0}.
\end{cases}
$$

Let Γ be the set of maxima of $u - \bar{w}$, which is non-empty and compact since $\bar{w}$ has compact sublevel sets. For each $x \in \Gamma$

$$
u(x) - \bar{w}(x) \geq u(\bar{x}) - \bar{w}(\bar{x}) > u^* - \gamma - \gamma_\varepsilon \geq u^* - 2\gamma
$$

$$
> \max_{B_{r_0}} u \geq \max_{B_{r_0}}(u - \bar{w}),
$$

(8.21)

hence $\Gamma \subset M \backslash \overline{B}_{r_0}$. From the first line in (8.21), we get

$$
u(x) \geq u^* - 2\gamma + \bar{w}(x) \geq u^* - 2\gamma > u^* - \eta,
$$

and thus $\Gamma \subset \left(M \backslash \overline{B}_{r_0}\right) \cap \{u > u^* - \eta\}$. Furthermore, for each $x \in \Gamma \backslash \mathrm{cut}(o)$ it holds

$$
|\nabla u(x)| = |\nabla \bar{w}(x)| = w'(r(x)) \leq \varepsilon.
$$

We claim that the same relation holds even for $x \in \Gamma \cap \mathrm{cut}(o)$. Let $\sigma : [0, r(x)] \to M$ be a unit speed minimizing geodesic from o to x, and for $0 < \tau << 1$ define $r_\tau(\cdot) = \tau + \mathrm{dist}(., \sigma(\tau))$. Then, $r_\tau \geq r$, with equality at x, and furthermore r_τ is smooth around x. Setting $\bar{w}_\tau(x) = w(r_\tau(x))$, $w' > 0$ implies that $\bar{w}_\tau \geq \bar{w}$, with equality at x. Hence x is a maximum for $u - \bar{w}_\tau$, which gives

$$|\nabla u(x)| = |\nabla w_\tau(x)| = w'\big(r_\tau(x)\big) = w'\big(r(x)\big) \leq \varepsilon.$$

We have therefore shown that $\Gamma \in \Omega_{\eta,\varepsilon}$. Equality $|\nabla u| = w'(r)$, combined with $w \in C^1$ and $w' > 0$ on $[r_0, \infty)$, guarantees the existence of $\delta > 0$ such that $\delta \leq |\nabla u| \leq \varepsilon$ on Γ. Using the continuity and positivity of l, we can fix a small open neighbourhood $V \in \Omega_{\eta,\varepsilon}$ of Γ of the form $\{u - \bar{w} > c\}$, c close enough to $\max\{u - \bar{w}\}$, such that $l(|\nabla u|) \geq 2^{-1} l(w'(r))$ on V. Consequently,

$$\begin{cases} \Delta_\varphi u \geq 2Kb(x)l(|\nabla u|) \geq Kb(x)l\big(w'(r(x))\big) \geq \Delta_\varphi \bar{w} = \Delta_\varphi(\bar{w} + c) & \text{on } V \\ u = \bar{w} + c & \text{on } \partial V. \end{cases}$$

By comparison, $u \leq \bar{w} + c$ on V, contradicting the very definition of V. $\square$

Remark 8.6 Unfortunately, Lemma 8.2 cannot be applied as above to prove Theorem 8.5 also in the range $\alpha > -2$ and $\chi = 0$. We recall that, for the p-Laplace operator, this corresponds to gradient terms with borderline growth $l(t) \asymp t^{p-1}$ for $t \in (0,1)$. In fact, for (8.20) to hold it is necessary that $\bar{\mu} \geq -\alpha/2$, but on the other hand, because of (8.18),

$$\text{if } \bar{\mu} < \chi + 1 = 1, \quad \text{then} \quad \liminf_{r \to \infty} \frac{\log \int_{r_0}^r v}{r^{\chi+1-\bar{\mu}}} = 0 \quad \text{iff} \quad \bar{\mu} < \chi - \frac{\alpha}{2} = -\frac{\alpha}{2}$$

$$\text{if } \bar{\mu} = \chi + 1 = 1, \quad \text{then} \quad \liminf_{r \to \infty} \frac{\log \int_{r_0}^r v}{\log r} \leq p \quad \text{does not hold for any } \alpha > -2.$$

Therefore, no choice of $\bar{\mu}$ is admissible for Lemma 8.2.

To better appreciate Theorem 8.5, we express it for the mean curvature operator, both in Euclidean space $\mathbb{R}^m$ and in the hyperbolic space $\mathbb{H}^m$.

Corollary 8.7 *Let $l \in C(\mathbb{R}_0^+)$ satisfy*

$$l(t) \geq C_1 \frac{t^{1-\chi}}{\sqrt{1+t^2}} \qquad \text{on } [0,1],$$

for some $\chi \in [0,1]$. Then, (SMP$_\infty$) holds for the operator

$$\big(1 + r(x)\big)^\mu l(|\nabla u|)^{-1} \mathrm{div}\left(\frac{\nabla u}{\sqrt{1 + |\nabla u|^2}}\right),$$

(i) in $\mathbb{R}^m$, provided that $\mu \leq \chi + 1$ and either $\chi > 0$ or $\chi = 0$ and $m = 2$;
(ii) in $\mathbb{H}^m$, provided that $\mu \leq \chi$ and $\chi > 0$.

Proof Let $\varphi(t) = t/\sqrt{1+t^2}$, and choose $p = 2$ in (8.3). To recover the Euclidean space set $\kappa = 0$, $\alpha = -2$, while for the hyperbolic space set $\kappa = 1$, $\alpha = 0$. The rest of the proof is a direct application of Theorem 8.5. □

As a further application of Theorem 8.5, in the next corollary we obtain a Liouville theorem for bounded solutions of

$$\Delta_p u \geq b(x) f(u) |\nabla u|^q - \bar{b}(x) \bar{f}(u) |\nabla u|^{\bar{q}}. \tag{8.22}$$

Corollary 8.8 *Let $(M, \langle , \rangle)$ be a m–dimensional complete manifold satisfying*

$$\mathrm{Ric}(\nabla r, \nabla r) \geq -(m-1)\kappa^2 (1+r^2)^{\alpha/2} \qquad on \ D_o,$$

for some $\kappa \geq 0$, $\alpha \geq -2$. Consider $b \in C(M)$ such that

$$b(x) \geq C_1 (1 + r(x))^{-\mu} \qquad on \ M,$$

for some $C_1 > 0$. Let $f, \bar{f} \in C(\mathbb{R})$, $\bar{b} \in C(M)$ and $C > 0$ be such that

$$\bar{b} \leq Cb \quad on \ M, \qquad \bar{f} \leq Cf \quad on \ \mathbb{R}. \tag{8.23}$$

Fix

$$p \in (1, \infty), \qquad q \in [0, p-1), \qquad \bar{q} > q$$

and consider a bounded above solution $u \in C^1(M)$ of

$$\Delta_p u \geq b(x) f(u) |\nabla u|^q - \bar{b}(x) \bar{f}(u) |\nabla u|^{\bar{q}}. \tag{8.24}$$

If

$$\mu \leq p - 1 - q - \frac{\alpha}{2}$$

and u is non-constant, then $f(u^*) \leq 0$. In particular, if $u \in C^1(M) \cap L^\infty(M)$ solves (8.22) with the equality sign, and

$$C^{-1} f \leq \bar{f} \leq Cf \qquad on \ \mathbb{R}, \tag{8.25}$$

then, u must be constant in each of the following cases:

(i) $f < 0$ on $(-\infty, t_0)$ and $f > 0$ on (t_0, ∞);
(ii) f has no zeroes.

Remark 8.9 If $q = 0$, under (i) above the only constant solution of (8.22) with the equality sign is $u \equiv t_0$, while, under (ii), (8.22) with the equality sign does not admit any constant solution.

Proof Suppose by contradiction that $f(u^*) = 4K > 0$, and pick $\eta < u^*$ such that $f(t) > 2K$ if $t > \eta$. Because of (8.24) and (8.23), on

$$\Omega_{\eta,\varepsilon} = \left\{x \in M \ : \ u(x) > \eta, \ |\nabla u(x)| < \varepsilon\right\}$$

we have

$$\Delta_p u \geq b(x) f(u) |\nabla u|^q \left(1 - C^2 |\nabla u|^{\bar{q}-q}\right) \geq 2K b(x) |\nabla u|^q \left(1 - C^2 \varepsilon^{\bar{q}-q}\right),$$

and since $\bar{q} > q$ we can choose $\varepsilon > 0$ small enough that

$$\Delta_p u \geq K b(x) |\nabla u|^q \qquad \text{on } \Omega_{\eta,\varepsilon}. \tag{8.26}$$

Set $\chi = p - 1 - q \in (0, p - 1]$. Then, in our assumptions, we can apply Theorem 8.5 to deduce that $(bl)^{-1} \Delta_p$ satisfies (SMP$_\infty$). Consequently, from (8.26) we get $K \leq 0$, contradiction.

Suppose now, by contradiction, that $u \in C^1(M) \cap L^\infty(M)$ is a non-constant solution of (8.22) with the equality sign. By the first part of the proof we get $f(u^*) \leq 0$. Next, observe that $\bar{u} = -u$ solves

$$\Delta_p \bar{u} = b(x) f_1(\bar{u}) |\nabla \bar{u}|^q - \bar{b}(x) \bar{f}_1(\bar{u}) |\nabla \bar{u}|^{\bar{q}},$$

with $f_1(t) = -f(-t)$ and $\bar{f}_1(t) = -\bar{f}(-t)$. In view of (8.25), applying again the first part we obtain $f_1(\bar{u}^*) \leq 0$, that is, $f(u_*) \geq 0$ with $u_* = \inf_M u$. From $f(u^*) \leq 0 \leq f(u_*)$ and using (i) or (ii), we deduce that u is necessarily constant, contradiction. □

Remark 8.10 Theorem 8.5 could be improved to include slowly growing solutions of $(P_\geq)$ as in (ii) of Theorem 7.5, provided that one is able to estimate from below the order of growth of a family of Khas'minskii potentials (8.1) in a way independent of the origin o and of $\eta, r_0, r_1, \varepsilon$. If this holds, repeating the proof verbatim one shows that any solution u of $(P_\geq)$ on M, or on some upper level set, is bounded from above and satisfies $f(u^*) \leq 0$

whenever

$$u_+(x) = o(\bar{w}(x)) \qquad \text{as } r(x) \to \infty,$$

with $\bar{w}$ being any of such Khas'minskii potentials. Growth estimates are achieved provided that one can explicitly exhibit $\bar{w}$, and this is the case when $l(0) > 0$. Indeed, when $l(0) > 0$ the first two of (8.3) are automatically satisfied, and to produce solutions of (8.1) we can consider radial solutions of

$$(v_g \varphi(w'))' = \sigma v_g \beta \qquad \text{on } [r_0, \infty),$$

for small enough $\sigma > 0$. Explicit integration with $w(r_0) = w'(r_0) = 0$ gives

$$w(r) = \int_{r_0}^r \varphi^{-1} \left(\frac{\sigma}{v_g(t)} \int_{r_0}^t v_g(s)\beta(s)ds \right) dt.$$

This approach has been developed in Section 6 of [182] and in [183, Thm. 18], the latter dealing with inequality

$$\Delta u \geq (1+r)^{-\mu} f(u)l(|\nabla u|)$$

on complete manifolds satisfying $\text{Ric} \geq -(m-1)\kappa^2 \langle\,,\,\rangle$, for some $\kappa > 0$, for increasing f and for $\mu \in [0, 1]$. The conclusion $f(u(x)) \leq 0$ on M is shown to hold provided that, as $r(x) \to \infty$,

$$u(x) = \begin{cases} o(r(x)^{1-\mu}) & \text{if } \mu \in [0, 1), \\ o(\log r(x)) & \text{if } \mu = 1. \end{cases}$$

Inspection shows that the case $\mu < 1$ well fits with (ii) of Theorem 7.5 (apply the theorem with $\varphi(t) = t$, $\sigma = 1 - \mu$, $\chi = 1$ and use Bishop–Gromov comparison to check the first of (7.9)). Case $\mu = 1$, on the other hand, has no analogue in Theorem 7.5.

8.2 Bernstein Theorems for Prescribed Mean Curvature Graphs

We now apply (SMP$_\infty$) to entire graphs with prescribed mean curvature in a warped product ambient manifold $\bar{M} = \mathbb{R} \times_h M$. We recall that the mean curvature of the totally umbilic slice $\{s = s_0\}$ of $\bar{M}$ in the upward direction ∂_s is

$$H_{\partial_s}(\{s = s_0\}) = -\frac{h'(s_0)}{h(s_0)}.$$

The next theorem gives an a priori estimate for entire graphs with prescribed mean curvature, and in particular it characterizes all constant mean curvature entire graphs. For simplicity, we state the result for warped products with $h(s) = \cosh s$, a class including the fibration $\mathbb{H}^{m+1} = \mathbb{R} \times_{\cosh s} \mathbb{H}^m$ by hyperspheres $\{s = s_0\}$ of constant mean curvature $H = -\tanh s_0 \in (-1, 1)$.

Theorem 8.11 *Let $\bar{M} = \mathbb{R} \times_{\cosh s} M$, for some complete manifold $(M^m, \langle \, , \, \rangle)$ whose Ricci tensor satisfies*

$$\mathrm{Ric}(\nabla r, \nabla r) \geq -(m-1)\kappa^2(1+r)^2 \qquad on \ D_o,$$

for some constant $\kappa > 0$. Fix a constant $H_0 \in (-1, 1)$, and consider an entire geodesic graph of $v : M \to \mathbb{R}$ with prescribed mean curvature $H(x) \geq -H_0$ in the upward direction. Then, v is bounded from above and satisfies

$$v^* \leq \mathrm{arctanh}(H_0).$$

In particular,

(i) there is no entire graph with prescribed mean curvature satisfying $|H(x)| \geq 1$ on M;
(ii) the only entire graph with constant mean curvature $H_0 \in (-1, 1)$ in the upward direction is the totally umbilic slice $\{s = \mathrm{arctanh}(H_0)\}$.

Proof Define t, $\lambda(t)$ and $u(x)$ as in (1.10)–(1.12) in Chap. 1, with the choice $h(s) = \cosh s$:

$$t(s) = \int_0^s \frac{d\sigma}{\cosh \sigma} = 2\arctan(e^s) - \frac{\pi}{2}, \quad \lambda(t) = h(s(t)), \quad u(x) = t(v(x)).$$

Note that $u : M \to \left(-\frac{\pi}{2}, \frac{\pi}{2}\right)$. Since $\lambda(u) = \cosh v$ and $\lambda_t(u)/\lambda(u) = \sinh v$, by (1.12) u satisfies

$$\mathrm{div}\left(\frac{\nabla u}{\sqrt{1 + |\nabla u|^2}}\right) = m \cosh v \left[H(x) + \tanh v \frac{1}{\sqrt{1 + |\nabla u|^2}}\right]$$

$$\geq m \cosh v \left[-H_0 + \tanh v \frac{1}{\sqrt{1 + |\nabla u|^2}}\right].$$

Suppose, by contradiction, that the following upper level set of v (hence, of u) is non-empty for some $\eta > 0$:

$$\Omega_\eta = \{\tanh v > H_0 + \eta\}.$$

Then,

$$\mathrm{div}\left(\frac{\nabla u}{\sqrt{1 + |\nabla u|^2}}\right) \geq \frac{m \cosh v}{\sqrt{1 + |\nabla u|^2}}\left[\eta - H_0(\sqrt{1 + |\nabla u|^2} - 1)\right] \quad \text{on } \Omega_\eta.$$

If $H_0 < 0$, from $\cosh v \geq 1$ we deduce

$$\mathrm{div}\left(\frac{\nabla u}{\sqrt{1 + |\nabla u|^2}}\right) \geq \frac{m\eta}{\sqrt{1 + |\nabla u|^2}} \quad \text{on } \Omega_\eta. \tag{8.27}$$

On the other hand, if $H_0 > 0$, for $\varepsilon > 0$ we consider the set $\Omega_{\eta,\varepsilon} = \Omega_\eta \cap \{|\nabla u| < \varepsilon\}$. Note that $\Omega_{\eta,\varepsilon}$ is non-empty by Ekeland's quasi-maximum principle, since M is complete. If ε is sufficiently small, the term in square brackets is less than $\eta/2$, and since $\cosh v \geq 1$ we deduce

$$\mathrm{div}\left(\frac{\nabla u}{\sqrt{1 + |\nabla u|^2}}\right) \geq \frac{m\eta}{2\sqrt{1 + |\nabla u|^2}} \quad \text{on } \Omega_{\eta,\varepsilon}. \tag{8.28}$$

We now apply Theorem 8.5 with the choices $\alpha = 2$, $\mu = 0$, $\chi = 1$ to deduce the validity of (SMP_∞) for the operator $l^{-1}\Delta_\varphi$, with

$$\varphi = \frac{t}{\sqrt{1 + t^2}}, \qquad l(t) = \frac{1}{\sqrt{1 + t^2}}.$$

Since u is bounded from above, applying (SMP_∞) to (8.28) (for $H_0 > 0$) or to (8.27) (for $H_0 < 0$) we reach the desired contradiction.

To prove (i), suppose that $|H(x)| \geq 1$ on M. Since $\cosh s$ is even, the graph of $-v$ has curvature $-H(x)$ in the upward direction. Thus, up to replacing v with $-v$ we can suppose that $H(x) \geq 1$. Applying the first part of the theorem to any $H_0 > -1$ we obtain $v^* \leq \mathrm{arctanh}(H_0)$, and the nonexistence of v follows by letting $H_0 \to -1$.

To prove (ii), let $H(x) = -H_0 \in (-1, 1)$ be the mean curvature of the graph of v in the upward direction. Then, Theorem 8.11 gives $\tanh v^* \leq H_0$. On the other hand, the graph of $-v$ has mean curvature $H(x) = H_0$ in the upward direction, and applying again Theorem 8.11 we deduce $\tanh[(-v)^*] \leq -H_0$, that is, $\tanh v_* \geq H_0$. Combining the two estimates gives $v \equiv \mathrm{arctanh}(H_0)$, as required. $\qquad\square$

Remark 8.12 If $H_0 < 0$, to conclude from (8.27) it is sufficient to require the validity of (WMP_∞).

Remark 8.13 Observe that (ii) generalizes item (ii) in Do Carmo–Lawson Theorem 1.3: it is sufficient to apply Theorem 8.11 to $\mathbb{H}^{m+1}$ with the warped product structure $\mathbb{R} \times_{\cosh r} \mathbb{H}^m$.

Remark 8.14 The above result can be generalized, with the same proof, to warped products $\mathbb{R} \times_h M$ for h satisfying

$$
\begin{cases}
h \quad \text{even}, \\
h^{-1} \in L^1(-\infty) \cap L^1(\infty), \\
(h'/h)' > 0 \quad \text{on } \mathbb{R}.
\end{cases}
$$

We leave the statement to the interested reader.

The Compact Support Principle   9

Consider the problem

$$\begin{cases} \Delta_\varphi u \geq b(x) f(u) l(|\nabla u|) & \text{on } \Omega \text{ end of } M. \\ u \geq 0, \qquad \lim_{x \in \Omega,\, x \to \infty} u(x) = 0. \end{cases} \tag{9.1}$$

We recall that an end $\Omega \subset M$ is a connected component with non-compact closure of $M \backslash K$, for some compact set K. In this section, we investigate the necessity and sufficiency of condition

$$\frac{1}{K^{-1} \circ F} \in L^1(0^+) \tag{KO$_0$}$$

for the validity of the compact support principle (CSP), that is, the statement that each u solving (9.1) has compact support. We assume the following:

$$\begin{cases} \varphi \in C(\mathbb{R}_0^+) \cap C^1(\mathbb{R}^+), & \varphi(0) = 0, \qquad \varphi' > 0 \text{ on } \mathbb{R}^+, \\ f \in C(\mathbb{R}), & f \geq 0 \text{ in } (0, \eta_0), \text{ for some } \eta_0 \in (0, \infty), \\ l \in C(\mathbb{R}_0^+), & l > 0 \text{ on } \mathbb{R}^+, \end{cases} \tag{9.2}$$

and moreover

$$\frac{t\varphi'(t)}{l(t)} \in L^1(0^+). \tag{9.3}$$

© The Author(s), under exclusive license to Springer Nature Switzerland AG 2021
B. Bianchini et al., *Geometric Analysis of Quasilinear Inequalities on Complete Manifolds*, Frontiers in Mathematics, https://doi.org/10.1007/978-3-030-62704-1_9

Having defined F, K as in (2.8) and (2.7), that is,

$$K(t) = \int_0^t \frac{s\varphi'(s)}{l(s)} ds, \qquad F(t) = \int_0^t f(s) ds, \qquad (9.4)$$

set $K_\infty = \lim_{t \to \infty} K(t) \in (0, \infty]$; since $\varphi' > 0$, the inverse $K^{-1} : [0, K_\infty) \to \mathbb{R}^+$ exists, and (KO$_0$) is meaningful. In most of the results, we also require

$$\begin{cases} f \text{ is } C\text{-increasing on } [0, \eta_0), \\ l \text{ is } C\text{-increasing on } [0, \xi), \text{ for some } \xi > 0. \end{cases} \qquad (9.5)$$

We underline that condition $f(0)l(0) = 0$ does not appear in (9.2). In fact, some of the next results do not need it. As usual, having fixed a relatively compact, smooth open set $\mathcal{O} \subset M$, we denote with $r(x) = \mathrm{dist}(x, \mathcal{O})$.

9.1   Necessity of (KO$_0$) for the Compact Support Principle

Suppose the failure, $(\neg$KO$_0)$, of the Keller–Osserman condition. Because of Theorem 6.8, under assumptions (9.2), (9.3) and $f(0)l(0) = 0$, each C^1 solution of (9.1) *with the equality sign* must satisfy (FMP), and consequently it cannot be compactly supported. However, finding solutions with the equality sign for (9.1), and especially proving their C^1-regularity, seems to be tricky in the generality of (9.2) and (9.3). For this reason, we follow a different path producing, on *each* complete manifold, radial solutions of inequality (9.1), which are positive on $\Omega = M \backslash B_{r_0}(\mathcal{O})$. The C^1-regularity will be therefore a consequence of the assumption that the radial function be smooth, that is, that the origin $\mathcal{O}$ be a pole of M.

   The key step is provided by the following theorem that considers the exterior Dirichlet problem. Fix $r_0 > 0$ and functions v and β satisfying

$$\begin{aligned} v \in C^1([r_0, \infty)), & \qquad v > 0, \ v' \ge 0 \quad \text{on } [r_0, \infty); \\ \beta \in C([r_0, \infty)), & \qquad \beta > 0 \quad \text{on } [r_0, \infty). \end{aligned} \qquad (9.6)$$

For $\eta, \xi > 0$, define f_η and l_ξ as in (5.6), that is,

$$f_\eta = \max_{[0,\eta]} f, \qquad l_\xi = \max_{[0,\xi]} l.$$

We are ready to state the following theorem.

Theorem 9.1 *Let φ, f, l satisfy (9.2) and*

$$\begin{cases} f \text{ is non-decreasing on } (0, \eta_0), \\ l \in \text{Lip}_{\text{loc}}(\mathbb{R}^+), \\ f(0)l(0) = 0. \end{cases}$$

Fix $r_0 > 0$, and let v and β be as in (9.6). Then, for each $R > 0$, $\xi \in (0, 1)$ and $\eta \in (0, \eta_0)$ (with η_0 as in (5.2)) satisfying

$$\frac{v(r_0 + R)}{v(r_0)} \varphi\left(\frac{\eta}{R}\right) + f_\eta l_\xi \left[\sup_{[r_0, r_0 + R)} \frac{1}{v(r)} \int_{r_0}^{r} v(s)\beta(s)ds \right] < \varphi(\xi), \tag{9.7}$$

there exists a solution $z \in C^1([r_0, \infty))$ of

$$\begin{cases} \left[v\varphi(z')\right]' = \beta v f(z)l(|z'|) & \text{on } [r_0, \infty) \\ z(r_0) = \eta, \quad -\xi < z' \leq 0 & \text{on } [r_0, \infty). \end{cases}$$

Furthermore, if

$$\varphi^{-1}\left(\frac{c}{v(r)}\right) \in L^1(\infty), \qquad \text{for some constant } c > 0, \tag{9.8}$$

there exists $\eta_1 = \eta_1(v, c, \varphi)$ such that, for each $\eta \in (0, \min\{\eta_0, \eta_1\})$ satisfying (9.7), $z(r) \to 0$ as $r \to \infty$.

Remark 9.2 Condition (9.8), to be meaningful, needs to be considered on an interval of integration of the type $[r_c, \infty)$ where the integrand is well defined, that is, because of the monotonicity of v, for $c < v(r_c)\varphi(\infty)$. The existence of such r_c is implicit since the validity of (9.8) and the monotonicity of v force $\lim_{r \to \infty} v(r) = \infty$.

Proof Set

$$h = \max\left\{1, 2\frac{\eta}{R}\right\}.$$

We define $\bar{\varphi}$ and $\bar{l}$ on $\mathbb{R}_0^+$ as follows:

$$\bar{\varphi}(t) = \varphi(t) \text{ on } [0, h], \qquad \bar{\varphi}(t) = \varphi(h) + (t - h) \text{ on } (h, \infty);$$

$$\bar{l} \in C(\mathbb{R}_0^+), \qquad \bar{l} = l \text{ on } [0, \xi], \qquad 0 < \bar{l} \leq l_\xi \text{ on } \mathbb{R}^+,$$

and we extend $\bar{\varphi}$ to an odd function on the entire $\mathbb{R}$. For each $j \in \mathbb{N}$, $j \geq 1$, set also

$$\wp_j(t) = v(r_0 + jR - t), \qquad a_j(t) = \beta(r_0 + jR - t),$$

and let w_j be a solution of the Dirichlet problem

$$
\begin{cases}
\left[\wp_j\bar{\varphi}(w_t)\right]_t = a_j\wp_j f(w)\bar{l}(|w_t|) & \text{on } [0, jR], \\
w(0) = 0, \qquad w(jR) = \eta, & \\
0 \leq w \leq \eta, \qquad w_t \geq 0 & \text{on } [0, jR],
\end{cases}
\tag{9.9}
$$

where the subscript t denotes differentiation in the t variable. We stress that w_j exists for each j. Indeed, we shall apply Theorem 5.1 with the parameter ξ replaced by some suitably chosen $\bar{\xi}$. Note that (5.7) is satisfied up to choosing $\bar{\xi}$ sufficiently large, because $\bar{\varphi}(\infty) = \infty$ and $\bar{l}_{\bar{\xi}} \leq l_{\xi}$. Observe that $\bar{\xi}$ might depend on j, but this does not affect the rest of the proof. From $\wp' \leq 0$, again by Theorem 5.1, we deduce

$$0 \leq (w_j)_t \leq \bar{\varphi}^{-1}\left(\frac{\wp(0)}{\wp(jR)}\bar{\varphi}\left(\frac{\eta}{jR}\right) + f_\eta l_\xi\left[\sup_{[0,jR]}\frac{1}{\wp_j(t)}\int_0^t \wp_j(s)a_j(s)\mathrm{d}s\right]\right). \tag{9.10}$$

Set $z_j(r) = w_j(r_0 + jR - r)$, and note that z_j solves

$$
\begin{cases}
\left[v\bar{\varphi}(z_j')\right]' = \beta v f(z_j)\bar{l}(|z_j'|) & \text{on } (r_0, r_0 + jR), \\
z_j(r_0) = \eta, \qquad z_j(r_0 + jR) = 0 & \\
0 \leq z_j \leq \eta, \qquad z_j' \leq 0 & \text{on } [r_0, r_0 + jR].
\end{cases}
\tag{9.11}
$$

Next, we estimate the derivative z_j' uniformly in j. First, observe that integrating on $[t_1, t_2]$ the inequality $[v\bar{\varphi}(z')]' \geq 0$ that follows from (9.11) and (5.2), we deduce

$$v(t_2)\left[\bar{\varphi}(z'(t_2)) - \bar{\varphi}(z'(t_1))\right] \geq \left[v(t_1) - v(t_2)\right]\bar{\varphi}(z'(t_1)) \geq 0.$$

Using $\bar{\varphi}(z') \leq 0$ and $v' \geq 0$, we conclude that $\bar{\varphi}(z')$, hence z', is increasing. In particular, since $z_j' \leq 0$, we have $|z_j'| \leq |z_j'(r_0)|$.
Claim: $\{z_j\}$ is a non-decreasing sequence.
We show that $z_j \leq z_{j+1}$ on $[r_0, r_0 + jR]$. Applying Lemma 5.2 to w_j and rephrasing for z_j, there exists $r_j \in (r_0, r_0 + jR]$ such that $z_j > 0$, $z_j' < 0$ on $[r_0, r_j)$, while $z_j = 0$ on

$[r_j, r_0 + jR]$. On (r_0, r_j), it holds

$$\begin{cases} \left[v\bar\varphi(z_j')\right]' = \beta v f(z_j)\bar l(|z_j'|) & \text{on } (r_0, r_j), \\ \left[v\bar\varphi(z_{j+1}')\right]' = \beta v f(z_{j+1})\bar l(|z_{j+1}'|) & \text{on } (r_0, r_j), \\ z_j(r_0) = z_{j+1}(r_0) = \eta, \qquad z_j(r_j) = 0 \le z_{j+1}(r_j). \end{cases}$$

The inequality $z_j \le z_{j+1}$ on $[r_0, r_j]$, hence on $[r_0, r_0 + jR]$, is then a consequence of the comparison result in Proposition 6.2 applied to the model manifold $M_g = [r_0, \infty) \times \mathbb{S}^{m-1}$ with the radially symmetric C^1-metric

$$dr^2 + g(r)^2 d\theta^2 \quad \text{with} \quad g(r) = \left(\frac{v(r)}{\omega_{m-1}}\right)^{\frac{1}{m-1}},$$

recall also Remark 6.4.

The convexity and monotonicity of z_j, together with the above claim, imply the uniform estimate $|z_j'| \le |z_1'(r_0)|$. Changing variables in (9.10) and exploiting (9.7), we deduce

$$|z_j'| \le |z_1'(r_0)|$$

$$\le \bar\varphi^{-1}\left(\frac{v(r_0 + R)}{v(r_0)}\bar\varphi\left(\frac{\eta}{R}\right) + f_\eta l_\xi \left[\sup_{[r_0, r_0+R]} \frac{1}{v(r)} \int_{r_0}^{r} v(s)\beta(s)ds\right]\right) < \xi,$$

where we used again the identity $\bar\varphi = \varphi$ on $[0, h]$, the definition of h and $\xi < 1$. Therefore, by the Ascoli–Arzelà Theorem, the sequence $\{z_j\}$ converges locally uniformly to a solution $z \in C^1([r_0, \infty))$ of

$$\begin{cases} \left[v\bar\varphi(z')\right]' = \beta v f(z)\bar l(|z'|) & \text{on } (r_0, \infty), \\ z(r_0) = \eta, \\ 0 \le z \le \eta, \qquad -\xi < z' \le 0 \quad \text{on } [r_0, \infty). \end{cases}$$

Since $\bar\varphi = \varphi$ and $\bar l = l$ on $(0, \xi) \subset (0, 1)$, z is the desired solution of (9.9). Suppose now (9.8), which in particular implies that $\lim_{r \to \infty} v(r) = \infty$, and we choose r_c such that

$$r_c \ge r_0, \qquad \frac{c}{v(r_c)} \le \varphi(1).$$

Define

$$\bar z(r) = \int_r^\infty \varphi^{-1}\left(\frac{c}{v(s)}\right) ds \qquad \text{on } [r_c, \infty).$$

From $\bar{\varphi} = \varphi$ on $[0, h] \supset [0, 1]$, $\bar{z}$ solves $0 = \left[v\varphi(\bar{z}')\right]' = \left[v\bar{\varphi}(\bar{z}')\right]' = 0$ on $[r_c, \infty)$. Choose now

$$\eta_1 = \int_{r_c}^{\infty} \varphi^{-1}\left(\frac{c}{v(s)}\right) ds,$$

and consider η satisfying the further restriction $\eta \in (0, \min\{\eta_0, \eta_1\})$. For j large enough, since $z_j(r_c) \leq z_j(r_0) = \eta < \eta_1 = \bar{z}(r_c)$, by the comparison Proposition 6.2 and the non-negativity of β, $f(z_j)$ and $l(|z'_j|)$, we get $\bar{z} \geq z_j$ on $[r_c, r_0 + jR]$, and thus $\bar{z} \geq z$ on $[r_c, \infty)$. The last claim of the theorem follows since $\bar{z} \to 0$ as $r \to \infty$. $\square$

We are ready to prove our main result, Theorem 2.37 in the Introduction, in the following more general form: it says, loosely speaking, that there is no geometric obstruction for (KO$_0$) to be necessary for the compact support principle.

Theorem 9.3 *Let $(M^m, \langle , \rangle)$ be a complete manifold, and let φ, f, l satisfy (9.2), (9.3), (9.5) and*

$$l \in \mathrm{Lip}_{\mathrm{loc}}\big((0, \xi_0)\big),$$

$$f(0)l(0) = 0.$$

Then, for each

origin $\mathcal{O} \subset M$ with associated distance $r(x) = \mathrm{dist}(x, \mathcal{O})$,
$r_0 > 0$, $\xi \in (0, \xi_0)$ (with ξ_0 as in (9.5)),
$0 < b \in C\big(M \backslash B_{r_0}(\mathcal{O})\big),$

there exist $\eta \in (0, \eta_0)$ sufficiently small and a radial solution $u \in \mathrm{Lip}(M \backslash B_{r_0}(\mathcal{O}))$ of

$$\begin{cases} \Delta_\varphi u \geq b(x)f(u)l(|\nabla u|) & \text{weakly on } M \backslash \overline{B_{r_0}(\mathcal{O})}, \\ 0 \leq u \leq \eta & \text{on } M \backslash B_{r_0}(\mathcal{O}), \\ u = \eta & \text{on } \partial B_{r_0}(\mathcal{O}), \quad u(x) \to 0 \quad \text{as } r(x) \to \infty, \\ |\nabla u| < \xi & \text{on } M \backslash B_{r_0}(\mathcal{O}). \end{cases}$$

Moreover, if $(\neg\mathrm{KO}_0)$ holds, then $u > 0$ on $M \backslash \mathcal{O}$. In particular, if $\mathcal{O}$ is a pole for M, $u \in C^1\big(M \backslash B_{r_0}(\mathcal{O})\big)$ and (KO$_0$) is necessary for the validity of the compact support principle (CSP).

Proof We choose $0 < \bar{g} \in C^\infty(\mathbb{R}_0^+)$ enjoying the following properties:

(*i*) if $H_{-\nabla r}$ is the mean curvature of $\partial \mathcal{O}$ with respect to the inward pointing unit normal $-\nabla r$,

$$(m-1)\frac{\bar{g}'(0)}{\bar{g}(0)} > \max\left\{0, \sup_{\partial \mathcal{O}} H_{-\nabla r}\right\}; \qquad (9.12)$$

(*ii*) setting $\bar{v}(r) = \omega_{m-1}\bar{g}(r)^{m-1}$,

$$\bar{v}' \geq 0 \qquad \text{on } \mathbb{R}^+, \qquad \bar{v}(0)^{-1} < \varphi(\infty).$$

$$\bar{v}(r) \geq \max\left\{1, \left[\varphi\left(\frac{1}{r^2}\right)\right]^{-1}\right\} \qquad \text{for } r \geq 1.$$

Note that, for $r \geq 1$,

$$\varphi^{-1}\left(\frac{1}{\bar{v}(r)}\right) \leq \varphi^{-1}\left(\varphi\left(\frac{1}{r^2}\right)\right) = \frac{1}{r^2} \in L^1(\infty). \qquad (9.13)$$

Next, choose $0 \leq \bar{G} \in C(\mathbb{R}_0^+)$ in such a way that

$$\mathrm{Ric} \geq -(m-1)\bar{G}(r)\langle\,,\,\rangle \qquad \text{on } M,$$

and define $G(r) = \max\{\bar{G}, \bar{g}''/\bar{g}\}$. Let $g \in C^2(\mathbb{R}_0^+)$ solve

$$\begin{cases} g'' - Gg = 0 \quad \text{on } \mathbb{R}^+ \\ g(0) = \bar{g}(0), \quad g'(0) = \bar{g}'(0), \end{cases} \qquad (9.14)$$

and let M_g be the model associated with g. By construction and by (9.13), setting $v_g(r) = \omega_{m-1}g(r)^{m-1}$, it holds

$$\mathrm{Ric} \geq -(m-1)G(r)\langle\,,\,\rangle \qquad \text{on } M,$$

$$v_g(r) \geq \bar{v}(r) \qquad \text{on } \mathbb{R}^+ \text{ by Sturm comparison}, \qquad (9.15)$$

$$\varphi^{-1}\left(\frac{1}{v_g(t)}\right) \in L^1(\infty).$$

We define

$$\bar{f}(t) = \sup_{[0,t]} f(s), \qquad \beta(r) = \sup_{\partial B_r} b$$

and note that, since f is C-increasing,

$$f(t) \le \bar{f}(t) \le Cf(t) \qquad \forall t \in [0, \eta_0). \tag{9.16}$$

Choose $\bar{l} \in \mathrm{Lip}_{\mathrm{loc}}(\mathbb{R}^+)$ such that

$$\bar{l} \ge l \quad \text{on } \mathbb{R}_0^+, \qquad \bar{l} \equiv l \quad \text{on } [0, \xi].$$

Eventually, fix $r_0 \ge 1$ and choose R small enough that

$$\bar{f}_{\eta_0} \bar{l}_1 \left[\sup_{[r_0, r_0+R)} \frac{1}{v_g(r)} \int_{r_0}^{r} v_g(s)\beta(s)\mathrm{d}s \right] < \frac{\varphi(\xi)}{2}.$$

Then, pick $\eta \in (0, \eta_0)$ small enough to enjoy

$$\frac{v_g(r_0 + R)}{v_g(r_0)} \varphi\left(\frac{\eta}{R}\right) < \frac{\varphi(\xi)}{2}.$$

Since $\bar{f}$ is increasing, $\eta \in (0, \eta_0)$ and $\xi \in (0, 1)$, the last two inequalities imply

$$\frac{v_g(r_0 + R)}{v_g(r_0)} \varphi\left(\frac{\eta}{R}\right) + \bar{f}_\eta \bar{l}_\xi \left[\sup_{[r_0, r_0+R)} \frac{1}{v_g(r)} \int_{r_0}^{r} v_g(s)\beta(s)\mathrm{d}s \right] < \varphi(\xi). \tag{9.17}$$

We are therefore in the position to apply Theorem 9.1 and infer the existence of η_1 such that, for each $\eta \in (0, \min\{\eta_0, \eta_1\})$ satisfying (9.17), there exists z solving

$$\begin{cases} \left[v_g \varphi(z')\right]' = \beta v_g \bar{f}(z) \bar{l}(|z'|) & \text{on } [r_0, \infty) \\ z(r_0) = \eta, \qquad -\xi < z' \le 0 & \text{on } [r_0, \infty), \\ z(r) \to 0 & \text{as } r \to \infty. \end{cases}$$

Define $u(x) = z(r(x))$ on $M \backslash B_{r_0}(\mathcal{O})$, with $r(x) = \mathrm{dist}(x, \mathcal{O})$. The first of (9.15) together with (9.14) and (9.12) implies, via the Laplacian comparison Theorem 3.8, the inequality

$$\Delta r \le \frac{v_g'(r)}{v_g(r)} \qquad \text{weakly on } M \backslash \mathcal{O},$$

and hence

$$\Delta_\varphi u \geq \varphi'(z')z'' + \varphi(z')\Delta r \geq \varphi'(z')z'' + \varphi(z')\frac{v_g'}{v_g}$$

$$= v_g^{-1}\big[v_g\varphi(z')\big]' = \beta\bar{f}(z)\bar{l}(|z'|) \geq b(x)f(z)l(|z'|)$$

$$= b(x)f(u)l(|\nabla u|)$$

weakly on $M\backslash B_{r_0}(\mathcal{O})$. This concludes the first part of the theorem.
Next, we prove that if ($\neg$KO$_0$) holds, then $z > 0$ on $[r_0, \infty)$. To see this,
we consider the radial function $v(x) = z(r(x))$ on the model M_g, which
satisfies

$$\Delta_\varphi v = v_g^{-1}\big[v_g\varphi(z')\big]' = \beta(r)\bar{f}(z)\bar{l}(|z'|) = \beta(r)\bar{f}(v)l(|\nabla v|),$$

where we used $|z'| < \xi$ and $l = \bar{l}$ on $[0, \xi]$. Moreover, $v = \eta$ on $\{r = r_0\}$.
Set

$$\bar{F}(t) = \int_0^t \bar{f}(s)\mathrm{d}s.$$

Because of (9.16), Lemma 5.6 guarantees that ($\neg$KO$_0$) is equivalent to

$$\frac{1}{K^{-1} \circ \bar{F}} \notin L^1(0^+).$$

Therefore, we can apply Theorem 6.8 on the set $\Omega = \{r > r_0\} \subset M_g$ to deduce that $v > 0$
on Ω, as claimed.

$\square$

9.2 Sufficiency of (KO$_0$) for the Compact Support Principle

General Operators and No Cut-Locus

As remarked in the Introduction, (KO$_0$) alone is not sufficient to prove (CSP) on complete
manifolds. Indeed, the influence of geometry is, for this property, particularly subtle, as
confirmed by the next refinement of Example 2.35.

Example 9.4 For $\delta \geq 0$, consider a model $M_\delta = (\mathbb{R}^m, ds_\delta^2)$ (cf. also Example 3.9) with

$$ds_\delta^2 = dt^2 + g_\delta(t)^2 d\theta^2, \qquad \text{where} \qquad \begin{cases} g_\delta \in C^2(\mathbb{R}_0^+) & g_\delta > 0 \quad \text{on } \mathbb{R}^+ \\ g_\delta(t) = t & \text{if } t \leq 1/4 \\ g_\delta(t) = \exp\{-t^\delta\} & \text{if } t \geq 1. \end{cases}$$

Define $\mathcal{O} = B_1(o)$, and let $r = t - 1$ be the distance from $\mathcal{O}$. We have

$$\text{II}_{-\nabla r} = -\delta ds_\delta^2,$$

$$K_{\mathrm{rad}} = -\delta\big[- (\delta - 1)(1 + r)^{\delta - 2} + \delta(1 + r)^{2\delta - 2}\big]$$

$$\leq -\delta^2(1 + r)^{\delta - 2}\big[(1 + r)^\delta - 1\big],$$

$$\Delta r = -(m - 1)\delta(1 + r)^{\delta - 1}.$$

Define $\alpha = 2\delta - 2 \geq -2$, and let $\mu, \chi, \omega \in \mathbb{R}$ satisfy

$$\mu > \chi - \frac{\alpha}{2}, \qquad \omega < \chi. \tag{9.18}$$

It is easy to show that, for

$$\sigma \in \left(0, \frac{\mu - \chi + \alpha/2}{\chi - \omega}\right],$$

and for each $p > 1$, the function $v(x) = (1 + r(x))^{-\sigma}$ is a bounded, positive solution of

$$\Delta_p v \geq C(1 + r)^{-\mu} v^\omega |\nabla v|^{p - 1 - \chi} \qquad \text{on } M_\delta \backslash \mathcal{O},$$

for some constant $C > 0$, and furthermore $v(x) \to 0^+$ as x diverges. However, defining $f(t) = t^\omega$ and $l(t) = t^{p - 1 - \chi}$, inequality $\omega < \chi$ implies

$$\frac{1}{K^{-1} \circ F} = C_1 s^{-\frac{\omega + 1}{\chi + 1}} \in L^1(0^+).$$

Thus, condition (KO$_0$) is met, but (CSP) fails on M_δ.

As in [194, 198], the proof of (CSP) will be achieved via the construction of a compactly supported C^1-supersolution $\bar{w}$ for (9.1). Under the above assumptions on φ, b, f, l, to produce $\bar{w}$, one could try to use the solution w of the related ODE (5.5) in a way analogous to the one in the proof of the finite maximum principle (Theorem 6.8). However, a direct use of w seems difficult, also because of the delicate interplay between the threshold η in (5.5) and the global behaviour of the constants $a_1, \wp_1, \wp_0, l_\xi$ in Theorem 5.1, depending on the interval $[0, T]$ under consideration. As we shall see, a certain independence between η and $a_1, \wp_1, \wp_0, l_\xi$ is key to conclude (CSP) from the existence of w. To overcome the problem, we will use a different technique: instead of solving a Dirichlet problem, we will exhibit an explicit, compactly supported supersolution by a direct use of (KO$_0$), an idea that is closer to the one in [194, 198, 208], which in turn are improvements of [196]. However, extending the method therein to non-constant b, l presents nontrivial hurdles and calls for new ideas. To this aim, we shall assume some further conditions that, although seemingly somewhat artificial, enable us to capture the right growths and achieve a sharp result.

To exhibit $\bar{w}$, we follow two slightly different constructions that need a (mildly) different set of assumptions. Besides (9.2), (9.3) and (9.5), for the first construction, we require

(C_1) there exists a constant $k_1 \geq 1$ such that

$$t K'(t) \leq k_1 K(t) \qquad \text{for each } t \in (0, 1];$$

(C_2) there exists a constant $k_2 \geq 1$ such that

$$K'(st) \leq k_2 K'(s) K'(t) \qquad \text{for each } s, t \in (0, 1];$$

(C_3) there exists a constant $\bar{C} \geq 1$ such that

$$\frac{t}{K^{-1}(t)} \text{ is } \bar{C}\text{-increasing,} \qquad \text{and} \qquad \frac{t}{K^{-1}(t)} \to 0 \quad \text{as } t \to 0^+.$$

(C_4) there exists a constant $c_F \geq 1$ such that

$$\frac{F(t)}{K^{-1}(F(t))} \leq c_F f(t) \qquad \text{for each } t \in (0, \min\{1, \eta_0\}).$$

Note that these requirements are all related to the behaviour of the various functions considered in a right neighbourhood of zero.

Example 9.5 Having fixed

$$p > 1, \qquad \chi \in [0, p-1], \qquad \omega > 0,$$

the prototype example of f, l, φ is given by

$$\varphi'(t) \asymp t^{p-2}, \qquad f(t) \asymp t^\omega, \qquad l(t) \asymp t^{p-1-\chi}$$

for $t \in (0, t_0)$. Then, $K(t) \asymp t^{\chi+1}$ satisfies (C_1) and (C_2), while (C_3) and (C_4) are met if and only if, respectively, $\chi > 0$ and $\omega \le \chi$.

Suppose that

$$b(x) \ge \beta\big(r(x)\big) \qquad \text{for } r(x) \ge r_0.$$

We express the relation between K and β in terms of an auxiliary weight $\bar{\beta}$, which is tied with K in the way expressed by the next two conditions. Later, (9.19) in Proposition 9.8 will relate β to $\bar{\beta}$.

(β_1) $0 < \beta \in C([r_0, \infty))$, $\quad \bar{\beta} \in C^1([r_0, \infty))$, $\quad 0 < \bar{\beta} < K_\infty$, $\bar{\beta}' \le 0$ on $[r_0, \infty)$;

(β_2) there exists a constant $c_\beta \ge 1$ such that

$$\frac{-\bar{\beta}'(t)}{K^{-1}(\bar{\beta}(t))} \le c_\beta \bar{\beta}(t) \qquad \text{for each } t \in [r_0, \infty).$$

Remark 9.6 Note that (β_2) is meaningful since $\bar{\beta} < K_\infty$ because of (β_1).

Example 9.7 Referring to Example 9.5, a borderline behaviour of $\bar{\beta}$ for the validity of (β_2) is

$$\bar{\beta}(t) = (1+t)^{-\chi-1}.$$

We first describe our main ODE result that should be compared to Lemma 4.1 in [209]. Here, we consider a different and (in some cases) weaker set of assumptions, and the proof that we present is considerably simpler. Below, we will describe in more detail the interplay between the two results.

Proposition 9.8 *Let φ, l, f satisfy (9.2), (9.3) and (9.5), and assume the validity of $(C_1), \ldots, (C_4)$ and $(\beta_1), (\beta_2)$. Having fixed a non-negative $\theta \in C([r_0, \infty))$, suppose that*

$$\max\left\{\frac{\bar{\beta}(s)}{\beta(s)}, \frac{\theta(s)\bar{\beta}(s)}{\beta(s)K^{-1}(\bar{\beta}(s))}\right\} \in L^\infty([r_0, \infty)). \tag{9.19}$$

If

$$\frac{1}{K^{-1} \circ F} \in L^1(0^+), \tag{KO_0}$$

then for each $\varepsilon > 0$, there exists $\lambda \in (0, 1)$ such that the following holds: for each $R \geq r_0$, there exist $R_1 > R$ and a function w with the following properties:

$$
\begin{cases}
w \in C^1([R, \infty)) \quad \text{and } C^2 \text{ except possibly at } R_1; \\
0 \leq w \leq \lambda, \qquad w(R) = \lambda, \qquad w \equiv 0 \quad \text{on } [R_1, \infty), \\
w' < 0 \quad \text{on } [R, R_1), \qquad |w'| \leq \varepsilon \quad \text{on } [R, \infty), \\
(\varphi(w'))' - \theta(r)\varphi(w') \leq \varepsilon \beta(r) f(w) l(|w'|) \qquad \text{on } [R, \infty).
\end{cases}
\tag{9.20}
$$

Remark 9.9 Inspecting the proof of Proposition 9.8, we can weaken the third in (2.5) to $l \in L^\infty_{loc}(\mathbb{R}^+_0) \cap C(\mathbb{R}^+)$ and $l > 0$ on $\mathbb{R}^+$, that is, the continuity of l at $t = 0$ is not needed. This will be used in the proof of Theorem 2.38.

Remark 9.10 Condition (9.19) relates β to $\bar{\beta}$. For instance, in Example 9.5, we can set $\bar{\beta}(t) = c(1 + t)^{-\chi-1}$ for a constant c small enough to satisfy (β_1) and choose $\beta(t) = (1 + t)^{-\mu}$, for some $\mu \in \mathbb{R}$. Then, (9.19) is met if and only if

$$\mu \leq \chi + 1 \qquad \text{and} \qquad \theta(s)s^{\mu-\chi} \in L^\infty([r_0, \infty)).$$

To begin the proof, we need a simple technical lemma.

Lemma 9.11 *Assume (9.5) and (9.3) and also (C_1), (C_2), (β_1), (β_2). Then, the following properties hold:*

(K_1) $K(st) \leq k_1 k_2 K(s) K(t)$ *for each* $s, t \in [0, 1]$;

(K_2) $K^{-1}(\tau) K^{-1}(\rho) \leq K^{-1}(k_1 k_2 \tau \rho)$ *for each* $\tau, \rho \in \left(0, \min\left\{1, \dfrac{K_\infty}{k_1 k_2}\right\}\right)$;

$(K\beta)$ *For each* $\sigma \in (0, 1)$,

$$K^{-1}(\sigma \bar{\beta}) \notin L^1(\infty).$$

Proof Property (K_1) follows immediately from (C_2) and (C_1) by integration:

$$K(st) = \int_0^{st} K'(\tau)d\tau = s \int_0^t K'(s\zeta)d\zeta \leq k_2 s K'(s)K(t) \leq k_2 k_1 K(s)K(t).$$

To show (K_2), use (K_1) with the choices $s = K^{-1}(\tau)$, $t = K^{-1}(\rho)$, and then apply K^{-1}. To prove $(K\beta)$, first observe that, because of (K_2), $K^{-1}(\sigma \bar{\beta}) \geq C_\sigma K^{-1}(\bar{\beta})$ for some $C_\sigma > 0$, and thus it is sufficient to restrict to $\sigma = 1$. Using (β_2), we deduce

$$c_\beta K^{-1}(\bar{\beta}) \geq -\frac{\bar{\beta}'}{\bar{\beta}}.
\tag{9.21}$$

If $\bar{\beta}$ is bounded from below by a positive constant, then $K^{-1}(\bar{\beta})$ is not infinitesimal and clearly $(K\beta)$ is met. Otherwise, from $\bar{\beta}' \leq 0$, we get $\bar{\beta}(r) \to 0$ as $r \to \infty$, and integrating (9.21) on $[r_1, r)$, we deduce

$$c_\beta \int_{r_1}^r K^{-1}(\bar{\beta}) \geq \log \bar{\beta}(r_1) - \log \bar{\beta}(r) \to \infty \quad \text{as } r \to \infty,$$

as claimed.

$\square$

Proof of Proposition 9.8 Let $\sigma, \lambda \in (0, \eta_0)$ to be specified later. Because of $(K\beta)$ in Lemma 9.11 and (KO_0), there exists $R_\sigma = R_\sigma(\sigma, \lambda, R) > R$ such that

$$\int_0^\lambda \frac{ds}{K^{-1}(F(s))} = \int_R^{R_\sigma} K^{-1}\big(\sigma\bar{\beta}(s)\big) ds. \tag{9.22}$$

We define implicitly a function α via the identity

$$\int_0^{\alpha(t)} \frac{ds}{K^{-1}(F(s))} = \int_{R_\sigma - t}^{R_\sigma} K^{-1}\big(\sigma\bar{\beta}(s)\big) ds.$$

Clearly, $\alpha(0) = 0$, $\alpha(R_\sigma - R) = \lambda$, $\alpha > 0$ on $(0, R_\sigma - R)$ and, differentiating,

$$\alpha'(t) = K^{-1}\big(F(\alpha(t))K^{-1}\big(\sigma\tilde{\beta}(t)\big) > 0 \quad \text{on } (0, R_\sigma - R),$$

$$\text{where} \quad \tilde{\beta}(t) = \bar{\beta}(R_\sigma - t), \quad \text{and} \quad \alpha'(0) = 0. \tag{9.23}$$

Note that $\tilde{\beta}$ is non-decreasing by (β_1). By construction, $\alpha \in [0, \lambda] \subset [0, \eta_0)$, and

$$0 < \alpha'(s) \leq K^{-1}\big(F(\alpha)\big)K^{-1}\big(\sigma\|\bar{\beta}\|_\infty\big)$$

$$\leq K^{-1}\big(F(\lambda)\big)K^{-1}\big(\|\bar{\beta}\|_\infty\big) \quad \text{on } (0, R_\sigma - R). \tag{9.24}$$

We can therefore reduce σ and λ, independently, in such a way that

$$K^{-1}(F(\alpha)) \leq \min\{1, \xi_0\}, \qquad K^{-1}\big(\sigma\tilde{\beta}\big) \leq 1$$

on $[0, R_\sigma)$. For convenience, we define

$$\rho = K^{-1}\big(F(\alpha(t)), \qquad \tau = K^{-1}\big(\sigma\tilde{\beta}(t)\big).$$

Applying K to (9.23) and differentiating, we obtain

$$\left(K(\alpha')\right)' = K'(\rho\tau)[\rho'\tau + \rho\tau'] \overset{(C_2)}{\le} k_2 K'(\rho)K'(\tau)\left[\frac{f(\alpha)\alpha'\tau}{K'(\rho)} + \frac{\rho\sigma\tilde{\beta}'}{K'(\tau)}\right]$$

$$\overset{(C_1)}{\le} k_2 k_1\left[f(\alpha)\alpha'K(\tau) + \sigma\tilde{\beta}'K(\rho)\right] \qquad (9.25)$$

$$= k_2 k_1\left[f(\alpha)\alpha'\sigma\tilde{\beta} + \sigma\tilde{\beta}'F(\alpha)\right].$$

However, by (C_4) and (β_2), together with (K_2) in Lemma 9.11 applied twice,

$$F(\alpha) \le c_F K^{-1}\left(F(\alpha)\right)f(\alpha),$$

$$\tilde{\beta}' \le c_\beta K^{-1}(\tilde{\beta})\tilde{\beta} \overset{(K_2)}{\le} \frac{\bar{c}_\beta}{K^{-1}(\sigma)}K^{-1}(\sigma\tilde{\beta})\tilde{\beta} \qquad (9.26)$$

for some $\bar{c}_\beta$ depending on c_β, K, k_1, k_2. Inserting into (9.25) and recalling (9.23), we get

$$\left(K(\alpha')\right)' \le k_2 k_1\sigma\left[f(\alpha)\alpha'\tilde{\beta} + \frac{c_F \bar{c}_\beta}{K^{-1}(\sigma)}\tilde{\beta}K^{-1}(F(\alpha))K^{-1}(\sigma\tilde{\beta})f(\alpha)\right]$$

$$= k_2 k_1\sigma\tilde{\beta}f(\alpha)\alpha'\left[1 + \frac{c_F \bar{c}_\beta}{K^{-1}(\sigma)}\right] \le \frac{C_1\sigma}{K^{-1}(\sigma)}\tilde{\beta}f(\alpha)\alpha',$$

for some constant C_1 depending on $k_1, k_2, \bar{c}_\beta, c_F$. Using the definition of K' and $\alpha' > 0$, we can simplify the inequality to deduce

$$\varphi'(\alpha')\alpha'' \le \frac{C_1\sigma}{K^{-1}(\sigma)}\tilde{\beta}f(\alpha)l(\alpha'). \qquad (9.27)$$

Moreover, observe that the first in (9.25) and the definition of ρ and τ imply $(K(\alpha'))' \ge 0$, and hence, expanding, $\alpha'' \ge 0$.
We now integrate (9.27) on $[0, t)$, and we use $\alpha'(0) = 0$ and $\varphi(0) = 0$ to obtain

$$\varphi(\alpha'(t)) \le \frac{C_1\sigma}{K^{-1}(\sigma)}\int_0^t \tilde{\beta}f(\alpha)l(\alpha'). \qquad (9.28)$$

Applying the third in (9.5) and noting that $\alpha'' \ge 0$, we get

$$\varphi(\alpha'(t)) \le \frac{C_2\sigma}{K^{-1}(\sigma)}f(\alpha(t))l(\alpha'(t))\int_0^t \tilde{\beta}$$

for some constant $C_2 > 0$. Next, we exploit (9.22) to estimate from above the integral term. To this end, we use (C_3), $\tilde{\beta}' \geq 0$ together with the second in (9.26) to get

$$\int_0^t \tilde{\beta} = \int_0^t \frac{\tilde{\beta}}{K^{-1}(\tilde{\beta})} K^{-1}(\tilde{\beta}) \overset{(C_3)}{\leq} \bar{C} \frac{\tilde{\beta}(t)}{K^{-1}(\tilde{\beta}(t))} \int_0^t K^{-1}(\tilde{\beta})$$

$$\overset{(K_2)}{\leq} \frac{\bar{C}\bar{c}_\beta}{K^{-1}(\sigma)} \frac{\tilde{\beta}(t)}{K^{-1}(\tilde{\beta}(t))} \int_R^{R_\sigma} K^{-1}(\sigma\tilde{\beta}) \tag{9.29}$$

$$= \frac{C_3}{K^{-1}(\sigma)} \frac{\tilde{\beta}(t)}{K^{-1}(\tilde{\beta}(t))} \int_0^\lambda \frac{ds}{K^{-1}(F(s))}.$$

Define $\tilde{\theta}(t) = \theta(R_\sigma - t)$. Putting together (9.27), (9.28) and (9.29), because of (9.19), we obtain

$$\left(\varphi(\alpha')\right)' + \tilde{\theta}(t)\varphi(\alpha')$$

$$\leq \beta(R_\sigma - t)f(\alpha)l(\alpha')\left[\frac{\tilde{\beta}(t)}{\beta(R_\sigma - t)} \frac{C_1\sigma}{K^{-1}(\sigma)}\right.$$

$$\left. + \frac{C_4\sigma}{[K^{-1}(\sigma)]^2} \frac{\tilde{\theta}(t)\tilde{\beta}(t)}{\beta(R_\sigma - t)K^{-1}(\tilde{\beta}(t))} \int_0^\lambda \frac{ds}{K^{-1}(F(s))}\right]$$

$$\leq \beta(t)f(\alpha)l(\alpha')\left[\left\|\frac{\tilde{\beta}}{\beta}\right\|_\infty \frac{C_1\sigma}{K^{-1}(\sigma)} + \frac{C_4\sigma}{[K^{-1}(\sigma)]^2}\left\|\frac{\theta\tilde{\beta}}{\beta K^{-1}(\tilde{\beta})}\right\|_\infty \int_0^\lambda \frac{ds}{K^{-1}(F(s))}\right].$$

Next, using the second in (C_3), we can choose σ sufficiently small to make the first term in square brackets smaller than $\varepsilon/2$. We can then choose $\lambda > 0$ small enough to make the second term smaller than $\varepsilon/2$. Eventually, define

$$w(r) = \alpha(R_\sigma - r) \qquad \text{and} \qquad R_1 = R_\sigma.$$

Because of (9.24)

$$|w'| \leq K^{-1}(F(\lambda))K^{-1}(\|\tilde{\beta}\|_\infty).$$

Up to further reducing λ, $|w'| \leq \varepsilon$ on $[R, R_1)$ (with ε as in the statement of the proposition). Using the definition of $\tilde{\theta}$ and $\tilde{\beta}$ and the fact that φ is odd on $\mathbb{R}$, it is immediate to check that w satisfies all the properties listed in (9.20). $\qquad\square$

Remark 9.12 A crucial feature of the above construction is that λ is independent of $R \geq r_0$. This will allow to construct compactly supported supersolutions attaining value λ on the boundary of any fixed geodesic sphere $\partial B_R(\mathcal{O})$ with $R \geq r_0$.

With this preparation, we can now prove our first main result for the compact support principle. We recall that a pole $\mathcal{O} \subset M$ is a smooth, relatively compact open set such that the normal exponential map realizes a diffeomorphism between $M \backslash \mathcal{O}$ and $\partial \mathcal{O} \times \mathbb{R}^+$. Let II$_{-\nabla r}$ be the second fundamental form of $\partial \mathcal{O}$ in the direction pointing towards $\mathcal{O}$.

Theorem 9.13 *Let $(M, \langle , \rangle)$ be a manifold possessing a pole $\mathcal{O}$, and let $r(x) =$ dist$(x, \mathcal{O})$. Suppose that*

$$\text{Sec}_{\text{rad}} \leq -\kappa^2 (1 + r)^\alpha \qquad \text{on } M \backslash \mathcal{O}, \quad \text{for some } \kappa \geq 0, \ \alpha \geq -2;$$

$$\text{II}_{-\nabla r} \geq \begin{cases} -\kappa \langle , \rangle & \text{if } \alpha \geq 0 \text{ or } \kappa = 0, \\ -\left[\dfrac{\alpha + \sqrt{\alpha^2 + 16\kappa^2}}{4} \right] \langle , \rangle & \text{otherwise,} \end{cases} \qquad (9.30)$$

where II$_{-\nabla r}$ *denotes the second fundamental form of $\partial \mathcal{O}$ in the inward pointing direction. Fix $0 < b \in C(M)$, and let $0 < \beta \in C([r_0, \infty))$ such that*

$$b(x) \geq \beta\big(r(x)\big) \qquad \text{for } r(x) \geq r_0.$$

Let φ, f, l satisfy (9.2), (9.3), (9.5) and $(C_1), \ldots, (C_4)$. Assume that, for some $\bar{\beta}$ matching (β_1) and (β_2), it holds

$$\max \left\{ \frac{\bar{\beta}(s)}{\beta(s)}, \frac{s^{\alpha/2} \bar{\beta}(s)}{\beta(s) K^{-1}(\bar{\beta}(s))} \right\} \in L^\infty([r_0, \infty)). \qquad (9.31)$$

Then,

$$(KO_0) \qquad \Longrightarrow \qquad (CSP) \text{ holds for } (9.1).$$

Moreover, if

$$l \in \text{Lip}_{\text{loc}}\big((0, \xi_0)\big), \qquad f(0)l(0) = 0,$$

then

$$(KO_0) \qquad \Longleftrightarrow \qquad (CSP) \text{ holds for } (9.1).$$

Proof We first prove implication $(KO_0) \Rightarrow (CSP)$. Note that it is enough to consider solutions of (9.1) when $\Omega = \Omega_{r_0}$ is a connected component of $M \setminus B_{r_0}(\mathcal{O})$. Let now u be a C^1 solution of (9.1) on Ω_{r_0}. By Proposition 3.12 applied with, respectively,

$$G(r) = \kappa^2 (1+r)^\alpha, \quad \begin{cases} \theta_* = 0, \quad D = D_-(\theta_*) = -1 & \text{if } \alpha \geq 0, \\[2ex] \theta_* = \dfrac{\alpha}{2\kappa}, \quad D = D_-(\theta_*) = -\dfrac{\alpha + \sqrt{\alpha^2 + 16\kappa^2}}{4\kappa} & \text{if } \alpha \in [-2, 0), \end{cases}$$

$$C = 1, \qquad \lambda = D\kappa,$$

we deduce that

$$g(t) = \exp\left\{ D \int_0^t \kappa(1+s)^{\alpha/2} ds \right\}$$

is a positive solution on $\mathbb{R}^+$ of $g'' - Gg \leq 0$, $g(0) = 1$, $g'(0) = D\kappa$, and thus by the Laplacian comparison theorem from below,

$$\Delta r \geq (m-1)\frac{g'(r)}{g(r)} = (m-1)D\kappa(1+r)^{\alpha/2} \qquad \text{for } r > 0. \tag{9.32}$$

Because of (9.31), we can apply Proposition 9.8 with

$$\theta(t) = (m-1)|D|\kappa(1+r)^{\alpha/2}, \qquad \varepsilon = 1/2C,$$

C the increasing constant in the third of (9.5), to infer the existence of λ sufficiently small such that, for each chosen $R \geq r_0$, there exists a solution w of

$$\begin{cases} w \in C^1([R, \infty)) \quad \text{and } C^2 \text{ except possibly at some } R_1 > R; \\[1ex] w \geq 0, \quad \text{on } [R, \infty), \qquad w \equiv 0 \quad \text{on } [R_1, \infty), \\[1ex] w(R) = \lambda, \qquad w' < 0 \quad \text{on } [R, R_1), \\[1ex] \left(\varphi(w')\right)' - \theta(r)\varphi(w') \leq \dfrac{1}{2C}\beta(r)f(w)l(|w'|) \qquad \text{on } [R, \infty). \end{cases}$$

We then specify $R \geq r_0$ large enough to satisfy

$$u(x) < \lambda \qquad \text{for } r(x) \geq R.$$

Defining $\bar{w}(x) = w(r(x))$ and using that $\mathcal{O}$ is a pole, we obtain

$$
\begin{cases}
\bar{w} \in C^1(M \setminus B_R(\mathcal{O})); \\[4pt]
\bar{w} \geq 0, \qquad \bar{w} \equiv 0 \quad \text{on } M \setminus B_{R_1}(\mathcal{O}), \\[4pt]
\bar{w} = \lambda \quad \text{on } \partial B_R(\mathcal{O}), \qquad |\nabla \bar{w}| > 0 \quad \text{on } B_{R_1}(\mathcal{O}) \setminus B_R(\mathcal{O})
\end{cases}
$$

and also, since $w' \leq 0$ and φ is odd, by (9.32),

$$
\Delta_\varphi \bar{w} = \big(\varphi(w')\big)' + \varphi(w')\Delta r \leq \big(\varphi(w')\big)' - \theta(r)\varphi(w')
$$

$$
\leq \frac{1}{2C}\beta(r)f(w)l(|w'|) = \frac{1}{2C}\beta(r)f(\bar{w})l(|\nabla \bar{w}|)
$$

weakly on $M \setminus B_R(\mathcal{O})$. Define $\Omega_R = \Omega \cap (M \setminus B_R(\mathcal{O}))$. By assumption, $u < \lambda = \bar{w}$ on $\partial \Omega_R$, and we are going to show that $u \leq \bar{w}$ on Ω_R. Once this is shown, then clearly u has compact support since $\bar{w}$ does, concluding our proof. We reason by contradiction, and we suppose that $c = \sup_{\Omega_R}(u - \bar{w}) > 0$. For $\delta \in (0, c)$, set $U_\delta = \{u - \bar{w} > \delta\} \neq \emptyset$. Note that $U_\delta \cap \partial \Omega_R = \emptyset$, and moreover U_δ is relatively compact since u vanishes at infinity and $\bar{w} \geq 0$. On the compact set $\Gamma = \{u - \bar{w} = c\}$, the identity $|\nabla \bar{w}| = |\nabla u|$ holds. We claim that $\inf_\Gamma |\nabla \bar{w}| > 0$. Suppose, by contradiction, that $|\nabla \bar{w}(x)| = 0$ for some $x \in \Gamma$. By the construction of $\bar{w}$, $x \in M \setminus B_{R_1}(\mathcal{O})$, and we examine two cases:

(i) $x \in M \setminus \overline{B_{R_1}(\mathcal{O})}$. In this case, $\bar{w} \equiv 0$ in a small neighbourhood V of x, and thus, by the definition of Γ and x is a local maximum of u on V. Since

$$
\Delta_\varphi u \geq b(x)f(u)l(|\nabla u|) \geq 0,
$$

applying the finite maximum principle in Theorem 6.8 to $c - u$, we deduce that $u \equiv c$ on V. Therefore, the set where $u = c$ is open, closed and non-empty in $\Omega_R \setminus B_{R_1}(\mathcal{O})$, and we conclude $u \equiv c$ on the connected component of $\Omega_R \setminus \overline{B_{R_1}(\mathcal{O})}$ containing x. Note then that $\Omega_R \setminus \overline{B_{R_1}(\mathcal{O})}$ is connected and unbounded; in fact, since $\mathcal{O}$ is a pole of M, the normal exponential map realizes a diffeomorphism between $M \setminus \mathcal{O}$ and $\partial \mathcal{O} \times \mathbb{R}^+$. In particular, Ω is diffeomorphic to $K \times (r_0, \infty)$ for some connected component $K \subset \partial \mathcal{O}$, and $\Omega_R \setminus \overline{B_{R_1}(\mathcal{O})}$ is diffeomorphic to the connected set $K \times (R_1, \infty)$. Concluding, $u \equiv c$ on $\Omega_R \setminus \overline{B_{R_1}(\mathcal{O})}$, a contradiction since u is assumed to vanish at infinity.

(ii) $x \in \partial B_{R_1}(\mathcal{O})$. In this case, $\nabla u(x) = \nabla \bar{w}(x) = 0$, and u on $\Omega_R \setminus B_{R_1}(\mathcal{O})$ has a boundary, global maximum at x. Moreover, by (i), the set Γ does not intersect $\Omega_R \setminus \overline{B_{R_1}(\mathcal{O})}$, and hence $u < c$ on $\Omega_R \setminus \overline{B_{R_1}(\mathcal{O})}$. Let $\gamma(t)$ be a ray from $\mathcal{O}$ with $\gamma(R_1) = x$. As in the proof of the finite maximum principle and the Hopf Lemma, Theorem 6.8, on a small enough annulus $E_\rho = B_{2\rho} \setminus B_\rho$ centered at $x_0 = \gamma(R_1 + 2\rho)$, we can construct a solution of

$$\Delta_\varphi v \geq 0, \qquad v = 0 \text{ on } \partial B_{2\rho}, \qquad v = \eta < c - \max_{\partial B_\rho} u \quad \text{on } \partial B_\rho$$

$$\langle \nabla v, \nabla r_{x_0} \rangle < 0 \quad \text{on } \partial B_{2\rho}, \qquad |\nabla v| > 0 \quad \text{on } \overline{E}_\rho,$$

where r_{x_0} is the distance from x_0. With the aid of Proposition 6.1, we can compare u with $c - v$ on $E_\rho \subset \Omega \backslash B_{R_1}(\mathcal{O})$ since

$$\begin{cases} \Delta_\varphi u \geq b(x) f(u) l(|\nabla u|) \geq 0 \geq \Delta_\varphi (c - v), \\[2mm] u - (c - v) = u - c \leq 0 \quad \text{on } \partial B_{2\rho}, \\[2mm] u - (c - v) \leq \max_{\partial B_\rho} u - c + \eta < 0 \quad \text{on } \partial B_\rho \end{cases}$$

to deduce $u \leq c - v$ on E_ρ. Since equality holds at $x \in \partial E_\rho$, we get

$$0 \leq \langle \nabla(u - c + v), \nabla r_{x_0} \rangle(x) = \langle \nabla v, \nabla r_{x_0} \rangle(x) < 0,$$

a contradiction.

We have therefore shown that $|\nabla \bar{w}| = |\nabla u| > 0$ on Γ, and we are in the position to conclude as usual: from $l > 0$ on $\mathbb{R}^+$, the quotient $l(|\nabla \bar{w}|)/l(|\nabla u|)$ is continuous and ≤ 2 on U_δ, for δ sufficiently close to c. By the C-increasing property of f,

$$f(\bar{w}(x)) l(|\nabla \bar{w}(x)|) \leq 2Cf(u(x)) l(|\nabla u(x)|) \qquad \forall x \in U_\delta,$$

and thus

$$\Delta_\varphi \bar{w} \leq \frac{1}{2C} b(x) f(\bar{w}) l(|\nabla \bar{w}|) \leq b(x) f(u) l(|\nabla u|) \leq \Delta_\varphi u$$

on U_δ, with $u = \bar{w} + \delta$ on ∂U_δ. By the comparison Proposition 6.1, $u \leq \bar{w} + \delta$ on U_δ, contradicting the very definition of U_δ and concluding the proof.

The reverse implication (CSP) $\Rightarrow$ (KO$_0$), under the further assumptions $l \in \mathrm{Lip}_{\mathrm{loc}}((0, \xi_0))$ and $f(0)l(0) = 0$, is a direct consequence of Theorem 9.3. $\square$

We next specialize Theorem 9.13, and we prove Theorem 2.38, that we rewrite for the reader's convenience.

Theorem 9.14 *Let $(M, \langle , \rangle)$ be a manifold with a pole $\mathcal{O}$ such that, setting $r(x) = \mathrm{dist}(x, \mathcal{O})$, (9.30) holds for some $\alpha \geq -2$, $\kappa \geq 0$. Consider φ, b, f, l satisfying (2.3),*

(2.5), (2.16) *and* (2.43). *Fix* $\chi, \mu \in \mathbb{R}$ *with*

$$\chi > 0, \qquad \mu \leq \chi - \frac{\alpha}{2}, \tag{9.33}$$

and assume that

$$l(t) \asymp t^{1-\chi} \varphi'(t) \qquad \text{for } t \in (0, 1),$$
$$b(x) \geq C_1 \big(1 + r(x)\big)^{-\mu} \qquad \text{for } r(x) \geq r_0, \tag{9.34}$$

for some constant $C_1 > 0$. *If there exists a constant* $c_F \geq 1$ *such that*

$$F(t)^{\frac{\chi}{\chi+1}} \leq c_F f(t) \qquad \text{for each } t \in (0, \eta_0), \tag{9.35}$$

then

$$\text{(CSP) holds for } (P_\geq) \qquad \Longleftrightarrow \qquad (KO_0).$$

Proof First, we note that requirements (2.3), (2.5), (2.16) and (2.43) on φ, f, l correspond to (9.2), (9.3), (9.5) and

$$f > 0 \text{ on } (0, \eta_0), \qquad l \in \mathrm{Lip}_{\mathrm{loc}}((0, \xi_0)), \qquad f(0)l(0) = 0.$$

We can therefore apply Theorem 9.3 to deduce the validity of implication (CSP) $\Rightarrow$ (KO_0). Vice versa, assume (KO_0) that in view of (9.34) is equivalent to

$$F^{-\frac{1}{\chi+1}} \in L^1(0^+). \tag{9.36}$$

We first observe that it is enough to prove (CSP) when

$$l(t) = C_2 t^{1-\chi} \varphi'(t) \qquad \text{if } t \in (0, 1),$$
$$b(x) = C_1 \big(1 + r(x)\big)^{-\mu} \qquad \text{if } r(x) \geq r_0,$$

for some positive constants C_1 and C_2. Indeed, the Keller–Osserman condition for $l(t) = C_2 t^{1-\chi} \varphi'(t)$ is still (9.36). As underlined in Remark 9.9, although the function $t^{1-\chi} \varphi'(t) \in L^\infty_{\mathrm{loc}}(\mathbb{R}_0^+)$ might fail to be continuous at $t = 0$, we can still apply Proposition 9.8, and thus Theorem 9.13, once we check the validity of the remaining assumptions: $\chi > 0$ implies both (9.3) and (C_3), (9.33) implies (C_1) and (C_2) and (9.35) is equivalent to (C_4). On the other hand, the function

$$\bar{\beta}(t) = c(1+t)^{-\chi-1}, \qquad \text{with} \quad c < K_\infty$$

satisfies (β_1) and (β_2). To conclude, note that (9.31) is equivalent to

$$\mu \leq \max\left\{\chi + 1, \chi - \frac{\alpha}{2}\right\} = \chi - \frac{\alpha}{2}.$$

$\square$

Remark 9.15 In the hypotheses of Theorem 9.14, set $v(r) = |\partial B_r(\mathcal{O})|$. Using the divergence theorem and coarea formula, we deduce

$$v'(r) = \int_{\partial B_r(\mathcal{O})} \Delta r \geq -C_1 r^{\chi - \mu} v(r),$$

where we used (9.32) with $\alpha/2 = \chi - \mu$. A further integration gives

$$\text{if } \mu < \chi + 1, \qquad v(r) \geq C_1 e^{-C_2 r^{\chi+1-\mu}}$$

$$\text{if } \mu = \chi + 1, \qquad v(r) \geq C_1 r^{-C_2}.$$

for some constants $C_1, C_2 > 1$. Assume that $|M| < \infty$. Integrating the above on (r, ∞), taking logarithms and recalling that

$$|M \backslash B_r(\mathcal{O})| = \int_r^\infty v(s)ds,$$

we deduce that

$$\text{if } \mu < \chi + 1, \qquad \limsup_{r \to \infty} \frac{-\log|M \backslash B_r(\mathcal{O})|}{r^{1+\chi-\mu}} < \infty;$$

$$\text{if } \mu = \chi + 1, \qquad \limsup_{r \to \infty} \frac{-\log|M \backslash B_r(\mathcal{O})|}{\log r} < \infty.$$

(9.37)

It is interesting to compare (9.37) with conditions in (7.6). In view of Theorem 7.5, one might wonder whether the (CSP) could be proved under a volume growth assumption like (9.37). Indeed, the problem seems to be quite hard, and one of the main reasons lies in the fact that a manifold satisfies (CSP) if and only if each of its ends does. This forces (9.37) to be satisfied on *each* end, otherwise an end with big volume would be sufficient for (9.37) to hold independently of the behaviour of the others. However, an approach via integral estimates like in Theorem 7.5, loosely speaking, seems unable to distinguish among different ends, and thus, at least, it needs to be complemented by new techniques.

Remark 9.16 Both (FMP) and (CSP) can be considered for more general inequalities, including the prototype ones

$$\Delta_p u = u^\omega \pm |\nabla u|^q$$

for some $\omega, q > 0$. In fact, the Keller–Osserman condition changes according to whether $q \geq p - 1$ or $q < p - 1$, see [196] and [88] for a detailed account. In particular, in [88], the authors propose suitable Keller–Osserman conditions for the case when the terms u^ω and $|\nabla u|^q$ strongly interact. The sharpness of these conditions for general nonlinearities in u and $|\nabla u|$ is, to our knowledge, still an open problem.

A Second ODE Lemma: Locating the Support

Proposition 9.8 guarantees the existence of $R_1 > R$ such that the supersolution w in (9.20) vanishes outside of B_{R_1}. However, the proof gives loose indication on the distance between R_1 and R. Although this further information is not needed in the results that we present here, we feel worth to underline that the construction of w can be modified in such a way to locate R_1, say to have $R_1 = 2R$. More importantly, this new method, which works under a set of assumptions which is skew with respect to that in Proposition 9.8, allows for weights $b(x)$ that may oscillate between two different polynomial type decays. In the sequel, we need

(C$_2$)′   there exist constants $d_1 > 0$ and $c_1 > 0$ such that

$$\varphi'(st) \leq d_1 \varphi'(s)\varphi'(t) \qquad \text{for each } s, t \in (0, 1];$$

$$l(s)l(t) \leq c_1 l(st) \qquad \text{for each } s, t \in (0, 1];$$

instead of the weaker (C$_2$) (cf. Lemma 9.34). On the other hand, we will not need (C$_3$). Regarding β, we assume that $\beta \equiv \bar{\beta}$ and that β vanishes at infinity, and a further condition (β_3), namely we require

(β_1)′ $0 < \beta \in C^1([r_0, \infty))$, $\beta' \leq 0$ for $t \geq r_0$, $\beta(t) \to 0$ as $t \to \infty$;
(β_2)′   there exists a constant $c_\beta \geq 1$ such that

$$\frac{-\beta'(t)}{K^{-1}(\beta(t))} \leq c_\beta \beta(t) \qquad \text{for each } t \in [r_0, \infty).$$

(β_3)   There exists a constant $\hat{c}_\beta > 0$ such that

$$\limsup_{t \to \infty} \frac{-t\beta'(t)}{\beta(t)} \geq \hat{c}_\beta.$$

Remark 9.17 Up to choosing r_0 large enough, by (β_1)′, we can assume $\beta(t) \in [0, K_\infty)$, and thus ($\beta_2$)′ is meaningful.

Example 9.18 Referring to Example 9.5, φ and l satisfy (C_1) and $(C_2)'$ for each $\chi \geq 0$, while (C_4) is met for $\omega \leq \chi$. If further $\beta(t) = (1+t)^{-\mu}$, $(\beta_1)'$ and (β_3) require $\mu > 0$ to be both satisfied, and $(\beta_2)'$ needs $\mu \leq \chi + 1$.

We are ready to state our second main ODE result, to be compared to Proposition 9.8. Its delicate proof originates from the paper [193], later refined in [198, 208] and [209], and to help readability, we postpone it to the end of this chapter.

Proposition 9.19 *Let* φ, f, l *satisfy* (9.5) *and* (9.3), *and assume the validity of* $(C_1), (C_2)', (C_4)$ *and* $(\beta_1)', (\beta_2)', (\beta_3),$ *for some* $r_0 > 0$. *Having fixed a non-negative* $\theta \in C([r_0, \infty)),$ *suppose that*

$$\limsup_{R \to \infty} K\left(\frac{1}{R K^{-1}(\beta(2R))}\right) R\theta(R) < \infty. \tag{9.38}$$

Then, there exists a diverging sequence $\{R_j\}$ *such that the following holds: if*

$$\frac{1}{K^{-1} \circ F} \in L^1(0^+), \tag{KO$_0$}$$

then for each $\epsilon \in (0, \xi_0)$, *there exist* $\lambda \in (0, \eta_0)$ *and, for each* $R \in \{R_j\}$, *a function* z *with the following properties:*

$$\begin{cases} z \in C^1([R, \infty)), \quad \text{and } C^2 \text{ except possibly at } 2R \\ 0 \leq z \leq \lambda, \quad z(R) = \lambda, \quad z \equiv 0 \quad \text{on } [2R, \infty), \\ z' < 0 \quad \text{on } [R, 2R), \quad |z'| \leq \epsilon \quad \text{on } [R, \infty), \\ (\varphi(z'))' - \theta(t)\varphi(z') \leq \epsilon\beta(t)f(z)l(|z'|) \quad \text{on } [R, \infty). \end{cases} \tag{9.39}$$

Remark 9.20 The two $\limsup$ in (β_3) and (9.38) could be simultaneously replaced by $\liminf$. Indeed, the sequence $\{R_j\}$ is just required to satisfy

$$\begin{cases} R_1 \geq 2r_0 \\ \dfrac{-R_j\beta'(R_j)}{\beta(R_j)} \geq \dfrac{\hat{c}_\beta}{2} \\ K\left(\dfrac{1}{R_j K^{-1}(\beta(2R_j))}\right) R_j\theta(R_j) \leq B_2, \end{cases} \tag{9.40}$$

for some $B_2 > 0$. Observe that, in the "double $\liminf$" case, the vanishing of β is automatic by integrating (β_3), and hence $(\beta_1)'$ coincides with (β_1).

As a direct corollary, we have the following result, whose proof follows verbatim that of Theorem 9.13 replacing, in the argument, Proposition 9.8 with Proposition 9.19.

Theorem 9.21 *Let $(M, \langle , \rangle)$ be a manifold possessing a pole $\mathcal{O}$, and let $r(x) = \text{dist}(x, \mathcal{O})$. Suppose that (9.30) holds, consider $0 < b \in C(M)$ and let $\beta \in C^1([r_0, \infty))$ such that*

$$b(x) \geq \beta(r(x)) \qquad \text{for } r(x) \geq r_0.$$

Let φ, f, l satisfy (9.2), (9.3) and (9.5), and assume the validity of $(C_1), (C_2)', (C_4)$ and $(\beta_1)', (\beta_2)', (\beta_3)$. Suppose that

$$\limsup_{R \to \infty} K \left(\frac{1}{RK^{-1}(\beta(2R))} \right) R^{1+\frac{\alpha}{2}} < \infty.$$

Then,

$$(\text{KO}_0) \qquad \Longrightarrow \qquad (\text{CSP}) \text{ holds for } (9.1).$$

Moreover, if

$$l \in \text{Lip}_{\text{loc}}((0, \xi_0)), \qquad f(0)l(0) = 0,$$

then

$$(\text{KO}_0) \qquad \Longleftrightarrow \qquad (\text{CSP}) \text{ holds for } (9.1).$$

Specified to power-like φ, f, l, Theorem 9.21 has the next corollary for general weights b. Note that here we use Remark 9.20.

Corollary 9.22 *Let M be a complete manifold with a pole $\mathcal{O}$ such that (9.30) holds. Suppose that φ, f, l satisfy (9.2), (9.3), (9.5) and $l \in \text{Lip}_{\text{loc}}((0, \xi_0))$. Moreover, assume that for some $p, \chi, \omega \in \mathbb{R}$ with*

$$p > 1, \qquad 0 < \chi \leq p - 1, \qquad \omega > 0,$$

it holds

$$\varphi'(t) \asymp t^{p-2}, \qquad l(t) \asymp t^{p-1-\chi}, \qquad f(t) \asymp t^{\omega}$$

for $t \in (0, 1)$. Suppose that there exist $r_0 > 0$ and $0 < \beta \in C^1([r_0, \infty))$ matching (β_1) and

$$-\beta'(t) \le B[\beta(t)]^{\frac{\chi+2}{\chi+1}} \qquad \text{on } [r_0, \infty);$$

$$\liminf_{t \to \infty} \frac{-t\beta'(t)}{\beta(t)} > 0, \qquad \liminf_{t \to \infty} \frac{t^{\frac{\alpha}{2}-\chi}}{\beta(t)} < \infty, \tag{9.41}$$

for some constant $B > 0$. If $0 < b \in C(M)$ satisfies

$$b(x) \ge \beta(r(x)) \qquad \text{for } r(x) \ge r_0,$$

then

$$(\text{CSP}) \text{ holds for } (9.1) \qquad \Longleftrightarrow \qquad \omega < \chi.$$

Remark 9.23 A careful analysis of (9.41) shows that the above corollary allows for bounds β that oscillate between the polynomial decays $t^{-1-\chi}$ and $t^{\alpha/2-\chi}$.

Non-Empty Cut-Locus: The p-Laplacian Case

With the help of Chap. 4, we now remove the pole condition in the particular case of the p-Laplace operator, exploiting the fake distance function ϱ. Let Ω be an end of M, and we consider a solution u of

$$\begin{cases} \Delta_p u \ge b(x) f(u) |\nabla u|^{p-1-\chi} & \text{on } \Omega, \\ u \ge 0, \qquad \lim_{x \in \Omega,\, x \to \infty} u(x) = 0, \end{cases} \tag{9.42}$$

where $0 \le \chi \le p - 1$. Since $\varphi(t) = t^{p-1}$ and $l(t) = t^{p-1-\chi}$, the function K in (9.4) automatically satisfies (C_1) and $(C_2)'$. Eventually, we assume (C_4) that in the present case can be written as follows:

(C_4) there exists $c_F \ge 1$ such that

$$c_F f(t) \ge F(t)^{\frac{\chi}{\chi+1}} \qquad \text{for } t \in [0, \eta_0).$$

Let $(M^m, \langle , \rangle)$ be a complete Riemannian manifold satisfying

$$\text{Ric} \ge -(m - 1)\kappa^2 \langle , \rangle \qquad \text{on } M,$$

for some constant $\kappa > 0$. Suppose that Δ_p is non-parabolic on M, and define the fake distance ϱ as in (4.10) associated to the hyperbolic space of curvature $-\kappa^2$ (that is, $g(r) = \kappa^{-1}\sinh(\kappa r)$). To be able to radialize with respect to ϱ, we shall assume

$$b(x) \geq \beta\big(\varrho(x)\big) \qquad \text{on } M, \tag{9.43}$$

for some function β matching the necessary assumptions to apply Proposition 9.8. In our case of interest, we can restrict to non-increasing β and to $\bar\beta = \beta^\gamma$, for some $\gamma \geq 1$. Then, (β_1) and (β_2) amount to the requirements:

(β_1) $0 < \beta \in C^1(\mathbb{R}_0^+)$, $\beta' \leq 0$ on $\mathbb{R}^+$.

(β_2) For some $\gamma \geq 1$,

$$-\beta'(t) \leq c_\beta \beta(t)^{\frac{\chi+1+\gamma}{\chi+1}} \qquad \text{on } [1, \infty).$$

The prototype example is given by the choice

$$\beta(t) = (1+t)^{-\mu} \qquad \text{with } \mu \in [0, \chi+1].$$

As explained in Chap. 2, we shall need a technical assumption, the weak Sard property $(\mathcal{WS})$ described in Definition 2.42. Written in terms of the fake distance ϱ, the property guarantees the existence of a diverging sequence $\{R_j\}$ such that the boundary of the fake ball

$$D_{R_j} \doteq \{\varrho < R_j\}$$

has the exterior ball condition. Here is our main result:

Theorem 9.24 *Let $(M^m, \langle\,,\,\rangle)$ be a complete Riemannian manifold such that, for some origin o,*

$$\mathrm{Ric} \geq -(m-1)\kappa^2\langle\,,\,\rangle \qquad \text{on } M,$$

for some constant $\kappa > 0$. Let $p \in (1, \infty)$; suppose that Δ_p is non-parabolic on M and that the minimal positive Green kernel $\mathcal{G}(x)$ with pole at o satisfies

$$\mathcal{G}(x) \to 0 \quad \text{as } x \text{ diverges.} \tag{9.44}$$

Let f satisfy

$$f \in C(\mathbb{R}), \qquad f > 0 \quad \text{and } C\text{-increasing on } (0, \eta_0) \tag{9.45}$$

for some $\eta_0 > 0$, and, for $\chi \in (0, p - 1]$, assume (C_4). Let $b \in C(M)$ satisfy (9.43), for some β matching (β_1) and (β_2) above. Suppose that either

(i) $\chi = p - 1$ and $f(0) = 0$, or
(ii) $\chi \in (0, p - 1)$, $p \in (1, 2)$ and property $(\mathcal{WS})$ holds (cf. Definition 2.42), or
(iii) $\chi \in (0, 1)$ and $p = 2$.

Then,

$$\text{(CSP) holds for (9.42)} \quad \Longleftrightarrow \quad F^{-\frac{1}{\chi+1}} \in L^1(0^+).$$

Remark 9.25 In view of Corollary 4.17, Theorem 4.24 and Examples 4.20, 4.22 and 4.23, the vanishing of $\mathcal{G}$ is granted provided either one of the next conditions holds:

(i) $p \in (1, \infty)$ and Ric ≥ 0 on M;
(ii) $p \in (1, \infty)$ and M supports the Sobolev inequality (4.31);
(iii) $p \in (1, m)$ and M is minimally immersed into a Cartan–Hadamard manifold;
(iv) $p \in (1, m)$,

$$\text{Ric} \geq -(m - 1)\kappa^2 \langle\,,\,\rangle, \qquad \inf_{x \in M} |B_1(x)| > 0$$

and M supports the Poincaré inequality (4.36);
(v) $p \in (1, m)$, M is roughly isometric to $\mathbb{R}^m$ and

$$\text{Ric} \geq -(m - 1)\kappa^2 \langle\,,\,\rangle, \qquad \text{inj}(M) > 0.$$

Proof Because of (9.44), the fake distance ϱ defined in (4.10) and associated with the hyperbolic space of curvature $-\kappa^2$ is proper. Furthermore, by (4.12) and Theorem 4.15,

$$|\nabla \varrho| \leq 1, \quad \Delta_p \varrho \geq 0 \qquad \text{on } M \backslash \{o\}. \tag{9.46}$$

For $s > 0$, write

$$D_s \doteq \{\varrho < s\}, \qquad \Omega_s \doteq \Omega \backslash \overline{D}_s.$$

We first prove the implication $(KO_0) \Rightarrow$ (CSP). Since (9.19) is automatically met for $\theta(t) \equiv 0$, for each fixed $\varepsilon > 0$, Proposition 9.8 guarantees the existence of $\lambda \in (0, \eta_0)$ sufficiently small and $r_0 > 1$ such that, for each $R > r_0$, we can find $R_1 > R$ and a solution w of

$$
\begin{cases}
w \in C^1([R, \infty)) \quad \text{and } C^2 \text{ except possibly at } R_1; \\
w \geq 0, \quad \text{on } [R, \infty), \quad w(R) = \lambda, \quad w \equiv 0 \quad \text{on } [R_1, \infty), \\
w' < 0 \quad \text{on } [R, R_1), \quad |w'| \leq \varepsilon \quad \text{on } [R, \infty), \\
\left(|w'|^{p-2} w'\right)' \leq \varepsilon \beta(r) f(w) |w'|^{p-1-\chi} \quad \text{on } [R, \infty).
\end{cases}
\tag{9.47}
$$

Choose r_0 such that $u < \lambda$ on Ω_{r_0}, and $R > r_0$. The value of ε will be specified later, depending just on the C-increasing constant. Define $\bar{w}(x) = w(\varrho(x))$. If we combine (4.13), $w' \leq 0$ and (4.12), by (9.46), we deduce

$$
\begin{cases}
\Delta_p \bar{w} = \left[\left(|w'|^{p-2} w'\right)' + \dfrac{v_g'}{v_g} |w'|^{p-2} w'\right] (\varrho) |\nabla \varrho|^p \\[2mm]
\qquad \leq \left(|w'|^{p-2} w'\right)' |\nabla \varrho|^p \leq \varepsilon \beta(\varrho) f(\bar{w}) |w'(\varrho)|^{p-1-\chi} |\nabla \varrho|^p \quad \text{on } M \backslash D_R, \\[2mm]
\bar{w} \in C^1(M \backslash D_R) \quad \text{and is in fact } C^{1,\alpha}_{\text{loc}} \text{ except possibly on } \partial D_{R_1}; \\[2mm]
\bar{w} \geq 0, \quad \text{on } M \backslash D_R, \quad \bar{w} \equiv 0 \quad \text{on } M \backslash D_{R_1}, \\[2mm]
\bar{w} = \lambda \quad \text{on } \partial D_R, \quad |\nabla \bar{w}| \leq \varepsilon \quad \text{on } D_{R_1} \backslash D_R.
\end{cases}
$$

To prove (CSP), we show, as before, that $u \leq \bar{w}$ on Ω_R: we proceed by contradiction, assuming $u > \bar{w}$ somewhere, and we look at the set

$$
\Gamma = \{u - \bar{w} = c\} \Subset \Omega_R, \quad \text{with} \quad c = \max_{\overline{\Omega}_R}(u - \bar{w}) > 0.
$$

We will then apply a comparison theorem on an upper level set

$$
U_\delta = \{u - \bar{w} > \delta\} \cap \Omega_R
$$

for δ close enough to c. To this aim, in Theorem 9.13, we obtained $\Delta_p u \geq \Delta_p \bar{w}$ on U_δ, crucially using $|\nabla \bar{w}| > 0$ on Γ. However, now $\bar{w}$ is radial with respect to the fake distance ϱ, which differently from r may possess stationary points. This forces us to use a different argument when $\chi < p - 1$, that is, in case (ii). On the other hand, in case (i), the gradient term disappears and the argument goes straightforwardly: first, choose ε so that $\varepsilon C \leq 1$, and then observe that $0 \leq u, \bar{w} \leq \lambda < \eta_0$ on Ω_R. Therefore, using that f is C-increasing, on U_δ, we have

$$
\Delta_p u \geq b(x) f(u) \geq \frac{1}{C} \beta(\varrho) f(\bar{w}) \geq \varepsilon \beta(\varrho) f(\bar{w}) |\nabla \varrho|^p \geq \Delta_p \bar{w},
$$

and by comparison, we conclude $u \leq \bar{w} + \delta$ on U_δ, a contradiction.

Cases (ii) and (iii) are more subtle. Although we cannot guarantee that $|\nabla \bar{w}| > 0$ on Γ, nevertheless, as a first step, we still claim that

$$\Gamma \Subset D_{R_1}. \tag{9.48}$$

Indeed, by contradiction, if there exists $x \in \Gamma \cap \Omega_{R_1}$, let V be the connected component of Ω_{R_1} containing x. By the second in (9.46) and $R_1 \geq 1$, we deduce that V is necessarily unbounded by the maximum principle, being a component of the upper level set $\{\varrho > R_1\}$ of a p-subharmonic function. Arguing as in the proof of Theorem 9.13, case (i), we deduce $u \equiv c$ on V, which contradicts the vanishing of u at infinity. Therefore, $u < c$ on Ω_{R_1}. Next, since R_1 depends continuously on ε and $R_1 \to \infty$ as $\varepsilon \to 0$, by the construction of w we can assume that R_1 is chosen such that ∂D_{R_1} has the exterior boundary condition. Indeed, in case (ii), the claim follows from property $(\mathcal{WS})$, while in case (iii), it follows by Sard's theorem. We can then apply the Hopf Lemma as in (ii) of Theorem 9.13 to get $u < c$ on $\partial \Omega_{R_1}$. This proves (9.48). As a consequence, from (9.47), the inequality $|w'(\varrho)| > 0$ holds on Γ. Coupling with $|\nabla u| = |\nabla \bar{w}| \leq \varepsilon$ on Γ, if $\varepsilon < 1/2$, we can choose δ_0 close enough to c in such a way that

$$|w'(\varrho)| > 0, \qquad |\nabla u| + |\nabla \bar{w}| < 1 \qquad \text{on } \overline{U}_{\delta_0}.$$

Note that δ_0 depends on ε. We come back to the differential inequality for $\bar{w}$, which on U_{δ_0} implies

$$\Delta_p \bar{w} \leq \varepsilon \beta(\varrho) f(\bar{w}) |w'(\varrho)|^{p-1-\chi} |\nabla \varrho|^p$$
$$= \varepsilon \beta(\varrho) |w'(\varrho)|^{-1-\chi} f(\bar{w}) |\nabla \bar{w}|^p. \tag{9.49}$$

For $\delta \in (\delta_0, c)$, we consider the open sets

$$E_\delta = U_\delta \cap \{|\nabla u| < |w'(\varrho)|\}, \qquad \hat{E}_\delta = U_\delta \cap \{|\nabla u| > |w'(\varrho)|/2\}.$$

On E_δ, using the C-increasing property of f and $u > \bar{w}$, whenever $\varepsilon \leq C^{-1}$, we deduce that u is a weak solution of

$$\Delta_p u \geq \beta(\varrho) f(u) |\nabla u|^p |\nabla u|^{-1-\chi} \geq \frac{\beta(\varrho)}{C} f(\bar{w}) |\nabla u|^p |w'(\varrho)|^{-1-\chi}$$
$$\geq \varepsilon \beta(\varrho) |w'(\varrho)|^{-1-\chi} f(\bar{w}) |\nabla u|^p. \tag{9.50}$$

On the other hand, on $\Gamma \cap \hat{E}_\delta$, we have

$$|\nabla u| = |\nabla \bar{w}| = |w'(\varrho)||\nabla \varrho|, \qquad \text{hence} \qquad |\nabla \varrho| > 1/2.$$

By continuity, we can therefore choose δ sufficiently close to c so that

$$\frac{|\nabla u|}{|w'(\varrho)|} \leq 2|\nabla\varrho| \leq 2 \qquad \text{on } \hat{E}_\delta.$$

As a consequence, if $\varepsilon \leq [C2^{\chi+1}]^{-1}$, on $\hat{E}_\delta$, the function u weakly solves

$$\Delta_p u \geq \beta(\varrho)f(u)|\nabla u|^p|\nabla u|^{-1-\chi} \geq \frac{\beta(\varrho)}{C}f(\bar{w})|\nabla u|^p|w'(\varrho)|^{-1-\chi}\frac{1}{2^{\chi+1}} \tag{9.51}$$

$$\geq \varepsilon\beta(\varrho)|w'(\varrho)|^{-1-\chi}f(\bar{w})|\nabla u|^p.$$

Putting together (9.49), (9.50) and (9.51), for ε small enough depending only on C, χ and setting $\bar{w}_\delta = \bar{w} + \delta$, the following inequalities hold on U_δ:

$$\begin{cases} \Delta_p u \geq \left[\varepsilon\beta(\varrho)|w'|^{-1-\chi}f(\bar{w})\right]|\nabla u|^p; \\[2mm] \Delta_p \bar{w}_\delta \leq \left[\varepsilon\beta(\varrho)|w'|^{-1-\chi}f(\bar{w})\right]|\nabla\bar{w}_\delta|^p, \\[2mm] u = \bar{w}_\delta \qquad \text{on } \partial U_\delta. \end{cases}$$

To conclude, we observe that, since $p \leq 2$, the p-Laplacian is non-degenerate elliptic. We claim that we can apply the comparison Theorem 5.3 in [11] (the manifold version of [194, Thm. 3.5.1]) with the choice

$$B(x, z, \xi) = \varepsilon\beta(\varrho)|w'(\varrho)|^{-1-\chi}f(\bar{w})|\xi|^p$$

to deduce $u \leq \bar{w}_\delta$ on U_δ, a contradiction. To ensure the applicability of the above theorem, we shall check that B is *regular*, in the sense specified in [11]: for each compact set $K \Subset \mathbb{R} \times TU_\delta$, there exists a constant $L_K > 0$ such that

$$\left|B(x, z, \xi) - B(x, z, \eta)\right| \leq L_K|\xi - \eta| \qquad \forall\, (x, z, \xi), (x, z, \eta) \in K.$$

Towards this end, let A be such that $|\xi| + |\eta| \leq A$ for $(x, z, \xi), (x, z, \eta) \in K$. From

$$\left||\xi|^p - |\eta|^p\right| = \left|\int_{|\eta|}^{|\xi|} pt^{p-1}dt\right| \leq pA^{p-1}\left|\int_{|\eta|}^{|\xi|} dt\right|$$

$$= pA^{p-1}\left||\xi| - |\eta|\right| \leq pA^{p-1}|\xi - \eta|,$$

we obtain

$$|B(x, z, \xi) - B(x, z, \eta)| \leq \varepsilon \|\beta(\varrho) w'(\varrho)^{-1-\chi} f(\bar{w})\|_\infty \big||\xi|^p - |\eta|^p\big|$$

$$\leq \bar{C}\varepsilon|\xi - \eta|,$$

for some $\bar{C} > 0$, as claimed. This concludes the proof that $u \leq \bar{w}$.

We next show that (CSP) $\Rightarrow$ (KO$_0$). Suppose the failure of (KO$_0$). Both cases (i) and (ii) imply $f(0)l(0) = 0$ since $l(t) = t^{p-1-\chi}$; thus, having fixed $r_0 > 0$ and $\xi \in (0, \xi_0)$, we can apply Theorem 9.3 to deduce the existence of η small enough and of a solution $u_1 \in \mathrm{Lip}(M \backslash B_{r_0})$ (B_{r_0} centered at some fixed origin o) of

$$\begin{cases} \Delta_p u_1 \geq b(x) f(u_1) |\nabla u_1|^{p-1-\chi} & \text{weakly on } M \backslash B_{r_0}, \\ 0 < u_1 \leq \eta & \text{on } M \backslash B_{r_0}, \\ u_1 = \eta & \text{on } \partial B_{r_0}, \quad u_1(x) \to 0 \quad \text{as } r(x) \to \infty, \\ |\nabla u_1| < \xi & \text{on } M \backslash B_{r_0}. \end{cases}$$

Set $u_2(x) = \mathcal{G}(x)$. Up to decreasing η, we can suppose that $u_1 \leq u_2$ on ∂B_{r_0}, and thus, by comparison (being $\Delta_p u_1 \geq 0$), we deduce $u_1 \leq u_2$ on $M \backslash B_{r_0}$. Since u_2 trivially solves $(P_\leq)$, by the subsolution–supersolution method (cf. [138, Thm. 4.4]), there exists a solution u of

$$\begin{cases} \Delta_p u = b(x) f(u) |\nabla u|^{p-1-\chi} & \text{on } M \backslash B_{r_0} \\ u_1 \leq u \leq u_2 & \text{on } M \backslash B_{r_0}. \end{cases}$$

The regularity theorem in [226, Thm. 1] ensures that $u \in C^{1,\beta}_{\mathrm{loc}}(M \backslash B_{r_0})$. Since, by assumption, u_2 vanishes at infinity, u shows the failure of (CSP), concluding our proof.

$\square$

To be able to handle the case of non-constant b, the bound (9.43) becomes effective provided that we know an upper bound on r in terms of ϱ. In view of Theorems 4.16 and 4.24 and Remark 4.18, this can be done in various cases of interest. By way of example, the following result applies to manifolds with non-negative Ricci curvature and maximal volume growth.

Theorem 9.26 *Let $(M^m, \langle\,,\,\rangle)$ be a complete Riemannian manifold of dimension $m \geq 3$ satisfying*

$$\mathrm{Ric} \geq 0, \qquad \lim_{r \to \infty} \frac{|B_r|}{r^m} > 0.$$

Fix

$$p \in (1, \infty), \qquad \chi \in (0, p-1], \qquad \mu \in [0, \chi + 1],$$

and let f satisfy

$$\begin{cases} f \in C(\mathbb{R}), \qquad f \text{ is positive and } C\text{-increasing on } (0, \eta_0), \\ \exists c_F \geq 1 \text{ such that } \quad c_F f(t) \geq F(t)^{\frac{\chi}{\chi+1}} \quad \text{for } t \in [0, \eta_0), \end{cases}$$

for some $\eta_0 > 0$. Assume that either

(i) $\chi = p - 1$ *and* $f(0) = 0$, *or*
(ii) $\chi \in (0, p-1)$, $p \in (1, 2)$ *and property* (WS) *holds (cf. Definition 2.42), or*
(iii) $\chi \in (0, 1)$ *and* $p = 2$.

Then, (CSP) *holds for solutions of*

$$\begin{cases} \Delta_p u \geq \big(1 + r(x)\big)^{-\mu} f(u) |\nabla u|^{p-1-\chi} \qquad \text{on } \Omega \text{ end of } M; \\ u \geq 0, \qquad \lim_{x \in \Omega, \, x \to \infty} u(x) = 0, \end{cases} \tag{9.52}$$

if and only if

$$F^{-\frac{1}{\chi+1}} \in L^1(0^+).$$

Proof Let ϱ be the fake distance associated with the model $\mathbb{R}^m$. From Theorem 4.24 and Example 4.21, in our assumptions, $\mathcal{G}(x) \leq C_1 r(x)^{-\frac{m-p}{p-1}}$, and hence

$$\varrho(x) \geq C_2 r(x) \qquad \forall x \in M,$$

for some constant $C_2 > 0$. Setting

$$b(x) = \big(1 + r(x)\big)^{-\mu} \geq C\big(1 + \varrho(x)\big)^{-\mu} \doteq \beta\big(\varrho(x)\big),$$

then (β_1) and (β_2) are matched because of our requirement on μ. The result then follows by applying Theorem 9.24. $\qquad \square$

Remark 9.27 In view of Remark 4.18, the above theorem applies more generally to manifolds satisfying Ric ≥ 0 and, for some origin o and some $C_1 > 0$, $b > 1$, the reverse doubling inequality

$$\forall t \geq s > 0, \qquad \frac{|B_t|}{|B_s|} \geq C_1 \left(\frac{t}{s}\right)^b.$$

Indeed, in this case,

$$\mathcal{G}(x) \leq C \int_{r(x)}^{\infty} \left(\frac{s}{|B_r|}\right)^{\frac{1}{p-1}} ds,$$

for some constant $C > 0$, and therefore

$$\varrho(x) \geq C_2 h(x) \doteq C_2 \left[\int_{r(x)}^{\infty} \left(\frac{s}{|B_s|}\right)^{\frac{1}{p-1}} ds\right]^{\frac{p-1}{p-m}}.$$

Consequently, Theorem 9.26 rephrases verbatim by replacing the weight $(1+r)^{-\mu}$ in (9.52) with the function

$$\left(1 + h(x)\right)^{-\mu}.$$

A Further Fake Distance and the Feller Property

When l is constant and Δ_φ is the Laplace–Beltrami operator, a different fake distance ς recently constructed in [20, Thm 2.1] turns out to be effective to improve Theorem 9.24.

Theorem 9.28 ([20]) *Let $(M, \langle\,,\,\rangle)$ be a complete manifold of dimension $m \geq 2$ satisfying*

$$\mathrm{Ric} \geq -(m-1)\kappa^2 \left(1+r^2\right)^{\alpha/2} \langle\,,\,\rangle \qquad on\ M, \tag{9.53}$$

for some $\kappa > 0$ and $\alpha \in [-2, 2]$. Fix an origin o with associated distance $r(x) = \mathrm{dist}(x, o)$. Then, there exist constants $C_1, C_2 > 0$ depending on m, κ, α, o and a function $\varsigma \in C^\infty(M)$ such that

$$C_1^{-1}\left(1+r(x)\right)^{1+\frac{\alpha}{2}} \leq \varsigma(x) \leq C_1\left(1+r(x)\right)^{1+\frac{\alpha}{2}} \quad if\ \alpha \in (-2, 2],$$

$$C_1^{-1} \log\left(2+r(x)\right) \leq \varsigma(x) \leq C_1 \log\left(2+r(x)\right) \quad if\ \alpha = -2, \tag{9.54}$$

$$\max\left\{|\nabla\varsigma|^2, |\Delta\varsigma|\right\} \leq C_2(1+r)^{\alpha} \qquad on\ M.$$

The proof of the existence of ς is delicate and inspired by that in [215], and we refer the reader to both references for details. Here, we show how to use ς to prove the compact support principle for solutions of

$$\Delta u \geq (1+r)^{-\mu} f(u)$$

under the only geometric requirement (9.53), for each $\alpha \in (-2, 2]$ and

$$\mu \leq 1 - \frac{\alpha}{2}, \tag{9.55}$$

provided that (KO$_0$) holds, i.e. if

$$\frac{1}{\sqrt{F(t)}} \in L^1(0^+).$$

The core is to construct the radial compactly supported supersolution $\bar{w} = w(\varsigma)$, for some w that we assume to be C^2, convex and strictly decreasing until it touches zero in a C^1 way. Using (9.54) and $w' < 0$, $w'' \geq 0$, for $\alpha > -2$, we deduce

$$\Delta \bar{w} = w'' |\nabla \varsigma|^2 + w' \Delta \varsigma \tag{9.56}$$

$$\leq C_2 (1+r)^\alpha \left\{ w'' - w' \right\} \leq C_3 \varsigma^{\alpha (1 + \frac{\alpha}{2})^{-1}} \left\{ w'' - w' \right\},$$

for some constant $C_3 > 0$. Now, let f satisfying (9.45), $f(0) = 0$ and

$$c_F f(t) \geq \sqrt{F(t)} \qquad \text{for } t \in [0, \eta_0),$$

for some constant $c_F > 0$. We apply Proposition 9.8 with the choices

$$\varphi(t) = t, \quad l(t) = 1, \quad \chi = 1, \quad \theta(t) = 1,$$

$$\beta(t) = t^{-(\mu + \alpha)(1 + \frac{\alpha}{2})^{-1}}, \qquad \bar{\beta}(t) = t^{-2}$$

to deduce the existence of w satisfying

$$\begin{cases} w \in C^1([R, \infty)) \quad \text{and } C^2 \text{ except possibly at } R_1; \\ 0 \leq w \leq \lambda, \qquad w(R) = \lambda, \qquad w \equiv 0 \quad \text{on } [R_1, \infty), \\ w' < 0 \quad \text{on } [R, R_1), \qquad |w'| \leq \varepsilon \quad \text{on } [R, \infty), \\ w'' - w' \leq \varepsilon t^{-(\mu + \alpha)(1 + \frac{\alpha}{2})^{-1}} f(w) \qquad \text{on } [R, \infty). \end{cases}$$

In fact, $K(t) \asymp t^2$ for $t \in (0, 1)$, and the growth requirement (9.19) is equivalent to (9.55). Plugging into (9.56) and using again (9.54), we get

$$\Delta \bar{w} \leq C_3 \varepsilon \varsigma^{\alpha\left(1+\frac{\alpha}{2}\right)^{-1}} \varsigma^{-(\mu+\alpha)\left(1+\frac{\alpha}{2}\right)^{-1}} f(w)$$

$$\leq C_4 \varepsilon (1+r)^{-\mu} f(\bar{w}),$$

for some $C_4 > 0$, and hence $\bar{w}$ is the desired supersolution. The proof of (CSP) now proceeds verbatim as in Theorem 9.24, case (i), leading to the following theorem.

Theorem 9.29 *Let M be a complete m-dimensional manifold satisfying (9.53), for some $\kappa > 0$ and $\alpha \in (-2, 2]$. Let f satisfy (9.45), $f(0) = 0$ and*

$$c_F f(t) \geq \sqrt{F(t)} \qquad \text{for } t \in [0, \eta_0),$$

for some constant $c_F > 0$. Fix $\mu \leq 1 - \frac{\alpha}{2}$. If

$$\frac{1}{\sqrt{F(t)}} \in L^1(0^+),$$

then (CSP) holds for solutions of

$$\begin{cases} \Delta u \geq \left(1 + r(x)\right)^{-\mu} f(u) & \text{on } \Omega \text{ end of } M; \\ u \geq 0, \qquad \lim\limits_{x \in \Omega, \, x \to \infty} u(x) = 0. \end{cases}$$

Remark 9.30 The same method directly applies to the p-Laplacian for each $p > 1$, provided that the corresponding of Theorem 9.28 holds. This is likely to be the case, but the construction of ς may reveal subtleties. In this respect, the gradient estimates in [235] should be useful.

We conclude this section by commenting on Theorem 9.29. Analogously to the link between (SMP$_\infty$) and (KO$_\infty$), it seems to us that the function-theoretic property that might describe how geometry relates to (CSP) is what is known in the literature as the Feller property.

Definition 9.31 We say that the *Feller property* (shortly (FE)) holds if, for every end Ω of M and every $\lambda \in \mathbb{R}^+$, the minimal positive solution h to

$$\begin{cases} \Delta h = \lambda h & \text{on } \Omega, \\ h = 1 & \text{on } \partial \Omega \end{cases}$$

satisfies $h(x) \to 0$ as x diverges in Ω.

The construction of h proceeds by fixing an exhaustion $\{\Omega_j\}$ of $\overline{\Omega}$ by smooth, relatively compact open sets containing $\partial\Omega$ and taking the limit of the solutions h_j to

$$\begin{cases} \Delta h_j = \lambda h_j & \text{on } \Omega_j, \\ h = 1 & \text{on } \partial\Omega, \\ h = 0 & \text{on } \partial\Omega_j. \end{cases}$$

By comparison, h is independent of the chosen exhaustion. Classically, the Feller property is introduced as the C_0 conservation property for the heat flow, that is, the fact that the heat semigroup P_t preserves the space $C_0(M)$ of functions on M that vanish at infinity:

$$\begin{array}{c} u(x) \to 0 \\ \text{as } x \text{ diverges} \end{array} \quad \Longrightarrow \quad \forall t > 0, \quad (P_t u)(x) \to 0 \text{ as } x \text{ diverges}.$$

Its equivalence with Definition 9.31 is shown by R. Azencott, cf. [13]. Various authors investigated the geometric conditions needed to guarantee the Feller property, notably [67, 78, 126, 145, 239] and the recent [178]. The most general criteria for its validity are, to the best of our knowledge, the following two. For $G \in C(\mathbb{R}_0^+)$, as usual, let $g \in C^2(\mathbb{R}_0^+)$ be the solution of Jacobi equation

$$\begin{cases} g'' - Gg = 0 & \text{on } \mathbb{R}^+ \\ g(0) = 0, \quad g'(0) = 1, \end{cases}$$

and set $v_g(r) = \omega_{m-1} g(r)^{m-1}$.

Theorem 9.32 *Let M be a complete manifold of dimension $m \geq 2$, fix $o \in M$ and let $r(x) = \text{dist}(x, o)$. Then, M is Feller provided that one of the following properties holds:*

(i) [178, Thms. 3.4 and 5.9] o is a pole,

$$\text{Sec}_{\text{rad}} \leq -G(r) \qquad \text{on } M\backslash\{o\},$$

for some $G \in C^\infty(\mathbb{R}_0^+)$, and

$$\text{either} \quad \frac{1}{v_g} \in L^1(\infty) \quad \text{or} \quad \frac{1}{v_g} \notin L^1(\infty), \quad \frac{\int_r^\infty v_g}{v_g(r)} \notin L^1(\infty), \tag{9.57}$$

where the last condition is intended to be trivially satisfied if $v_g \notin L^1(\infty)$.

(ii) [126, 127] the Ricci curvature satisfies

$$\mathrm{Ric} \geq -G(r)\langle\,,\,\rangle \qquad \text{on } M,$$

for some $G \in C^\infty(\mathbb{R}_0^+)$ matching

$$G > 0, \quad G' \geq 0 \quad \text{on } \mathbb{R}_0^+, \quad \text{and} \qquad \frac{1}{\sqrt{G}} \notin L^1(\infty).$$

In view of [178, Thm. 3.4], (i) and (ii) are sharp for the Feller property, and indeed (9.57) is both necessary and sufficient for the model manifold M_g to be Feller. Observe that the inequalities in (ii) coincide with those appearing in (2.18) and (2.19) to guarantee the (SMP$_\infty$), and that the limit polynomial threshold for both (i) and (ii) is

$$G(r) \asymp 1 + r^2 \qquad \text{on } \mathbb{R}^+,$$

that is, $\alpha = 2$. Setting $l \equiv 1$ and $b \equiv 1$ and restricting to the Laplace–Beltrami operator, cases (i) and (ii) match, respectively, with the geometric conditions (9.30) in Theorem 9.13 and (9.53) in Theorem 9.29. In view of these remarks, we feel interesting to suggest the following:

Problem 5 Investigate the validity of the implication

$$(\mathrm{KO}_0) \quad + \quad (\mathrm{FE}) \quad \Longrightarrow \quad (\mathrm{CSP})$$

on a (complete) Riemannian manifold, possibly restricting to the inequality

$$\Delta_p u \geq f(u)|\nabla u|^{p-1-\chi}.$$

Can we obtain, for (CSP), a result analogous to Theorem 10.5 below?

Proof of Proposition 9.19

We report the statement to help readability.

Proposition 9.33 *Let φ, f, l satisfy (9.5) and (9.3), and assume the validity of (C_1), $(C_2)'$, (C_4) and $(\beta_1)'$, $(\beta_2)'$, (β_3), for some $r_0 > 0$. Having fixed a non-negative*

$\theta \in C([r_0, \infty))$, *suppose that*

$$\limsup_{R \to \infty} K \left(\frac{1}{RK^{-1}(\beta(2R))} \right) R\theta(R) < \infty. \tag{9.58}$$

Then, there exists a diverging sequence $\{R_j\}$ such that the following holds: if

$$\frac{1}{K^{-1} \circ F} \in L^1(0^+), \tag{KO$_0$}$$

then for each $\epsilon \in (0, \xi_0)$, there exist $\lambda \in (0, \eta_0)$ and, for each $R \in \{R_j\}$, a function z with the following properties:

$$\begin{cases} z \in C^1([R, \infty)), & \text{and } C^2 \text{ except possibly at } 2R \\ 0 \le z \le \lambda, & z(R) = \lambda, \qquad z \equiv 0 \quad \text{on } [2R, \infty), \\ z' < 0 \quad \text{on } [R, 2R), & |z'| \le \epsilon \quad \text{on } [R, \infty), \\ (\varphi(z'))' - \theta(t)\varphi(z') \le \epsilon\beta(t)f(z)l(|z'|) & \text{on } [R, \infty). \end{cases}$$

We first need the next simple result, whose proof is by direct integration.

Lemma 9.34 *If (C_1) and $(C_2)'$ hold, then (C_2) holds, and also*

$(K_4) \qquad \varphi(st) \le d_1 K'(t)l(t)\varphi(s) = d_1 t \varphi'(t)\varphi(s) \quad$ *for each $s, t \in (0, 1]$*

Proof of Proposition 9.19 Because of (β_3) and (9.58), we can choose a sequence $\{R_j\}$ satisfying (9.40), for some $B_2 > 0$. Let $\lambda \in (0, \eta_0)$ to be specified later. Using (KO$_0$), the quantity

$$C_\lambda = \int_0^\lambda \frac{ds}{K^{-1}(F(s))}$$

is well defined, increasing on $(0, \eta_0)$ and $C_\lambda \downarrow 0$ as $\lambda \to 0^+$. For each fixed $j \in \mathbb{N}$, we set $R = R_j$ and choose $T = T(R, \lambda)$ small enough that

$$T \le R, \qquad \int_{2R-T}^{2R} K^{-1}(\beta(s))ds \le C_\lambda. \tag{9.59}$$

We also set

$$D = D(\lambda, T, R) = \frac{C_\lambda}{\int_{2R-T}^{2R} K^{-1}(\beta(s))ds}$$

and note that $D \geq 1$. Next, we implicitly define $\alpha : [0, TD] \to [0, \lambda]$ by the formula

$$\int_0^{\alpha(s)} \frac{d\tau}{K^{-1}(F(\tau))} = D \int_{2R-\frac{s}{D}}^{2R} K^{-1}(\beta(\tau)) d\tau.$$

Then, α is increasing and $\alpha(0) = 0$, $\alpha(TD) = \lambda$. Hereafter, the subscript s denotes differentiation in the s variable. From

$$\alpha_s(s) = K^{-1}(F(\alpha(s))) K^{-1}(\beta(2R - s/D)) > 0 \qquad \text{on } (0, TD], \qquad (9.60)$$

we deduce $\alpha_s(0) = 0$ and $0 \leq \alpha(s) \leq \alpha(DT) = \lambda$. Using $K(0) = 0$, we choose λ small enough that

$$K^{-1}(F(\alpha(s))) \leq 1 \qquad \text{for } s \in [0, DT].$$

Furthermore, since $2R - s/D \geq 2R - T \geq R$ in view of (9.59), by $(\beta_1)'$ we can choose $R_2 \geq R_1$ large enough that

$$K^{-1}(\beta(2R - s/D)) \leq 1 \qquad \text{for } s \in [0, DT],$$

whence

$$|\alpha_s(s)| \leq 1 \qquad \text{for each } s \in [0, DT]. \qquad (9.61)$$

To simplify the writing, set

$$\tilde{\beta}(s) = \beta(2R - s/D), \qquad \rho(s) = K^{-1}(F(\alpha(s))), \qquad \tau(s) = K^{-1}(\tilde{\beta}(s)), \qquad (9.62)$$

and note that $(\beta_2)'$ can be rewritten as

$$\frac{\tilde{\beta}_s(s)}{K^{-1}(\tilde{\beta}(s))} \leq \frac{c_\beta}{D} \tilde{\beta}(s) \leq c_\beta \tilde{\beta}(s),$$

the last inequality being a consequence of $D \geq 1$. Equation (9.60) becomes $\alpha_s = \rho\tau$, and therefore

$$K(\alpha_s) = K(\rho\tau).$$

Differentiating this latter and using, in the order, (9.62), (C_1) together with (C_2)' (and so (C_2) by Lemma 9.34), (C_1), (β_2)', (C_4) and (9.60), we obtain

$$
\begin{aligned}
\left(K(\alpha_s)\right)_s &= K'(\rho\tau)\left[\frac{f(\alpha)\alpha_s\tau}{K'(\rho)} + \frac{\tilde{\beta}_s\rho}{K'(\tau)}\right] \\
&\overset{(C_2)}{\leq} k_2 K'(\rho)K'(\tau)\left[\frac{f(\alpha)\alpha_s\tau}{K'(\rho)} + \frac{\tilde{\beta}_s\rho}{K'(\tau)}\right] \\
&= k_2\left[f(\alpha)\alpha_s\tau K'(\tau) + \tilde{\beta}_s\rho K'(\rho)\right] \\
&\overset{(C_1)}{\leq} k_2 k_1\left[f(\alpha)\alpha_s K(\tau) + \tilde{\beta}_s K(\rho)\right] \\
&= k_2 k_1\left[f(\alpha)\alpha_s\tilde{\beta} + \tilde{\beta}_s F(\alpha)\right] \\
&\overset{(\beta_2)'}{\leq} k_2 k_1\left[f(\alpha)\alpha_s\tilde{\beta} + c_\beta\tilde{\beta}K^{-1}(\tilde{\beta})F(\alpha)\right] \\
&\overset{(C_4)}{\leq} k_2 k_1\left[f(\alpha)\alpha_s\tilde{\beta} + c_F c_\beta\tilde{\beta}K^{-1}(\tilde{\beta})f(\alpha)K^{-1}(F(\alpha))\right] \\
&\overset{(9.60)}{=} k_2 k_1\left[1 + c_F c_\beta\right]\tilde{\beta}f(\alpha)\alpha_s \qquad \text{for each } s \in (0, DT).
\end{aligned}
$$
(9.63)

Therefore, differentiating K, we deduce

$$
\frac{\alpha_s\varphi_s(\alpha_s)}{l(\alpha_s)}\alpha_{ss} = \left(K(\alpha_s)\right)_s \leq k_2 k_1\left[1 + c_F c_\beta\right]\tilde{\beta}f(\alpha)\alpha_s,
$$

and since $\alpha_s > 0$ and $T \leq R$, by the monotonicity of $\tilde{\beta}$ we obtain

$$
\begin{aligned}
\left(\varphi(\alpha_s)\right)_s &\leq k_2 k_1\left[1 + c_F c_\beta\right]\tilde{\beta}f(\alpha)l(\alpha_s) \\
&\leq k_2 k_1\left[1 + c_F c_\beta\right]\tilde{\beta}\left(\tfrac{Rs}{T}\right)f(\alpha)l(\alpha_s) \qquad \text{on } (0, DT).
\end{aligned}
$$
(9.64)

We also note that $\left(K(\alpha_s)\right)_s \geq 0$ follows from the first line in (9.63) and the fact that $\tilde{\beta}_s \geq 0$, and hence $\alpha_{ss} \geq 0$. Integrating (9.64) on $(0, s]$, $s \in (0, DT]$ and using $K^{-1}(0) = 0$, the monotonicity of $\tilde{\beta}_s$ and α_s, (9.5) and $T \leq R$, we get

$$
\varphi(\alpha_s(s)) \leq k_2 k_1\left[1 + c_F c_\beta\right]C^2 T D\tilde{\beta}\left(\frac{Rs}{T}\right)f(\alpha(s))l(\alpha_s(s)).
$$
(9.65)

Next, we define the function $z : [R, \infty) \to [0, \lambda)$ by setting

$$
z(t) = \begin{cases} \alpha(s), & s = DT\left(2 - \frac{t}{R}\right) & \text{if } t \in [R, 2R]; \\ 0 & & \text{if } t > 2R. \end{cases}
$$

Then, $z \in C^1([R, \infty))$ (actually, C^2 with a possible exception at $t = 2R$), and z is non-increasing. Furthermore,

$$
z(R) = \alpha(DT) = \lambda, \qquad z(2R) = z'(2R) = 0, \qquad z'(t) = -\frac{DT}{R}\alpha_s(s). \tag{9.66}
$$

We pause for a moment to estimate the quotient DT/R. By definition, and since β is decreasing,

$$
\frac{DT}{R} = \frac{C_\lambda T}{R \int_{2R-T}^{2R} K^{-1}(\beta(\tau))d\tau} \leq \frac{2C_\lambda}{2RK^{-1}(\beta(2R))}. \tag{9.67}
$$

Using (β_3) and recalling that $R = R_j$ satisfy (9.40),

$$
\frac{-2R\beta'(2R)}{\beta(2R)} \geq \frac{\hat{c}_\beta}{2},
$$

whence applying $(\beta_2)'$,

$$
\frac{1}{2RK^{-1}(\beta(2R))} \leq \frac{c_\beta \beta(2R)}{-2R\beta'(2R)} \leq \frac{2c_\beta}{\hat{c}_\beta},
$$

and inserting into (9.67),

$$
\frac{DT}{R} \leq \frac{4c_\beta C_\lambda}{\hat{c}_\beta}.
$$

Up to a further reduction of λ, we can guarantee that $DT/R \leq \epsilon$, and consequently by (9.61) and (9.66),

$$
|z'| \leq \epsilon.
$$

This shows the third relation in (9.39). Next, since $|\alpha_s| \leq 1$ on $(0, DT)$, applying (K_4) in Lemma 9.34, (9.65), $(C_2)'$ and since

$$
\tilde{\beta}\left(\frac{R}{T}s\right) = \tilde{\beta}(2DR - Dt) = \bar{\beta}(t),
$$

we infer

$$-\varphi\big(z'(t)\big) = \varphi\big(-z'(t)\big) = \varphi\left(\frac{DT}{R}\alpha_s(s)\right)$$

$$\overset{(K_4)}{\leq} d_1 K'\left(\frac{DT}{R}\right) l\left(\frac{DT}{R}\right)\varphi\big(\alpha_s(s)\big)$$

$$\overset{(9.65)}{\leq} d_1 k_1 k_2\big[1+c_F c_\beta\big]C^2 DT K'\left(\frac{DT}{R}\right) l\left(\frac{DT}{R}\right)\tilde{\beta}\left(\frac{Rs}{T}\right) f\big(\alpha(s)\big)l\big(\alpha_s(s)\big)$$

$$\overset{(C_2)'}{\leq} c_1 d_1 k_1 k_2\big[1+c_F c_\beta\big]C^2 DT K'\left(\frac{DT}{R}\right)\beta(t)f\big(\alpha(s)\big)l\big(\frac{DT}{R}\alpha_s(s)\big)$$

$$= c_1 d_1 k_1 k_2\big[1+c_F c_\beta\big]C^2 DT K'\left(\frac{DT}{R}\right)\beta(t)f\big(z(t)\big)l\big(|z'(t)|\big)$$

$$\leq c_1 d_1 k_1^2 k_2\big[1+c_F c_\beta\big]C^2 RK\left(\frac{DT}{R}\right)\beta(t)f\big(z(t)\big)l\big(|z'(t)|\big).$$

$$(9.68)$$

We next investigate $(\varphi(z'))'$. By definition, and because of $(C_2)'$, $\alpha_{ss} \geq 0$, (9.64) and (C_1), we obtain

$$(\varphi(z'))' = \varphi'\left(\frac{DT}{R}\alpha_s(s)\right)\frac{D^2 T^2}{R^2}\alpha_{ss}(s)$$

$$\overset{(C_2)'}{\leq} d_1\varphi'\left(\frac{DT}{R}\right)\varphi'(\alpha_s(s))\frac{D^2 T^2}{R^2}\alpha_{ss}(s)$$

$$= d_1\varphi'\left(\frac{DT}{R}\right)\frac{D^2 T^2}{R^2}\big(\varphi(\alpha_s(s))\big)_s$$

$$\overset{(9.64)}{\leq} k_2 k_1\big[1+c_F c_\beta\big]d_1\varphi'\left(\frac{DT}{R}\right)\frac{D^2 T^2}{R^2}\beta(t)f\big(\alpha(s)\big)l\big(\alpha_s(s)\big) \qquad (9.69)$$

$$= k_2 k_1\big[1+c_F c_\beta\big]d_1\varphi'\left(\frac{DT}{R}\right)\frac{D^2 T^2}{R^2}\beta(t)f\big(z(t)\big)l\left(\frac{R}{DT}|z'(t)|\right)$$

$$= k_2 k_1\big[1+c_F c_\beta\big]d_1\frac{DT}{R}K'\left(\frac{DT}{R}\right)l\left(\frac{DT}{R}\right)\beta(t)f\big(z(t)\big)l\left(\frac{R}{DT}|z'(t)|\right)$$

$$\overset{(C_1),\,(C_2)'}{\leq} c_1 k_2 k_1^2\big[1+c_F c_\beta\big]d_1 K\left(\frac{DT}{R}\right)\beta(t)f\big(z(t)\big)l\big(|z'(t)|\big).$$

Combining (9.68) and (9.69) and using $z' \le 0$, for $t \in [R, 2R]$, we get

$$\left(\varphi(z')\right)' - \theta(t)\varphi(z') = \left(\varphi(z')\right)' + \theta(t)\varphi(-z')$$

$$\le c_1 d_1 k_1^2 k_2 [1 + c_F c_\beta] K\left(\frac{DT}{R}\right)\left[1 + C^2 R\theta(R)\right]\beta(t)f(z)l\left(|z'|\right). \tag{9.70}$$

Observe now that properties (C_1) and $(C_2)'$ (hence, (C_2) by Lemma 9.34) guarantee the validity of (K_1) in Lemma 9.11. Possibly reducing λ in such a way that

$$\max\left(\sqrt{C_\lambda},\, K\left(\frac{\sqrt{C_\lambda}4c_\beta}{\hat{c}_\beta}\right)\right) \le 1$$

and using (9.67), we get

$$K\left(\frac{DT}{R}\right) \le K\left(\frac{C_\lambda}{RK^{-1}(\beta(2R))}\right)$$

$$\overset{(K_1)}{\le} k_1 k_2 K(\sqrt{C_\lambda}) K\left(\frac{\sqrt{C_\lambda}}{RK^{-1}(\beta(2R))}\right)$$

$$\le k_1 k_2 K(\sqrt{C_\lambda}) \min\left[K\left(\frac{1}{RK^{-1}(\beta(2R))}\right), 1\right].$$

Inserting into (9.70), and using $C \ge 1$ and (9.58), we obtain

$$\left(\varphi(z')\right)' - \theta(t)\varphi(z') \le c_1 d_1 k_1^3 k_2^2 [1 + c_F c_\beta] C^2 K\left(\sqrt{C_\lambda}\right).$$

$$\times \left[1 + K\left(\frac{1}{RK^{-1}(\beta(2R))}\right) R\theta(R)\right]\beta(t)f(z)l\left(|z'|\right)$$

$$\le (1 + B_2) c_1 d_1 k_1^3 k_2^2 [1 + c_F c_\beta] C^2 K\left(\sqrt{C_\lambda}\right)\beta(t)f(z)l\left(|z'|\right)$$

for $t \ge R$. With a possible smaller choice of λ, still independent of R, we can ensure that

$$(1 + B_2) c_1 d_1 k_1^3 k_2^2 [1 + c_F c_\beta] C^2 K(\sqrt{C_\lambda}) \le \epsilon,$$

concluding the proof. $\square$

Keller–Osserman, A Priori Estimates and the (SL) Property

In this section, we relate the Keller–Osserman condition

$$\frac{1}{K^{-1} \circ F} \in L^1(\infty) \tag{KO$_\infty$}$$

to the strong Liouville property (SL) for solutions of $(P_\geq)$. It is particularly interesting to see how geometry comes into play via the validity of the weak or the strong maximum principle for $(bl)^{-1}\Delta_\varphi$. Hereafter, we require

$$\begin{cases} \varphi \in C(\mathbb{R}_0^+) \cap C^1(\mathbb{R}^+), \quad \varphi(0) = 0, \quad \varphi' > 0 \text{ on } \mathbb{R}^+, \\[2mm] l \in C(\mathbb{R}_0^+), \quad l > 0 \text{ on } \mathbb{R}^+, \\[2mm] f \in C(\mathbb{R}), \\[2mm] f > 0 \quad \text{and } C\text{-increasing on } (\bar{\eta}_0, \infty), \text{ for some } \bar{\eta}_0 \geq 0, \end{cases} \tag{10.1}$$

and moreover

$$\frac{t\varphi'(t)}{l(t)} \in L^1(0^+)\backslash L^1(\infty). \tag{10.2}$$

Having defined K as in (2.7), by (10.2) K is a homeomorphism of $\mathbb{R}_0^+$ onto itself, thus (KO_∞) is well defined with

$$F(t) = \int_{\bar{\eta}_0}^{t} f(s)\mathrm{d}s. \tag{10.3}$$

B. Bianchini et al., *Geometric Analysis of Quasilinear Inequalities on Complete Manifolds*, Frontiers in Mathematics, https://doi.org/10.1007/978-3-030-62704-1_10

Note that, for the mean curvature operator, if $l \equiv 1$, then K is not surjective on $\mathbb{R}^+$, that is, (10.2) does not hold. Indeed, for operators of mean curvature type one is able to guarantee property (SL) *without* the need of the Keller–Osserman condition, at least in some instances. In this respect, the following result of Y. Naito and H. Usami [168] is illustrative:[1]

Theorem 10.1 ([168, Thms. 1, 2, 3]) *Let* φ, f *satisfy* (10.1) *and*

$$f(0) = 0, \qquad f > 0 \quad \text{and non-decreasing on } \mathbb{R}^+.$$

Consider a non-negative solution $u \in C^1(\mathbb{R}^m)$ *of* $\Delta_\varphi u \geq f(u)$ *on* $\mathbb{R}^m$.

(i) *If* $\varphi(\infty) < \infty$, *then* $u \equiv 0$ *on* $\mathbb{R}^m$.
(ii) *If* $\varphi(\infty) = \infty$, *then the only non-negative solution is* $u \equiv 0$ *if and only if* (KO_∞) *holds.*

Remark 10.2 As observed in [168], when $l \equiv 1$ condition $\varphi(\infty) = \infty$ implies $K_\infty = \infty$ and thus (KO_∞) is meaningful. This follows from the next inequalities:

$$K(t) + \int_0^1 \varphi(s)ds = \int_0^t s\varphi'(s)ds + \int_0^1 \varphi(s)ds$$

$$= t\varphi(t) - \int_1^t \varphi(s)ds \geq \varphi(t).$$

In fact, in Appendix of [168] it is also proved that if $t\varphi(t) \asymp t^p$ in a neighbourhood of infinity, for some $p > 1$, then also $K(t) \asymp t^p$.

In the last subsection, we will discuss in detail the case of mean curvature type operators, for which we describe appropriate Keller–Osserman conditions (that are necessary in some cases!) for the validity of (SL). First, we focus on those operators for which (10.2) holds, and begin with considering the implication (SL) $\Rightarrow$ (KO_∞).

10.1   Necessity of (KO_∞) for the (SL) Property

The main result of this section is Theorem 10.3 below: under the failure of (KO_∞), we exhibit a non-constant, non-negative solution $u \in C^1(M)$ of $(P_\geq)$ on any complete

---

[1]The statement reported here is slightly different from the original one in Theorems 1, 2, 3 of [168]. However, the two are equivalent in view of Lemma 5.6. Moreover, their notion of solution needs the further condition $|\nabla u|^{-1}\varphi(|\nabla u|)\nabla u \in C^1(\mathbb{R}^m)$, and $\Delta_\varphi u \geq f(u)$ is meant in the pointwise sense.

manifold with a pole o and satisfying the mild curvature restriction

$$\text{Sec}_{\text{rad}} \leq -G(r) \qquad \text{on } M \backslash \{o\}, \tag{10.4}$$

for some $G \in C(\mathbb{R}_0^+)$ with the property that

$$t \int_t^\infty G_-(s)ds \leq \frac{1}{4} \qquad \forall t \in \mathbb{R}^+ \tag{10.5}$$

with $G_- = -\min\{G, 0\}$. In view of Remark 3.4, (10.5) guarantees both $g > 0$ and $g' > 0$ on $\mathbb{R}^+$. In fact, (10.4) and (10.5) are only used to guarantee that the model to be compared to M is defined on the entire $\mathbb{R}^m$, and seems to be merely a technical requirement (although, probably challenging to remove). Loosely speaking, this would suggest that there is no geometric obstruction to ensure that (KO$_\infty$) be necessary for (SL).

Theorem 10.3 *Let M^m be a complete Riemannian manifold with a pole $o \in M$, and assume (10.4) for some $G \in C(\mathbb{R}_0^+)$ enjoying (10.5). Let φ, f, l satisfy (10.1) and (10.2), and*

$$f(0) = 0, \qquad f > 0 \qquad \text{on } \mathbb{R}^+.$$

If

$$\frac{1}{K^{-1} \circ F} \notin L^1(\infty), \tag{$\neg$KO$_\infty$}$$

then for each $b \in C(M)$, $b > 0$ on M there exists a non-constant, non-negative $u \in C^1(M)$ satisfying $(P_\geq)$, that is,

$$\Delta_\varphi u \geq b(x) f(u) l(|\nabla u|) \qquad \text{on } M.$$

In particular, (SL) does not hold on M.

Proof Let $g \in C^2(\mathbb{R}_0^+)$ be the solution of

$$\begin{cases} g'' - Gg = 0 & \text{on } \mathbb{R}^+, \\ g(0) > 0, \quad g'(0) = 1, \end{cases} \tag{10.6}$$

that is positive and increasing in view of Remark 3.4. Let $v_g(r) = \omega_{m-1} g(r)^{m-1}$ be the volume growth of spheres centred at the origin of the model M_g. Define $a \in C(\mathbb{R}_0^+)$ and $\bar{l} \in C(\mathbb{R}_0^+)$ in such a way that

$$b(x) \le a\big(r(x)\big) \qquad \forall x \in M,$$

$$\bar{l} \ge l \quad \text{on } \mathbb{R}_0^+, \qquad \bar{l}(0) > 0, \qquad \bar{l} = l \quad \text{on } [1, \infty).$$

Then, all of the assumptions in Proposition 5.13 are satisfied with $\bar{l}$ replacing l and $\wp = v_g$, and there exists a non-constant function $w \in C^1(\mathbb{R}_0^+)$ solving

$$\begin{cases} \big[v_g \varphi(w')\big]' = a v_g f(w) \bar{l}(|w'|) & \text{on } \mathbb{R}^+, \\ w'(0) = 0, \quad w \ge 0, \quad w' \ge 0 & \text{on } \mathbb{R}^+. \end{cases}$$

Set $u(x) = w\big(r(x)\big)$. By the Laplacian comparison theorem from below (i.e. taking traces in (3.7) of Theorem 3.6), and using that $\bar{l} \ge l$, $w' \ge 0$ and $\varphi(w') \in C^1(\mathbb{R}_0^+)$, we have

$$\Delta_\varphi u = \big[\varphi(w')\big]' + \varphi(w') \Delta r \ge \big[\varphi(w')\big]' + \varphi(w') \frac{\wp'}{v_g} = \wp^{-1}\big[v_g \varphi(w')\big]'$$

$$= a f(w) \bar{l}(|w'|) \ge b(x) f(u) l(|\nabla u|). \tag{10.7}$$

Since o is a pole and $w'(0) = 0$, $u \in C^1(M)$ and provides the desired solution.   □

Remark 10.4 Evidently, (10.5) can be replaced by the only requirement that the solution g of the Jacobi equation (10.6) is positive and non-decreasing on $\mathbb{R}^+$.

The use of the mixed Dirichlet–Neumann problem to prove Theorem 10.3 is inspired by the very recent [33]: in Theorem 1.1 therein, the authors prove existence under $(\neg KO_\infty)$ in the setting of the Heisenberg group, for each $l \in C(\mathbb{R}_0^+)$ with $l > 0$ on $\mathbb{R}_0^+$. In particular, no C-monotonicity of l is needed. To the best of our knowledge, in the literature the existence of entire solutions of $(P_\ge)$ under the failure of (KO_∞) has been shown just in a few further special cases, see, for instance, [157, p. 694–696], [147, Thm. 1.3], [95, Cor. 1.1]. Differently from [33], in the constructions in [95, 147, 157] the radial function u is defined implicitly by a direct use of (KO_∞), in a way analogous to that in Proposition 9.8. This method needs various structural assumptions on φ, b, l, that considerably restrict the range of the operators. On the contrary, the use of the Dirichlet–Neumann problem allows to make a clean, simpler proof of the existence of solutions of $(P_\ge)$ and, at the same time, to remove the unnecessary conditions on φ, f, l: in particular, neither we assume a controlled growth of b nor the C-monotonicity (increasing or decreasing) of l. However, the presence of a pole, intimately related to the use of the comparison theorem from below, is unavoidable for our method to work, as well as for those in the above references. It would be interesting to investigate the following

Question Can one prove the necessity of (KO$_\infty$) for property (SL) on a general complete manifold, at least for some classes of φ and l?

Note that, in the p-Laplacian case, a direct use of the fake distance ϱ as in the proof of Theorem 2.43 is not enough to conclude. Indeed, from (4.13) and taking into account (10.7), for u to solve ($P_\geq$) we need a *global lower bound* for $|\nabla \varrho|$. Despite the fact that lower bounds for $|\nabla \varrho|$ seem very difficult to achieve, their existence coupled with the properness of ϱ would still force, by Morse theory, topological restrictions on M.

10.2 Sufficiency of (KO$_\infty$) for the (SL) Property

The investigation of the sufficiency of (KO$_\infty$) for (SL) in a manifold setting began with the pioneering [50, 239], for the prototype semilinear example $\Delta u \geq f(u)$. There, geometry is taken into account via a constant lower bound on the Ricci tensor of M. The Liouville theorems therein proved to be remarkably effective in a wealth of different geometric problems. Among them, we stress a striking proof of the generalized Schwarz-Pick Lemma for maps between Kahler manifolds in [237], and the Bernstein theorem for maximal hypersurfaces in Minkowski space in [51]. Since then, various authors studied possible useful generalizations, notably [166] for the inequality

$$\Delta u \geq \varphi(u, |\nabla u|).$$

The topic has first been considered from a general perspective in [182], that also contains a detailed account of the previous literature, and later more specifically in [157] for quasilinear equations including ($P_\geq$) with non-constant l. As usual, the geometric requirements range from a control on the Ricci to a growth estimate for the volume of geodesic balls. In the next subsections, we will describe improvements of the results therein, as well as new theorems, and discuss their sharpness. Of particular interest for us is the case of mean curvature type operators, for which interesting specific phenomena appear. Typically, but not exclusively, we will consider a gradient nonlinearity of the type

$$l(t) \asymp \frac{\varphi(t)}{t^\chi} \qquad \text{on } \mathbb{R}^+,$$

that, for the mean curvature operator, vanishes both as $t \to 0^+$ and as $t \to \infty$ when $\chi \in (0,1)$. When l is allowed to vanish both at $t = 0$ and at $t = \infty$ we cannot rely on the existing literature, because all of the results that we know require a C-monotonicity of l, either increasing (cf. [157]) or decreasing (cf. [33]).

We begin with the following result that considers homogeneous operators and a power-like gradient dependence. In this case, we can give a very simple proof of the next implication:

$$(\text{SMP}_\infty) \quad + \quad (\text{KO}_\infty) \implies (\text{SL}).$$
(10.8)

The argument naturally splits into two steps. First, the combination of (SMP_∞) and (KO_∞) guarantees that each solution of $(P_\geq)$ is, in fact, bounded from above; in the second step we are left to prove the validity of (L).

Theorem 10.5 *Let M be a Riemannian manifold. Fix $p > 1$, $\chi \in (-1, p-1]$, and let $f \in C(\mathbb{R})$ satisfying*

$$\begin{cases} f > 0 \quad \text{on } (\bar{\eta}_0, \infty) \\ F(t)^{\frac{\chi}{\chi+1}} \leq c_F f(t), \quad \text{on } (\bar{\eta}_0, \infty), \text{ for some constant } c_F > 0. \end{cases}$$
(10.9)

Let $u \in C^1(M)$ solve $\Delta_p u \geq f(u)|\nabla u|^{p-1-\chi}$ on M. If

1. *$l^{-1}\Delta_p$ satisfies (SMP_∞) with $l(t) = t^{p-1-\chi}$, and*
2. *the Keller–Osserman condition (KO_∞) holds, that is,*

$$F(t)^{-\frac{1}{\chi+1}} \in L^1(\infty),$$

then

$$u^* = \sup_M u < \infty \quad \text{and} \quad f(u^*) \leq 0.$$

Remark 10.6 Note that the second in (10.9) is the equivalent, at infinity, of condition (C_4), repeatedly used in the proof of the compact support principle. It is easy to see that the condition is, for general f, unrelated to (KO_∞).

Proof In our assumptions, the function K defined in (2.7) satisfies $K(t) \asymp t^{\chi+1}$ on, say, $[1, \infty)$, thus $K_\infty = \infty$ and (KO_∞) is meaningful. Let $g \in C^2(\mathbb{R})$ be such that

$$g' > 0 \quad \text{on } \mathbb{R}, \quad g(t) = \int_{\bar{\eta}_0}^t \frac{ds}{K^{-1}(F(s))} \quad \text{for } t \geq \bar{\eta}_0 + 1.$$

Suppose by contradiction that $u^* = \infty$, so that $\Omega_{\bar\eta_0+1} = \{x \in M : u(x) > \bar\eta_0 + 1\}$ is non-empty. Set $h(x) = g(u(x))$. Then, $h \in C^1(M)$ and $h^* < \infty$ because of (KO$_\infty$). Computations show that, on $\Omega_{\bar\eta_0+1}$,

$$\nabla h = g'(u)\nabla u = \frac{\nabla u}{K^{-1}(F(u))}$$

$$\Delta_p h = (p-1)(g')^{p-2}g''|\nabla u|^p + (g')^{p-1}\Delta_p u$$

$$\geq \left[-|\nabla h|^p K^{-1}(F(u))^{p-1-\chi} + |\nabla u|^{p-1-\chi}\right]\frac{f(u)}{K^{-1}(F(u))^{p-1}}$$

$$= \left[1 - |\nabla h|^{\chi+1}\right]\frac{f(u)|\nabla h|^{p-1-\chi}}{K^{-1}(F(u))^\chi}$$

in the weak sense. Next, by (10.9),

$$\frac{f(u)}{K^{-1}(F(u))^\chi} \geq \hat{c}_F > 0 \qquad \text{on } (\bar\eta_0, \infty),$$

for some constant $\hat{c}_F > 0$. Then, for each $\varepsilon \in (0, 1)$ on the open, non-empty set

$$\Omega_{\eta,\varepsilon} = \{x \in M : h(x) > \eta \text{ and } |\nabla h(x)| < \varepsilon\}$$

the function h solves

$$\Delta_p h \geq \hat{c}_F(1 - \varepsilon^p)|\nabla h|^{p-1-\chi},$$

contradicting the validity of (SMP$_\infty$) for $l^{-1}\Delta_p$. Therefore, $u^* < \infty$ and, since (SMP$_\infty$) is in force, we can apply Proposition 7.4 to deduce that $f(u^*) \leq 0$. This concludes the proof.   □

Corollary 10.7 *In the assumptions of Theorem 10.5, (SL) holds for each $f > 0$ on $\mathbb{R}^+$ satisfying (10.9).*

Remark 10.8 Theorem 10.5 should be compared with [182, Thm. 1.31], that improves on previous results of Cheng and Yau [50] and Motomiya [166]. In [182], the authors consider solutions of

$$\Delta u \geq \varphi(u, |\nabla u|) \qquad \text{on } M$$

under suitable assumptions on φ that are skew with those in Theorem 10.5, and infer the bound $f(u^*) \leq 0$ under the validity of (SMP$_\infty$) for the Laplace–Beltrami operator Δ.

Remark 10.9 We stress that Theorem 10.5 also holds for $l(t) = t^{p-1-\chi}$ when $\chi \in (-1, 0)$. However, this range is not included in Theorem 7.5 and, indeed, currently we do not know which geometric conditions ensure (SMP$_\infty$) or even (WMP$_\infty$) for $l^{-1}\Delta_p$ and negative χ. For further related comments, we refer to Remark 10.42 below.

As a prototype example of applicability of the above theorem, we give a quick proof of the following classical result. The first part of Theorem 10.10 below is due to R. Osserman who introduced (KO$_\infty$), as we have mentioned in the Introduction, to prove his result; the second is a restatement of the classical Schwarz-Pick Lemma from complex analysis.

Theorem 10.10 ([175]) *Let (M, g) be a complete, non-compact, simply connected Riemann surface with Sec ≤ -1. Then, M is conformally equivalent to the Poincaré disk $(\mathbb{D}, g_{\mathbb{H}})$ via some conformal diffeomorphism $\varphi : \mathbb{D} \to M$ satisfying $\varphi^* g \leq g_{\mathbb{H}}$.*

Proof We recall that, if $g = e^{2u}h$ is a conformal deformation of a metric h on a surface M, then u turns out to satisfy the Yamabe equation

$$\Delta u = -K_g e^{2u} + K_h,$$

where we denoted with K_g and K_h the Gaussian curvatures of g and h, and Δ is the Laplace–Beltrami operator of h. By the Riemann-Köbe uniformization theorem, M is conformal either to the Euclidean plane $\mathbb{R}^2$ with its flat metric $g_{\mathbb{R}}$, or to the Poincaré disk $(\mathbb{D}, g_{\mathbb{H}})$ with the hyperbolic metric $g_{\mathbb{H}}$ of, say, sectional curvature -1. Both on $\mathbb{R}^2$ and on $\mathbb{D}$, the operator Δ satisfies (SMP$_\infty$) because of Theorem 8.5. Suppose by contradiction that $g = e^{2u}g_{\mathbb{R}}$. Then, u satisfies

$$\Delta u = -K_g(x)e^{2u} \geq e^{2u} \qquad \text{on } \mathbb{R}^2.$$

We therefore apply Theorem 10.5 with $f(t) = e^{2t}$ to deduce that u is bounded above and $f(u^*) \leq 0$, that is clearly impossible. Hence, M is conformally the Poincaré disk and setting $g = e^{2u}g_{\mathbb{H}}$, from $K_{g_{\mathbb{H}}} = -1$ and $K_g \leq -1$ the function u satisfies

$$\Delta u = -K_g e^{2u} - 1 \geq e^{2u} - 1.$$

Setting $f(t) = e^{2t} - 1$, f still satisfies (KO$_\infty$) and (10.9). Furthermore, on the Poincaré disk Δ satisfies (SMP$_\infty$), and again by Theorem 10.5 we get $f(u^*) \leq 0$, that is $u^* \leq 0$. Therefore, $g = e^{2u}g_{\mathbb{H}} \leq g_{\mathbb{H}}$.

□

We pause for a moment to comment on the necessity to require (SMP$_\infty$) in Theorem 10.5. The validity of (L) for $f > 0$ on $\mathbb{R}^+$ is granted under the sole assumption (WMP$_\infty$) by Proposition 7.4. Thus, one might wonder whether the implication

$$(\text{SMP}_\infty) \quad + \quad (\text{KO}_\infty) \quad \Longrightarrow \quad (\text{SL})$$

for $f > 0$ on $\mathbb{R}^+$ satisfying (10.9) could be improved to

$$(\text{WMP}_\infty) \quad + \quad (\text{KO}_\infty) \quad \Longrightarrow \quad (\text{SL}).$$

This amounts to showing that the combination (WMP$_\infty$) + (KO$_\infty$) guarantees global L^∞-estimates for solutions of $\Delta_p u \ge f(u)$. This is generally false, as the following example shows.

Example 10.11 Consider the punctured Euclidean space $\mathbb{R}^m \backslash \{0\}$ with its flat metric. Then, it is easy to see that Δ satisfies (WMP$_\infty$). Indeed, define a function $w \in C^2(\mathbb{R}^m \backslash \{0\})$ as follows:

$$w(x) = -\log|x| + |x|^2 \quad \text{if } m = 2,$$
$$w(x) = |x|^{2-m} + |x|^2 \quad \text{if } m \ge 3,$$

then w is an exhaustion on $\mathbb{R}^m \backslash \{0\}$, i.e. it has relatively compact sublevel sets, and $\Delta w = 2m \le w$ outside a compact set. The function w is therefore a good Khasminskii potential, whose existence implies (WMP$_\infty$), see, for instance, [107] and [182, Prop. 3.2]. In fact, the existence of such a w is equivalent to (WMP$_\infty$), cf. [154, 155] for details. Now, consider

$$\sigma \in \left(1, \frac{m+2}{m}\right), \quad \beta = \frac{2}{\sigma - 1}, \quad u(x) = |x|^{-\beta}.$$

Then, a computation shows that

$$\Delta u = \beta(\beta - m) u^\sigma.$$

Hence, $\Delta u \ge c u^\sigma$ for some constant $c > 0$ in our range on σ. On the other hand, it is easy to check that the function $f(t) = ct^\sigma$ satisfies $f > 0$ on $\mathbb{R}^+$ together with (10.9) and (KO$_\infty$). Therefore, (WMP$_\infty$) is not enough to conclude (SL) even for functions matching (KO$_\infty$).

Remark 10.12 Example 10.11 is on a geodesically incomplete manifold. As suggested in the Introduction, it would be very interesting to find an analogous phenomenon for a complete manifold. It is likely that the technique in [32] to construct a complete manifold such that Δ satisfies (WMP$_\infty$), but not (SMP$_\infty$), be useful to produce a complete example.

10.3 Ricci Curvature and (SL)

When the operator is not of p-Laplacian type, the straightforward method described in Theorem 10.5 does not work, and we cannot directly infer implication (10.8). Nevertheless, in what follows we will describe how to obtain a sharp result for a much larger class of operators including various of geometric interest. The geometric assumption is given in term of a control on the Ricci curvature at infinity, the same as the one in Theorem 8.5 to guarantee the validity of the (SMP$_\infty$). Similarly, the approach is inspired by the Phrágmen-Lindelöff method. To construct the relevant supersolutions, we need some mild conditions relating φ and l. We assume (10.1) and (10.2), and furthermore that

$$l \text{ is } C\text{-increasing on } \mathbb{R}_0^+. \tag{10.10}$$

Moreover, we require the existence of $\chi_1, \chi_2 \in \mathbb{R}$ such that

$$(\chi_1) \quad t \mapsto \frac{\varphi'(t)}{l(t)} t^{1-\chi_1} \quad \text{is } C\text{-increasing on } \mathbb{R}^+,$$

$$(\chi_2) \quad t \mapsto \frac{\varphi(t)}{l(t)} t^{-\chi_2} \quad \text{is } C\text{-increasing on } \mathbb{R}^+.$$

Concerning β and $\bar\beta$, we require

$$(\beta\bar\beta) \quad \beta \in C([r_0, \infty)), \quad \bar\beta \in C^1([r_0, \infty)),$$

$$\bar\beta' \le 0, \quad \bar\beta \notin L^1(\infty).$$

Example 10.13 If $\varphi(t) = t^{p-1}$ and $l(t) \asymp t^{p-1-\chi}$ on $\mathbb{R}^+$, then $(\chi_1), (\chi_2)$ are both satisfied provided that

$$\chi_j \le \chi \quad \text{for each } j \in \{1, 2\}.$$

If $\varphi(t) = t/\sqrt{1+t^2}$ is the mean curvature operator, and $l(t) \asymp \varphi(t)/t^\chi$ on $\mathbb{R}^+$ for some $\chi \in \mathbb{R}$, then $(\chi_1), (\chi_2)$ hold provided that

$$\chi_1 \le \chi - 2, \quad \chi_2 \le \chi.$$

Finally, if $\varphi(t) = te^{t^2}$ is the operator of exponentially harmonic functions and $l(t) \asymp t^q$ on $\mathbb{R}^+$, (χ_1) and (χ_2) hold whenever

$$\max\{\chi_1, \chi_2\} \le 1 - q.$$

We shall first deduce some useful properties from the validity of (χ_1) and (χ_2).

Lemma 10.14 *Assume that φ and l satisfy* (10.1). *Then,* (χ_1) *with* $\chi_1 > -1$ *implies* (10.2). *Moreover,* (χ_2) *with* $\chi_2 \geq 0$ *implies*

$$t^{-\chi_2} \frac{\varphi(t)}{l(t)} \in L^\infty((0, 1)). \tag{10.11}$$

In particular, $\varphi(t)/l(t) \to 0$ *as* $t \to 0^+$ *whenever* $\chi_2 > 0$.

Proof Assume (χ_1). By definition, there exists $C \geq 1$ such that

$$0 < s^{1-\chi_1} \frac{\varphi'(st)}{l(st)} \leq C \frac{\varphi'(t)}{l(t)} \quad \forall t \in \mathbb{R}^+ \ s \in (0, 1],$$

or, equivalently,

$$s^{1-\chi_1} \frac{\varphi'(st)}{l(st)} \geq C^{-1} \frac{\varphi'(t)}{l(t)} \quad \forall t \in \mathbb{R}^+ \ s \in [1, \infty). \tag{10.12}$$

Setting $t = 1$, if $\chi_1 > -1$ we deduce that $\frac{s\varphi'(s)}{l(s)} \in L^1(0+) \setminus L^1(\infty)$, that is, (10.2). Similarly, if (χ_2) holds then

$$\frac{\varphi(st)}{l(st)} s^{-\chi_2} \leq C \frac{\varphi(t)}{l(t)} \quad \forall t \in \mathbb{R}^+ \ s \in (0, 1],$$

and (10.11) follows by setting $t = 1$. $\qquad\square$

Lemma 10.15 *Assume that φ and l satisfy* (10.1), *and let F be a positive function defined on* $(\bar\eta_0, \infty)$. *If* (χ_1) *holds with* $\chi_1 > -1$, *then there exists a constant* $B \geq 1$ *such that, for every* $\sigma \leq 1$,

$$\frac{\sigma^{\frac{1}{\chi_1+1}}}{K^{-1}(\sigma F(t))} \leq \frac{B}{K^{-1}(F(t))} \quad on \ (\bar\eta_0, \infty).$$

Proof According to Lemma 10.14, (χ_1) with $\chi_1 > -1$ implies (10.2), so K^{-1} is well defined on $\mathbb{R}_0^+$. Changing variables in the definition of K, and using (10.12) above, for every $\lambda \geq 1$ and $t \in \mathbb{R}^+$ we have

$$K(\lambda t) = \int_0^{\lambda t} s \frac{\varphi'(s)}{l(s)} ds = \lambda^2 \int_0^t s \frac{\varphi'(\lambda s)}{l(\lambda s)} ds$$

$$\geq C^{-1} \lambda^{\chi_1+1} \int_0^t s \frac{\varphi'(s)}{l(s)} ds = C^{-1} \lambda^{\chi_1+1} K(t),$$

where $C \geq 1$ is the constant in (χ_1). Applying K^{-1} to both sides of the above inequality, and setting $t = K^{-1}(\sigma F(s))$ we deduce

$$\lambda K^{-1}(\sigma F(s)) \geq K^{-1}(\lambda^{\chi_1+1}\sigma C^{-1}F(s)),$$

whence, setting $\lambda = (C/\sigma)^{1/(\chi_1+1)} \geq 1$, the required conclusion follows with $B = C^{1/(\chi_1+1)}$.

$\square$

We are ready to construct blowing-up supersolutions that remain close to the constant $\bar{\eta}_0$ in (10.1) on an arbitrarily fixed annulus. The construction is an improvement of the one in [157, Prop. 3.4], and at the same a simplification of it. We recall that F is defined as in (10.3).

Proposition 10.16 *Assume* (10.1) *and* (10.10). *Suppose further* (χ_1) *and* (χ_2) *with*

$$\chi_1 > 0, \qquad \chi_2 > 0,$$

and let $\beta, \bar{\beta}$ *satisfy* $(\beta\bar{\beta})$. *Fix* $\theta \in C([r_0, \infty))$ *with the property that*

$$\begin{cases} \dfrac{\bar{\beta}(r)^{\chi_1+1}}{\beta(r)} \in L^\infty([r_0, \infty)), \\[2mm] \dfrac{\theta(r)\bar{\beta}(r)^{\chi_2}}{\beta(r)} \in L^\infty\big([r_0, \infty)\big). \end{cases} \tag{10.13}$$

If (KO$_\infty$) *holds, then for each* $\varepsilon > 0$, $0 < \delta < \lambda$ *and* $r_1 > r_0$, *there exist* $R_1 > r_1$ *and a* C^1 *function* $w : [r_0, R_1) \to [\bar{\eta}_0 + \delta, \infty)$ *solving*

$$\begin{cases} \big(\varphi(w')\big)' + \theta(r)\varphi(w') \leq \varepsilon\beta(r)f(w)l(w') & \text{on } [r_0, R_1) \\[2mm] w' > 0 \ \text{ on } [r_0, R_1), \qquad w(r) \to \infty \ \text{ as } r \to R_1^- \\[2mm] \bar{\eta}_0 + \delta \leq w \leq \bar{\eta}_0 + \lambda \quad \text{on } [r_0, r_1]. \end{cases} \tag{10.14}$$

Proof Note first of all that (KO$_\infty$) is meaningful because, by Lemma 10.14, (χ_1) with $\chi_1 > -1$ implies (10.2). Since $\bar{\beta}$ is bounded on $[r_0, \infty)$, by rescaling we can assume that $\bar{\beta} \leq 1$. For a given $\sigma \in (0, 1]$ to be specified later, set

$$C_\sigma = \int_{\bar{\eta}_0+\delta}^{\infty} \frac{ds}{K^{-1}(\sigma F(s))},$$

which is well defined in view of (KO$_\infty$), (10.10) and Lemma 5.6. Since $\bar{\beta} \notin L^1(\infty)$, there exists $R_\sigma > r_0$ such that

$$C_\sigma = \int_{r_0}^{R_\sigma} \bar{\beta}(s) \, ds.$$

We note that, by monotone convergence, $C_\sigma \to \infty$ as $\sigma \to 0^+$, and we can therefore choose $\sigma > 0$ small enough that $R_\sigma > r_1$. We let $w : [r_0, R_\sigma) \to [\bar{\eta}_0 + \delta, \infty)$ be implicitly defined by the equation

$$\int_r^{R_\sigma} \bar{\beta}(s) \, ds = \int_{w(r)}^{\infty} \frac{ds}{K^{-1}(\sigma F(s))}, \tag{10.15}$$

so that, by definition,

$$w(r_0) = \bar{\eta}_0 + \delta, \qquad w(r) \to \infty \text{ as } r \to R_\sigma^-.$$

Differentiating (10.15) yields

$$w'(r) = \bar{\beta}(r) K^{-1}\big(\sigma F(w(r))\big), \tag{10.16}$$

so that $w' > 0$ on $[r_0, R_\sigma)$, and

$$\sigma F(w) = K(w'/\bar{\beta}).$$

Differentiating once more, using the definition of K and (10.16), we obtain

$$\sigma f(w) w' = K'(w'/\bar{\beta})(w'/\bar{\beta})' = \frac{w'}{\bar{\beta}} \frac{\varphi'(w'/\bar{\beta})}{l(w'/\bar{\beta})} \left(\frac{w'}{\bar{\beta}}\right)',$$

that is,

$$\varphi'\left(\frac{w'}{\bar{\beta}}\right)\left(\frac{w'}{\bar{\beta}}\right)' = \sigma \bar{\beta} f(w) l\left(\frac{w'}{\bar{\beta}}\right) \qquad \text{on } [r_0, R_\sigma). \tag{10.17}$$

Since $f > 0$ on $(\bar{\eta}_0, \infty)$ and $w' > 0$, we infer that $w'/\bar{\beta}$ is non-decreasing. Moreover, from $\bar{\beta}' \leq 0$ we deduce

$$\left(\frac{w'}{\bar{\beta}}\right)' = \frac{w''}{\bar{\beta}} - \frac{w'\bar{\beta}'}{\bar{\beta}^2} \geq \frac{w''}{\bar{\beta}}.$$

Inserting this into (10.17), using $\bar{\beta} \le 1$ and (χ_1) (in the form of (10.12)), and rearranging we obtain

$$\varphi'(w')w'' \le \left\{ C\sigma \frac{\bar{\beta}^{\chi_1+1}}{\beta} \right\} \beta f(w)l(w') \qquad \text{on } [r_0, R_\sigma). \tag{10.18}$$

We next integrate (10.17) on $[r_0, r]$, and we use (10.10) coupled with the monotonicity of both w and $w/\bar{\beta}$ to deduce

$$\varphi\left(\frac{w'}{\bar{\beta}}\right) \le \varphi\left(\frac{w'}{\bar{\beta}}\right)(r_0) + C^2\sigma f(w)l\left(\frac{w'}{\bar{\beta}}\right) \int_{r_0}^r \bar{\beta}ds, \tag{10.19}$$

and thus using (χ_2), $\bar{\beta} \le 1$ and $w(r_0) = \bar{\eta}_0 + \delta$ we get

$$\frac{\varphi(w')}{l(w')} \overset{(\chi_2)}{\le} C^2 \bar{\beta}^{\chi_2} \frac{\varphi(w'/\bar{\beta})}{l(w'/\bar{\beta})}$$

$$\le C^2 \bar{\beta}^{\chi_2} \left[\frac{\varphi(w'/\bar{\beta})(r_0)}{l(w'/\bar{\beta})} + C^2\sigma f(w) \int_{r_0}^r \bar{\beta}ds \right]$$

$$\overset{(10.1)+(10.10)}{\le} C^2 \frac{\bar{\beta}^{\chi_2}}{\beta} \left[C^2 \frac{\varphi(w'/\bar{\beta})(r_0)}{f(\bar{\eta}_0+\delta)l(w'/\bar{\beta})(r_0)} + C^2\sigma \int_{r_0}^r \bar{\beta}ds \right] \beta f(w)$$

$$\overset{(10.15)}{\le} C^4 \frac{\bar{\beta}^{\chi_2}}{\beta} \left[\frac{\varphi(w'/\bar{\beta})(r_0)}{f(\bar{\eta}_0+\delta)l(w'/\bar{\beta})(r_0)} + \sigma \int_{\bar{\eta}_0+\delta}^\infty \frac{ds}{K^{-1}(\sigma F(s))} \right] \beta f(w).$$

Next, we use (10.16) with $r = r_0$ and the fact that, by Lemma 10.15,

$$\sigma \int_{\bar{\eta}_0+\delta}^\infty \frac{ds}{K^{-1}(\sigma F(s))} \le B\sigma^{\frac{\chi_1}{\chi_1+1}} \int_{\bar{\eta}_0+\delta}^\infty \frac{ds}{K^{-1}(F(s))}$$

to obtain

$$\varphi(w') \le C_1 \frac{\bar{\beta}^{\chi_2}}{\beta} \left[\frac{\varphi(K^{-1}(\sigma F(\bar{\eta}_0+\delta)))}{f(\bar{\eta}_0+\delta)l(K^{-1}(\sigma F(\bar{\eta}_0+\delta)))} \right.$$

$$\left. + \sigma^{\frac{\chi_1}{\chi_1+1}} \int_{\bar{\eta}_0+\delta}^\infty \frac{ds}{K^{-1}(F(s))} \right] \beta f(w)l(w') \tag{10.20}$$

$$= C_1 \hat{C}_\sigma \frac{\bar{\beta}^{\chi_2}}{\beta} \beta f(w)l(w')$$

for some $C_1 > 0$, and where we set

$$\hat{C}_\sigma = \frac{\varphi(K^{-1}(\sigma F(\bar{\eta}_0+\delta)))}{f(\bar{\eta}_0+\delta)l(K^{-1}(\sigma F(\bar{\eta}_0+\delta)))} + \sigma^{\frac{\chi_1}{\chi_1+1}} \int_{\bar{\eta}_0+\delta}^\infty \frac{ds}{K^{-1}(F(s))}.$$

Putting together (10.18) and (10.20), and using (10.13), we obtain

$$(\varphi(w'))' + \theta(r)\varphi(w')$$

$$\leq \left\{ C_\sigma \left\| \frac{\bar{\beta}^{\chi_1+1}}{\beta} \right\|_{L^\infty([r_0,\infty))} + C_1 \hat{C}_\sigma \left\| \frac{\theta \bar{\beta}^{\chi_2}}{\beta} \right\|_{L^\infty([r_0,\infty))} \right\} \beta f(w) l(w').$$

Because of (χ_2) with $\chi_2 > 0$ and Lemma 10.14, $\hat{C}_\sigma \to 0$ as $\sigma \to 0$, and we can choose σ small enough such that the differential inequality in (10.14) is satisfied. To prove the last condition in (10.14), simply observe that by (10.15)

$$\int_{r_0}^{r_1} \bar{\beta}(s)ds = \int_{\bar{\eta}_0+\delta}^{w(r_1)} \frac{ds}{K^{-1}(\sigma F(s))},$$

thus $w(r_1) \to \bar{\eta}_0 + \delta$ as $\sigma \to 0$. Since w is increasing, it is enough to choose σ in such a way that

$$\int_{r_0}^{r_1} \bar{\beta}(s)ds < \int_{\bar{\eta}_0+\delta}^{\bar{\eta}_0+\lambda} \frac{ds}{K^{-1}(\sigma F(s))}.$$

$\square$

Remark 10.17 We can weaken our assumptions on χ_1, χ_2 to

$$\chi_1 \geq 0, \quad \chi_2 \geq 0 \quad \text{and} \quad \chi_1\chi_2 = 0,$$

assuming the further condition

$$\limsup_{t \to +\infty} f(t) = \infty.$$

Indeed, one can still guarantee the existence of a *divergent* sequence $\delta_j \to \infty$ such that, for each $\delta \in \{\delta_j\}$ and $\lambda > \delta$, there exists w solving (10.14). All is needed to conclude the proof is that $\hat{C}_\sigma$ can be made arbitrarily small for suitable σ and δ. First, using (χ_2) and Lemma 10.14, choose δ_j so that

$$\frac{1}{f(\bar{\eta}_0 + \delta_j)} \left\| \frac{\varphi}{l} \right\|_{L^\infty([0,1])} + \int_{\bar{\eta}_0+\delta_j}^{\infty} \frac{ds}{K^{-1}(F(s))}$$

is small enough, and then choose σ in such a way that $K^{-1}(\sigma F(\bar{\eta}_0+\delta)) \leq 1$. The coupling of these two conditions guarantee the smallness of $\hat{C}_\sigma$.

Remark 10.18 It is interesting to compare Proposition 10.16 with the corresponding Proposition 9.8 for the compact support principle. Although their underlying idea is the same, the two constructions are *not specular*, neither are the conditions on φ, f, l. The reason is that, while the monotonicity of the supersolution changes, the weight $\bar{\beta}$ is still decreasing. To grasp the core of the technical problem, we invite the interested reader to try to prove Proposition 9.8 by following the estimates in Proposition 10.16, suitably replacing (χ_1) and (χ_2) in a neighbourhood of zero.

We can now investigate the validity of (SL).

Theorem 10.19 *Let M^m be a complete Riemannian manifold of dimension $m \geq 2$ such that, for some origin $o \in M$, the distance function $r(x)$ from o satisfies*

$$\mathrm{Ric}(\nabla r, \nabla r) \geq -(m-1)\kappa^2\big(1+r^2\big)^{\alpha/2} \qquad on\ D_o, \tag{10.21}$$

for some $\kappa \geq 0$ and $\alpha \geq -2$. Let φ, f, l meet (10.1), (10.10), and assume (χ_1) and (χ_2) with

$$\chi_1 > 0, \qquad \chi_2 > 0. \tag{10.22}$$

Consider $b \in C(M)$ such that

$$b(x) \geq C_1\big(1+r(x)\big)^{-\mu} \qquad on\ M,$$

for some constants $C_1 > 0$, $\mu \in \mathbb{R}$ satisfying

$$\mu \leq \min\left\{\chi_1 + 1,\ \chi_2 - \frac{\alpha}{2}\right\}. \tag{10.23}$$

If (KO_∞) holds, then any non-constant solution $u \in C^1(M)$ of $(P_\geq)$ is bounded above and $f(u^*) \leq 0$. In particular, if $f > 0$ on $\mathbb{R}^+$ then (SL) holds for C^1 solutions.

Proof We first prove that u is bounded above. By contradiction, let $u^* = \infty$ and consider a geodesic ball B_{r_0} centred at o, with $r_0 > 0$. Fix $0 < \delta < \lambda$ and $\bar{x} \notin B_{r_0}$ enjoying

$$u_0^* = \max_{B_{r_0}} u \leq \bar{\eta}_0 + \delta, \qquad u(\bar{x}) > \bar{\eta}_0 + \lambda,$$

and choose r_1 in such a way that $\bar{x} \in B_{r_1}$. From (10.21) and the Laplacian comparison theorem,

$$\Delta r \leq A r^{\alpha/2} \qquad weakly\ on\ M \backslash B_{\frac{r_0}{2}}, \tag{10.24}$$

for some constant $A = A(r_0, \alpha, \kappa, m) > 0$. Applying Proposition 10.16 with

$$\theta(r) = Ar^{\alpha/2}, \qquad \bar{\beta}(r) = (1+r)^{-1}, \qquad \beta(r) = C_1(1+r)^{-\mu}, \qquad \varepsilon = \frac{1}{2C}$$

we deduce the existence of w satisfying (10.14) (note that (10.23) guarantees (10.13)). Setting $\bar{w}(x) = w(r(x))$ and taking into account $w' > 0$, (10.24) and $b \geq \beta(r)$, $\bar{w}$ solves

$$\begin{cases} \Delta_\varphi \bar{w} \leq \frac{1}{2C} b(x) f(\bar{w}) l(w'(r)) & \text{on } B_{R_1} \backslash B_{r_0} \\ w'(r) > 0 \text{ on } B_{R_1} \backslash B_{r_0}, \qquad \bar{w} \to \infty \text{ as } x \to \partial B_{R_1} \\ \bar{\eta}_0 + \delta \leq \bar{w} \leq \bar{\eta}_0 + \lambda & \text{on } B_{r_1} \backslash B_{r_0}. \end{cases} \qquad (10.25)$$

We compare u and $\bar{w}$ on $B_{R_1} \backslash B_{r_0}$. By construction, $u \leq \bar{w}$ on ∂B_{r_0} and $u - \bar{w} \to -\infty$ approaching ∂B_{R_1}. On the other hand,

$$u(\bar{x}) > \bar{\eta}_0 + \lambda \geq \bar{w}(\bar{x}),$$

and thus $c = \max\{u - \bar{w}\}$ is positive and attained on some compact set $\Gamma = \{u - \bar{w} = c\} \Subset B_{R_1} \backslash B_{r_0}$. For $\eta \in (0, c)$, consider $U_\eta = \{u - \bar{w} > \eta\} \Subset B_{R_1} \backslash B_{r_0}$. If $x \in \Gamma \backslash \mathrm{cut}(o)$, then $\bar{w} \in C^1$ around x and therefore

$$\nabla u(x) = \nabla \bar{w}(x) = w'(r(x)).$$

The same relation also holds if $x \in \mathrm{cut}(o)$ by using Calabi's trick (see the proof of Theorem 8.5). Since both ∇u and $w'(r)$ are continuous, for η close enough to c the inequality $|\nabla u| \geq w'(r)/2$ holds on U_η. From

$$\Delta_\varphi u \geq b(x) f(u) l(|\nabla u|) \geq \frac{1}{2C} b(x) f(\bar{w}) l(w'(r)) \geq \Delta_\varphi \bar{w} = \Delta_\varphi(\bar{w} + \eta),$$

we deduce by comparison that $u \leq \bar{w} + \eta$ on U_η, contradiction.

It remains to prove that $f(u^*) \leq 0$. If $f(u^*) > K > 0$, then choose an upper level set Ω_η with $\eta < u^*$ in such a way that $f(u) > K$ on Ω_η, and a continuous function $\bar{f} \leq f$ with

$$\bar{f}(\eta) = 0, \qquad \bar{f} \text{ is positive and } C\text{-increasing on } (\eta, \infty),$$

$$\bar{f} \leq f \quad \text{on } (\eta, u^*), \qquad \bar{f} = f \quad \text{on } (\max\{\bar{\eta}_0, u^*\}, \infty).$$

Then, $\bar{u} = \max\{u, \eta\}$ satisfies

$$\Delta_\varphi \bar{u} \geq b(x) \bar{f}(\bar{u}) l(|\nabla \bar{u}|) \quad \text{on } M$$

(observe that $\bar{f}(\eta)l(0) = 0$). Let $\bar{u}_0^* = \sup_{B_{r_0}} \bar{u}$, and note that $\bar{u}_0^* < \bar{u}^*$ in view of the finite maximum principle applied to $\bar{u}^* - \bar{u}$. Fix $\lambda > 0$ such that $\bar{u}_0^* + 2\lambda < \bar{u}^* - 2\lambda$, set $\delta = \lambda/2$, let $\bar{x}$ satisfy $\bar{u}(\bar{x}) > \bar{u}^* - \lambda$ and choose r_1 big enough that $\bar{x} \in B_{r_1}$. With our choices, $\bar{x}$ belongs to the relatively compact set $\{\bar{u} > \bar{w}\}$ and, consequently, the desired contradiction is achieved by proceeding verbatim as in the case $u^* = \infty$, with $\bar{u}$ replacing u, $\bar{u}_0^*$ replacing u_0^* and with the same function w. $\square$

Once the bound $u^* < \infty$ is shown, an alternative way to conclude $f(u^*) \le 0$ in Theorem 10.19 is to use Theorem 2.22 to ensure the validity of (WMP_∞). However, we should require the extra condition (2.27), that is avoided in the above argument. Nevertheless, this second approach is needed to deal with the relevant, borderline case when either χ_1 or χ_2 vanish, that is considered in the next

Theorem 10.20 *In the assumptions of Theorem 10.19, suppose that (10.22) is replaced by*

$$\chi_1 \ge 0, \qquad \chi_2 \ge 0, \qquad \chi_1 \chi_2 = 0$$

and the validity of

$$\limsup_{t \to +\infty} f(t) = \infty. \tag{10.26}$$

Assume that

$$\begin{cases} l(t) \ge C_1 \dfrac{\varphi(t)}{t^{\chi_2}} & \text{on } \mathbb{R}^+, \text{ for some } C_1 > 0, \\[2mm] \varphi(t) \le C_2 t^{p-1} & \text{on } [0, 1], \text{ for some } C_2 > 0, \ p > 1, \\[2mm] \varphi(t) \le \bar{C}_2 t^{\bar{p}-1} & \text{on } [1, \infty), \text{ for some } \bar{C}_2 > 0, \ \bar{p} > 1, \end{cases} \tag{10.27}$$

and that, besides (10.23), one of the following conditions is met:

$$\begin{cases} \alpha \ge -2, & \chi_2 > 0, & or \\ \alpha \ge -2, & \chi_2 = 0, & \mu < -\frac{\alpha}{2}, & or \\ \alpha > -2, & \chi_2 = 0, & \mu = -\frac{\alpha}{2}, & V_\infty = 0, & or \\ \alpha = -2, & \chi_2 = 0, & \mu = -\frac{\alpha}{2}, & V_\infty \le p, \end{cases} \tag{10.28}$$

where

$$
V_\infty =
\begin{cases}
\displaystyle\liminf_{r\to\infty} \frac{\log|B_r|}{r^{1+\alpha/2}} & \text{if } \alpha \geq -2, \\[3mm]
\displaystyle\liminf_{r\to\infty} \frac{\log|B_r|}{\log r} & \text{if } \alpha = -2.
\end{cases}
$$

Then, if (KO_∞) *holds, any non-constant solution* $u \in C^1(M)$ *of* $(P_\geq)$ *on M is bounded above and* $f(u^*) \leq 0$. *In particular, if* $f > 0$ *on* $\mathbb{R}^+$, *(SL) holds for* C^1 *solutions.*

Proof Because of Remark 10.17, for δ large and suitably chosen we can still produce a solution w of (10.14), that gives rise to $\bar{w}$ solving (10.25). Following the proof of Theorem 10.19 we obtain $u^* < \infty$. To conclude, we observe that we are in the position to apply Theorem 2.22 with the choice $\chi = \chi_2$: in this respect, note that the first requirement in (2.27) corresponds to the first in (10.27). The conclusion $f(u^*) \leq 0$ follows from the validity of (WMP_∞) and Proposition 2.19. $\qquad\square$

Remark 10.21 By the Bishop–Gromov theorem (Theorem 3.11 and the subsequent remarks in Sect. 3.2), (10.21) implies $V_\infty < \infty$ and in particular, if $\alpha = -2$,

$$
V_\infty \leq (m-1)\bar{\kappa} + 1 \qquad \text{with} \qquad \bar{\kappa} = \frac{1 + \sqrt{1 + 4\kappa^2}}{2}. \tag{10.29}
$$

Therefore, condition $V_\infty \leq p$ in the last of (10.28) is implied by $\bar{\kappa} \leq \frac{p-1}{m-1}$.

Remark 10.22 One could alternatively use Theorem 2.17 to conclude $f(u^*) \leq 0$. Doing so, on the one hand (10.27) would weaken to (2.22), requiring φ, l only on $[0, 1]$, but on the other hand the conclusion in the case $(\chi_2 =) \chi = 0$ in (2.24) is only possible under the Euclidean type behaviour $\alpha = -2$.

We conclude this section with some comments on Theorem 10.19. We begin with the following corollary for the p-Laplace operator, a slight improvement of [157, Cor. A1].

Corollary 10.23 *Let* M^m *be a complete Riemannian manifold satisfying*

$$
\mathrm{Ric}(\nabla r, \nabla r) \geq -(m-1)\kappa^2 (1+r^2)^{\alpha/2} \qquad \text{on } D_o,
$$

for some $\kappa \geq 0$, $\alpha \geq -2$ *and some origin o. Fix* $p > 1$ *and* $\chi \in (0, p-1]$. *Consider* $b \in C(M)$ *and* $f \in C(\mathbb{R})$ *such that*

$$
b(x) \geq C_1(1 + r(x))^{-\mu} \qquad \text{on } M,
$$

$$
f(0) = 0, \qquad f > 0 \text{ and } C\text{-increasing on } \mathbb{R}^+,
$$

for some constants $C, C_1 > 0$ *and*

$$\mu \leq \chi - \frac{\alpha}{2}.$$

If the Keller–Osserman condition

$$F(t)^{-\frac{1}{\chi+1}} \in L^1(\infty)$$

is met, then (SL) *holds for* C^1 *solutions of*

$$\Delta_p u \geq b(x) f(u) |\nabla u|^{p-1-\chi}.$$

The same conclusion holds if $\chi = 0$, *provided that* M *satisfies one of the next further conditions: either*

$$\alpha > -2, \qquad \liminf_{r \to \infty} \frac{\log |B_r|}{r^{1+\alpha/2}} = 0, \qquad or$$

$$\alpha = -2, \qquad \bar{\kappa} \leq \frac{p-1}{m-1},$$

with $\bar{\kappa}$ *as in* (10.29).

Proof It is a direct application of Theorems 10.19, 10.20 and Remark 10.21. If $\chi = 0$, note that (10.26) follows from $F^{-\frac{1}{\chi+1}} \in L^1(\infty)$.   □

Dealing with the mean curvature operator, a substantial problem arises: in view of (10.10), l is bounded from below in a neighbourhood of infinity, and thus $K_\infty < \infty$ and (KO$_\infty$) is meaningless. To overcome the problem and to be able to include inequalities of the type

$$\text{div}\left(\frac{\nabla u}{\sqrt{1 + |\nabla u|^2}}\right) \geq b(x) f(u) |\nabla u|^q,$$

taking into account that the mean curvature operator satisfies

$$t\varphi'(t) \leq C\varphi(t) \qquad \text{on } \mathbb{R}^+ \tag{10.30}$$

for some $C > 0$, the authors in [157, Sect. 4] propose to replace $t\varphi'(t)$ with $\varphi(t)$ in the definition of K in (2.7). In this way, the corresponding (KO$_\infty$) makes sense for some classes of C-increasing l. As we shall see in a moment, this seemingly "rough" replacement allows indeed to obtain a sharp result, but in the course of the proof in

[157] the authors lose optimality in some inequalities, and consequently their main result (Corollary A2 therein) is not sharp. We now describe how to achieve the optimal range of parameters. Clearly, the bulk is to get an analogue of Proposition 10.16 for mean curvature type operators and *not requiring that l be C-increasing*. Note that the *C*-monotonicity of *l* is essential to obtain inequality (10.19). We restrict to consider the relevant case of operators satisfying (10.30) and gradient terms *l* of the type

$$l(t) = \frac{\varphi(t)}{t^\chi},$$

for $\chi > 0$ small enough to make *l* continuous at $t = 0$. Observe that φ may vanish both at $t = 0$ and at infinity. Following the idea in [157], we set

$$K(t) = \int_0^t \frac{\varphi(s)}{l(s)} ds \asymp t^{\chi+1} \qquad \text{on } [1, \infty),$$

and the Keller–Osserman condition becomes $F^{-\frac{1}{\chi+1}} \in L^1(\infty)$, with *F* as in (10.3).

Proposition 10.24 *Let φ satisfy (10.1) and (10.30). Fix $\chi > 0$, $f \in C(\mathbb{R})$ satisfying*

$$f > 0 \quad \text{and C-increasing on } (\bar{\eta}_0, \infty), \text{ for some } \bar{\eta}_0 > 0,$$

and $\beta, \bar{\beta}$ satisfying $(\beta\bar{\beta})$. Let $\theta \in C([r_0, \infty))$ with the property that

$$\frac{\max\{\bar{\beta}(r), \theta(r)\} \cdot \bar{\beta}(r)^\chi}{\beta(r)} \in L^\infty([r_0, \infty)). \tag{10.31}$$

If the Keller–Osserman condition

$$F^{-\frac{1}{\chi+1}} \in L^1(\infty)$$

holds, then for each $\varepsilon > 0$, $0 < \delta < \lambda$ and $r_1 > r_0$, there exist $R_1 > r_1$ and a C^1 function $w : [r_0, R_1) \to [\bar{\eta}_0 + \delta, \infty)$ solving

$$\begin{cases} (\varphi(w'))' + \theta(r)\varphi(w') \le \varepsilon\beta(r)f(w)\dfrac{\varphi(w')}{[w']^\chi} & \text{on } [r_0, R_1) \\[2mm] w' > 0 \ \text{ on } [r_0, R_1), \qquad w(r) \to \infty \ \text{ as } r \to R_1^- \\[2mm] \bar{\eta}_0 + \delta \le w \le \bar{\eta}_0 + \lambda & \text{on } [r_0, r_1]. \end{cases}$$

Proof We proceed as in Proposition (10.16), so we skip some of the details and just concentrate on the main differences. For $\sigma \in (0, 1]$ to be specified later, set

$$C_\sigma = \int_{\bar\eta_0+\delta}^{\infty} [\sigma F(s)]^{-\frac{1}{\chi+1}} ds,$$

and since $\bar\beta \notin L^1(\infty)$, pick $R_\sigma > r_0$ such that

$$C_\sigma = \int_{r_0}^{R_\sigma} \bar\beta(s) ds.$$

We can choose $\sigma > 0$ small enough that $R_\sigma > r_1$. We let $w : [r_0, R_\sigma) \to [\bar\eta_0 + \delta, \infty)$ be implicitly defined by the equation

$$\int_r^{R_\sigma} \bar\beta(s) ds = \int_{w(r)}^{\infty} [\sigma F(s)]^{-\frac{1}{\chi+1}} ds. \tag{10.32}$$

Differentiating and rearranging,

$$\sigma F(w) = [w'/\bar\beta]^{\chi+1}.$$

Set $l(t) = \varphi(t)/t^\chi$. A second differentiation gives

$$\sigma f(w)w' = (\chi + 1)(w'/\bar\beta)^\chi (w'/\bar\beta)' = (\chi + 1)\frac{\varphi(w')}{l(w')}\bar\beta^{-\chi}(w'/\bar\beta)'$$

$$\geq (\chi + 1)\frac{\varphi(w')}{l(w')}w''\bar\beta^{-\chi-1}, \tag{10.33}$$

where we used that $w'/\bar\beta$ is increasing by the first equality in (10.33), and $\bar\beta' \leq 0$ by $(\beta\bar\beta)$. We next use (10.30) and simplify to deduce

$$\varphi'(w')w'' \leq \left\{c_1\sigma\frac{\bar\beta^{\chi+1}}{\beta}\right\} \beta f(w)l(w'), \tag{10.34}$$

for some constant $c_1 > 0$ independent of σ. On the other hand, from the first equality in (10.33) we deduce

$$\sigma f(w)\bar\beta = (\chi + 1)(w'/\bar\beta)^{\chi-1}(w'/\bar\beta)',$$

thus integrating on $[r_0, r]$ and using the C-monotonicity of f we get

$$\frac{\chi + 1}{\chi}(w'/\bar{\beta})^\chi = \frac{\chi + 1}{\chi}(w'/\bar{\beta})(r_0)^\chi + \sigma \int_{r_0}^r f(w)\bar{\beta}$$

$$\leq \frac{\chi + 1}{\chi}(w'/\bar{\beta})(r_0)^\chi + C\sigma f(w(r)) \int_{r_0}^r \bar{\beta}.$$

Up to rescaling, we can always assume that $\bar{\beta} \leq 1$, hence

$$\frac{\varphi(w')}{l(w')} = [w']^\chi \leq \bar{\beta}^\chi (w'/\bar{\beta})(r_0)^\chi + c_2\sigma\bar{\beta}^\chi f(w) \int_{r_0}^r \bar{\beta},$$

for some constant $c_2 > 0$ independent of σ. Rearranging, by (10.32) and the C-monotonicity of f we deduce

$$\varphi(w') \leq \frac{\bar{\beta}^\chi}{\beta}\left\{C\frac{(w'/\bar{\beta})(r_0)^\chi}{f(\bar{\eta}_0 + \delta)} + c_2\sigma \int_{r_0}^r \bar{\beta}\right\}\beta f(w)l(w')$$

$$\leq \frac{\bar{\beta}^\chi}{\beta}\left\{C\frac{(w'/\bar{\beta})(r_0)^\chi}{f(\bar{\eta}_0 + \delta)} + c_2\sigma \int_{\bar{\eta}_0+\delta}^\infty [\sigma F(s)]^{-\frac{1}{\chi+1}}ds\right\}\beta f(w)l(w').$$

Coupling with (10.34) and using (10.31), we finally infer

$$\varphi'(w')w'' + \theta(r)\varphi(w')$$

$$\leq \beta f(w)l(w')\left\{c_1\sigma \left\|\frac{\bar{\beta}^{\chi+1}}{\beta}\right\|_{L^\infty([r_0,\infty))}\right.$$

$$\left. + \left[C\frac{(w'/\bar{\beta})(r_0)^\chi}{f(\bar{\eta}_0 + \delta)} + c_2\sigma^{\frac{\chi}{\chi+1}} \int_{\bar{\eta}_0+\delta}^\infty [F(s)]^{-\frac{1}{\chi+1}}ds\right]\left\|\frac{\theta\bar{\beta}^\chi}{\beta}\right\|_{L^\infty([r_0,\infty))}\right\}.$$

The desired conclusions now follow verbatim from the arguments in Proposition 10.16.

□

Remark 10.25 We point out that $l(t) = \varphi(t)/t^\chi$ can be singular at $t = 0$. Indeed, by construction $w' > 0$ on $[r_0, R_1)$ and the continuity of l at $t = 0$ is not needed.

Once Proposition 10.24 is established, we proceed as in Theorem 10.19 to obtain the next result, that also applies to mean curvature type operators. In particular, a direct application of the following theorem yields Theorem 2.29 in the Introduction.

Theorem 10.26 *Let M^m be a complete Riemannian manifold satisfying*

$$\text{Ric}(\nabla r, \nabla r) \geq -(m-1)\kappa^2(1+r^2)^{\alpha/2} \qquad on \ D_o, \tag{10.35}$$

for some $\kappa \geq 0$, $\alpha \geq -2$ and some fixed origin o. Let φ, l meet (10.1) and

$$t\varphi'(t) \leq C_2\varphi(t), \qquad l(t) \geq C_1\frac{\varphi(t)}{t^\chi} \qquad on \ \mathbb{R}^+, \tag{10.36}$$

for some constants $C_1, C_2 > 0$ and $\chi > 0$. Consider $b \in C(M)$ such that

$$b(x) \geq C_3(1+r(x))^{-\mu} \qquad on \ M,$$

for some constants $C_3 > 0$, $\mu \in \mathbb{R}$ satisfying

$$\mu \leq \chi - \frac{\alpha}{2}. \tag{10.37}$$

Then, under the validity of the Keller–Osserman condition

$$F^{-\frac{1}{\chi+1}} \in L^1(\infty) \tag{10.38}$$

with F as in (10.3), any non-constant solution $u \in C^1(M)$ of $(P_\geq)$ on M is bounded above and $f(u^*) \leq 0$. In particular, if $f > 0$ on $\mathbb{R}^+$ then (SL) holds for C^1-solutions.

Remark 10.27 The continuity of l at $t = 0$ forces an upper bound on χ by (10.36). If we appropriately define solutions of $(P_\geq)$ when l has a singularity, it is likely that the upper bound on χ be removable or, at least, weakened. We will not pursue this issue here, and we leave it to the interested reader.

Remark 10.28 The above proof of Proposition 10.24 fails if $\chi = 0$, and thus, in this borderline case the possible validity of an analogous of Remark 10.17 and of Theorem 10.26 is yet to be investigated.

10.4 Sharpness

We conclude this section by discussing the sharpness of Theorem 10.26. Consider the polynomial case $f(t) = t^\omega$, for some $\omega \geq 0$. Then, (10.38) becomes

$$\omega > \chi. \tag{10.39}$$

We are going to contradict (SL) under the failure of (10.39), on a suitable manifold and for $\varphi, l, b, \chi, \mu, \alpha$ meeting all of the remaining requirements in Theorem 10.26. Let (M, ds_g^2) be a model manifold as in Sect. 7.4, and suppose further that $\varphi' \geq 0$ on $\mathbb{R}^+$. Note that, because of (7.77) and the asymptotic behaviour $\Delta r \sim (m-1)/r$ as $r \to 0$,

$$\Delta r \geq c(1+r^2)^{\alpha/4} \qquad \text{on } M, \tag{10.40}$$

for some constant $c > 0$. For $\sigma > 1$ define the smooth function $u = w(r) = (1+r^2)^\sigma$. A direct computation using $\alpha \geq -2$ and $\varphi' \geq 0$, w', $w'' \geq 0$ gives

$$\Delta_\varphi u = \varphi'(w')w'' + \varphi(w')\Delta r \geq c(1+r^2)^{\alpha/4}\varphi(w').$$

Therefore, u solves

$$\Delta_\varphi u \geq C\big(1+r(x)\big)^{-\mu} u^\omega \frac{\varphi(|\nabla u|)}{|\nabla u|^\chi} \qquad \text{on } M, \tag{10.41}$$

for some $C > 0$, if and only if

$$2\sigma(\omega - \chi) \leq \frac{\alpha}{2} + \mu - \chi. \tag{10.42}$$

Since the right-hand side is non-positive because of (10.37), (10.42) is always satisfied for some σ large enough if and only if

$$\begin{cases} \omega < \chi, & \text{for each } \mu \leq \chi - \frac{\alpha}{2}, \quad \text{or} \\ \omega = \chi & \text{and} \quad \mu = \chi - \frac{\alpha}{2}, \end{cases} \tag{10.43}$$

that proves the sharpness of (10.39). Also, the last restriction in the second of (10.43) is optimal: in fact, in the Euclidean setting $\alpha = -2$, if $\omega = \chi$ and $\mu < \chi + 1$ then entire solutions of (10.41), with the equality sign, are constant whenever they have polynomial growth, see [87, Thm. 12] and also Example 4 at p. 4402 therein.

Remark 10.29 Differently from Theorem 10.26, the above counterexample also works if $\chi = 0$ and $\omega = 0$.

10.5 Volume Growth and (SL)

In this section, we study property (SL) for solutions u of $(P_\geq)$ when the condition on the Ricci curvature is replaced by a volume growth requirement, in the particular case when

$f(t) \asymp t^\omega$ on $[t_0, \infty)$ and

$$l(t) \asymp \frac{\varphi(t)}{t^\chi} \qquad \text{on } \mathbb{R}^+.$$

In this setting, Theorem 10.26 and the subsequent remarks show that a sharp Keller–Osserman condition to guarantee the boundedness of u and $f(u^*) \leq 0$ is (10.38), that is, $\omega > \chi$. The condition is optimal also for the mean curvature operator. However, a quite interesting phenomenon happens in this case: we begin by commenting on the following Liouville theorem for solutions of $(P_\geq)$, specific to mean curvature type operators and polynomial volume growths, where *no Keller–Osserman condition* is needed on f nor growth requirements are imposed on u. The result considers $(P_\geq)$ with a borderline gradient dependence $l(t) \geq C_2\varphi(t)$ on $\mathbb{R}^+$. Its proof is inspired by the original argument of Tkachev in [225] for $b \equiv 1, l \equiv 1$, later extended in [219].

Theorem 10.30 *Let M be a complete Riemannian manifold, and consider*

$$\begin{cases} \varphi \in C(\mathbb{R}_0^+), & 0 \leq \varphi \leq C_1 \quad \text{on } \mathbb{R}_0^+; \\ f \in C(\mathbb{R}), & f \text{ non-decreasing on } \mathbb{R}; \\ l \in C(\mathbb{R}_0^+), & l(t) \geq C_2\varphi(t) \quad \text{on } \mathbb{R}_0^+, \end{cases}$$

for some constant $C_1, C_2 > 0$. Fix $b \in C(M)$ satisfying

$$b(x) \geq C\big(1 + r(x)\big)^{-\mu} \qquad \text{on } M,$$

for some constants $C > 0$, $\mu < 1$. Let $u \in \mathrm{Lip}_{\mathrm{loc}}(M)$ be a non-constant solution of

$$\Delta_\varphi u \geq b(x)f(u)l(|\nabla u|) \qquad \text{on } M. \tag{10.44}$$

If

$$\liminf_{r \to \infty} \frac{\log|B_r|}{\log r} < \infty, \tag{10.45}$$

then

$$f(u)\varphi(|\nabla u|) \leq 0 \qquad \text{on } M.$$

In particular, if $\varphi > 0$ on $\mathbb{R}^+$, then $f(u) \leq 0$ on M.

Furthermore, under the same assumptions, if $u \in \mathrm{Lip}_{loc}(M)$ is a non-constant solution of $(P_=)$ then

$$f(u)\varphi(|\nabla u|) \equiv 0 \qquad on\ M,$$

and $f(u) \equiv 0$ provided that $\varphi > 0$ on $\mathbb{R}^+$.

Proof Let $\{f_k\}$ be a sequence of locally Lipschitz functions on $\mathbb{R}$ converging pointwise to f from below: for instance, one can choose

$$f_k(t) = \inf_{y \in [t-1, t+1]} \left\{ f(y) + k|t - y| \right\}.$$

Since f is increasing, up to replacing f_k with $\bar{f}_k(t) = \sup_{(-\infty, t)} f_k$ we can further suppose that f_k is increasing for each k. From (10.44) we deduce

$$\Delta_\varphi u \geq b(x) f_k(u) l(|\nabla u|) \qquad on\ M. \tag{10.46}$$

Fix a divergent sequence $\{R_j\}$ such that $\{2R_j\}$ realizes the liminf in (10.45), and let d_0, C be positive constants such that

$$|B_{2R_j}| \leq C R_j^{d_0} \qquad for\ each\ j. \tag{10.47}$$

Suppose that the set $U = \{x : f(u(x)) > 0\}$ is non-empty, otherwise there is nothing to prove. We are going to prove that $f(u)\varphi(|\nabla u|) = 0$ on U. Fix a cut-off function $0 \leq \psi \in \mathrm{Lip}_c(M)$ whose support intersects U, and to test the weak formulation of (10.46) choose the function

$$\phi = \left(f_k(u)\right)_+^{\alpha-1} \psi,$$

with $(f_k)_+ = \max\{f_k, 0\}$ the positive part of f_k, and with α a fixed real number satisfying

$$\alpha > \max\left\{4, d_0, \frac{d_0 - \mu}{1 - \mu}\right\}.$$

Define the open set $U_k = \{x : f_k(u(x)) > 0\}$ and note that $U_k \uparrow U$ by the monotone convergence of f_k, thus $U_k \neq \emptyset$ and $\phi \not\equiv 0$ on U_k for large k. Using $l(t) \geq C_2 \varphi(t)$, we obtain

$$C_2 \int_{U_k} b|f_k(u)|^\alpha \varphi(|\nabla u|)\psi \leq \int_{U_k} b|f_k(u)|^\alpha l(|\nabla u|)\psi$$

$$\leq -\int_{U_k} f_k(u)^{\alpha-1} \langle \frac{\varphi(|\nabla u|)}{|\nabla u|} \nabla u, \nabla \psi \rangle$$

$$- (\alpha - 1) \int_{U_k} f_k(u)^{\alpha-2} f_k'(u) \psi \varphi(|\nabla u|) |\nabla u|$$

$$\leq - \int_{U_k} f_k(u)^{\alpha-1} \langle \frac{\varphi(|\nabla u|)}{|\nabla u|} \nabla u, \nabla \psi \rangle,$$

where we used $f_k' \geq 0$, $\psi \geq 0$ and $\varphi \geq 0$ to get rid of the second integral on the right-hand side. Thus, applying the Cauchy–Schwarz and Hölder inequalities we get

$$C_2 \int_{U_k} b |f_k(u)|^{\alpha} \varphi(|\nabla u|) \psi \leq \int_{U_k} |f_k(u)|^{\alpha-1} \varphi(|\nabla u|) |\nabla \psi|$$

$$\leq \left\{ \int_{U_k} b |f_k(u)|^{\alpha} \varphi(|\nabla u|) \psi \right\}^{\frac{\alpha-1}{\alpha}} \left\{ \int_{U_k} \varphi(|\nabla u|) b^{1-\alpha} \frac{|\nabla \psi|^{\alpha}}{\psi^{\alpha-1}} \right\}^{1/\alpha},$$

whence, rearranging and using the boundedness of φ,

$$\int_{U_k} b |f_k(u)|^{\alpha} \varphi(|\nabla u|) \psi \leq C_3 \int_{U_k} b^{1-\alpha} \frac{|\nabla \psi|^{\alpha}}{\psi^{\alpha-1}} \leq C_3 \int_M b^{1-\alpha} \frac{|\nabla \psi|^{\alpha}}{\psi^{\alpha-1}}$$

for some constant $C_3 > 0$ depending on α. Let $\psi(x) = \psi_j(x) = \gamma(r(x)/R_j)$, where $\gamma \in \mathrm{Lip}(\mathbb{R})$ is such that

$$\gamma(t) = \begin{cases} 1 & \text{on } [0, 1] \\ (2 - t)^{\alpha} & \text{on } [1, 2) \\ 0 & \text{on } (2, \infty). \end{cases}$$

Note that $\psi_j \to 1$ locally uniformly on M, and that $|\gamma'|^{\alpha}/\gamma^{\alpha-1} = \alpha^{\alpha}$ is bounded on $[1, 2]$. Using our bounds on b, the coarea formula and integrating by parts, we deduce

$$\int_{U_k} b |f_k(u)|^{\alpha} \varphi(|\nabla u|) \psi_j \leq \frac{C_4}{R_j^{\alpha}} \int_{R_j}^{2R_j} |\partial B_t| (1 + t)^{\mu(\alpha-1)} dt$$

$$= \frac{C_4}{R_j^{\alpha}} \left\{ \left[|B_t| (1 + t)^{\mu(\alpha-1)} \right]_{R_j}^{2R_j} \right.$$

$$\left. - \mu(\alpha - 1) \int_{R_j}^{2R_j} |B_t| (1 + t)^{\mu(\alpha-1)-1} dt \right\}.$$

for some constant $C_4 > 0$. From (10.47) and the above we eventually obtain

$$\int_{U_k} b|f_k(u)|^\alpha \varphi(|\nabla u|)\psi_j$$

$$\le \frac{C_4}{R_j^\alpha}\left\{C_5 R_j^{d_0+\mu(\alpha-1)} - \mu(\alpha-1)\int_{R_j}^{2R_j}|B_t|(1+t)^{\mu(\alpha-1)-1}dt\right\}. \tag{10.48}$$

If $\mu \ge 0$ we get rid of the integral in brackets, while if $\mu < 0$ we use inequality $|B_t| \le |B_{2R_j}|$, integrate $(1+t)^{\mu(\alpha-1)-1}$ and exploit (10.47). In both of the cases, from (10.48) we infer the existence of a constant $C_6 > 0$ such that

$$\int_{U_k} b|f_k(u)|^\alpha \varphi(|\nabla u|)\psi_j \le C_6 R_j^{d_0+\mu(\alpha-1)-\alpha},$$

and letting $k \to \infty$ we get

$$\int_U b|f(u)|^\alpha \varphi(|\nabla u|)\psi_j \le C_6 R_j^{d_0+\mu(\alpha-1)-\alpha}.$$

Because of our choice of α, the exponent of R_j is negative. Letting $j \to \infty$ and using $b > 0$, $\psi_j \to 1$ we deduce $f(u)\varphi(|\nabla u|) \equiv 0$ on U, as claimed. Moreover, from $f(u) > 0$ on U we get $\varphi(|\nabla u|) = 0$ on U. Next, if $\varphi > 0$ on $\mathbb{R}^+$ then $\nabla u = 0$ on U, that is, u is constant on the connected components of U. We claim that this is impossible unless U is empty. Indeed, if $\partial U = \varnothing$ we deduce that u must be globally constant, contradicting our assumption. On the other hand, if $\partial U \ne \varnothing$ then by continuity $f(u) = 0$ on ∂U, and thus $f(u) = 0$ on the entire U, contradicting the very definition of U. In conclusion, if $\varphi > 0$ on $\mathbb{R}^+$ then U is empty, that is, $f(u) \le 0$ on M.

If u solves $(P_=)$ and is non-constant, we apply the first part of Theorem 10.30 both to u and to $v = -u$, which solves

$$\Delta_\varphi v \ge b(x)\bar{f}(v)l(|\nabla v|) \qquad \text{with } \bar{f}(t) = -f(-t),$$

to deduce both $f(u)\varphi(|\nabla u|) \le 0$ and $\bar{f}(v)\varphi(|\nabla v|) \le 0$ on M. The conclusion follows since $\bar{f}(v) = -f(u)$.   □

Remark 10.31 Since φ is bounded, choosing $l \equiv 1$ in Theorem 10.30 we include solutions of

$$\Delta_\varphi u \ge b(x)f(u) \qquad \text{on } M. \tag{10.49}$$

However, a minor modification of the above proof shows that, in fact, if u solves (10.49) then the stronger $f(u) \leq 0$ holds on M, regardless of the behaviour of φ. With the equality sign, (10.49) has been considered in [225], see also [168], while in [219] the author investigated more general equalities of the type

$$\mathrm{div}\mathbf{A}(x, u, \nabla u) = b(x)f(u),$$

where $\mathbf{A}(x, u, \nabla u) \leq Cr(x)^{\lambda}$, cf. also [63, 86, 87].

It is instructive to compare Theorem 10.30 with Theorem 10.26 and Corollary 7.9. First, we observe that if $f \leq 0$ on $\mathbb{R}$ the conclusion of Theorem 10.30 is straightforward. Otherwise, since f is increasing, there exists a constant $C > 0$ such that $f(t) \geq C > 0$ for $t \gg 1$. Hence, Theorem 10.30 considers the range

$$\omega = \chi = 0,$$

that is not covered by Theorem 10.26 (cf. Remark 10.28). The sharpness of $\mu < 1$ in Theorem 10.30 follows from the counterexample in Sect. 10.4: otherwise, if $\mu = 1$, we can choose $\alpha = -2$ (hence, M of polynomial growth) and $\chi = \omega = 0$ (by Remark 10.29) to produce a non-constant smooth solution of

$$\mathrm{div}\left(\frac{\nabla u}{\sqrt{1 + |\nabla u|^2}}\right) \geq C\left(1 + r(x)\right)^{-1} \frac{|\nabla u|}{\sqrt{1 + |\nabla u|^2}} \qquad \text{on } M.$$

Also, Theorem 10.30 is specific to operators of mean curvature type, that is, those satisfying $\varphi \leq C_1$ on $\mathbb{R}$. To see it, suppose that φ is unbounded, more precisely that

$$t\varphi'(t) \geq c_1 \varphi(t) \qquad \text{on } \mathbb{R}^+, \tag{10.50}$$

for some constant $c_1 > 0$. Note that, by integration, $\varphi(t) \geq c_2 t^{c_1}$ for some positive c_2. For such φ, we are going to produce

(i) a manifold M satisfying (10.35), for any chosen $\alpha \geq -2$ (in particular, for $\alpha = -2$, geodesic balls in M grow polynomially), and
(ii) for each $\mu \in \mathbb{R}$, a $\mathrm{Lip}_{\mathrm{loc}}$, non-negative unbounded solution u of

$$\Delta_{\varphi} u \geq C\left(1 + r(x)\right)^{-\mu} \varphi(|\nabla u|) \qquad \text{on } M, \tag{10.51}$$

for some constant $C > 0$. The combination of (i) and (ii) with $\alpha = -2$ show the failure of Theorem 10.30 for operators satisfying (10.50). Consider the model manifold in Sect. 10.4. For a smooth, radial function $u = w(r)$ with w convex and strictly increasing, by (10.40) we compute

$$\Delta_\varphi u = \varphi'(w')w'' + \varphi(w')\Delta r \geq \left[c_1 \frac{w''}{w'} + \Delta r \right] \varphi(w')$$

$$\geq \left[c_1 \frac{w''}{w'} + c(1 + r^2)^{\alpha/4} \right] \varphi(w'),$$

for some constant $c > 0$. Therefore, choosing

$$w(r) = \int_0^r \exp\left\{ (1 + t^2)^\sigma \right\} dt,$$

we obtain

$$\Delta_\varphi u \geq c_3 (1 + r^2)^{\max\{\sigma - 1, \alpha/4\}} \varphi(|\nabla u|),$$

and u solves (10.51) whenever $\sigma \geq 1 - \mu/2$, as claimed.

In the next result, we show how the technique in Theorem 10.30 can be adapted to handle $(P_\geq)$ with a more general gradient term $l(|\nabla u|)$ that is not necessary borderline, and with no bound on the decay of b. In this case, however, a slow volume growth is needed.

Theorem 10.32 *Let M be a complete Riemannian manifold, consider*

$$\begin{cases} \varphi \in C(\mathbb{R}_0^+), & 0 \leq \varphi \leq C_1 \quad on \ \mathbb{R}_0^+; \\ f \in C(\mathbb{R}), & f \ non\text{-}decreasing \ on \ \mathbb{R}; \\ l \in C(\mathbb{R}_0^+), & l \geq 0 \quad on \ \mathbb{R}^+; \\ b \in C(M), & b > 0 \quad on \ M, \end{cases}$$

for some constant $C_1 > 0$. Let $u \in \mathrm{Lip}_{\mathrm{loc}}(M)$ be a non-constant solution of

$$\Delta_\varphi u \geq b(x) f(u) l(|\nabla u|) \qquad on \ M.$$

If

$$\liminf_{r \to \infty} \frac{|B_r|}{r} = 0, \tag{10.52}$$

then $f(u)l(|\nabla u|) \le 0$ on M. If u is non-constant and solves $(P_=)$, then

$$f(u)l(|\nabla u|) \equiv 0 \qquad on\ M.$$

Proof We proceed as in the proof of Theorem 10.30: let $\{f_k\}$ be a sequence of increasing, locally Lipschitz functions converging to f from below, and note that u solves

$$\Delta_\varphi u \ge b(x) f_k(u) l(|\nabla u|) \qquad on\ M. \tag{10.53}$$

For $\varepsilon > 0$ we define

$$\eta_\varepsilon(t) = \frac{\big(f_k(t)\big)_+}{\sqrt{\big(f_k(t)\big)_+^2 + \varepsilon^2}}.$$

The monotonicity of f_k implies that $\eta_\varepsilon' \ge 0$. Define $U_k = \{f_k(u) > 0\}$ and $U = \{f(u) > 0\}$, and assume that $U \ne \emptyset$, otherwise the conclusion is immediate. Fix a cut-off function $\psi \in \mathrm{Lip}_c(M)$, to be chosen later, and insert

$$\phi = \eta_\varepsilon(u)\psi \in \mathrm{Lip}_c(M)$$

in the weak definition of (10.53). Then, apply Cauchy–Schwarz inequality to deduce

$$\int b\eta_\varepsilon(u) f_k(u) l(|\nabla u|)\psi \le -\int \eta_\varepsilon(u) \langle \frac{\varphi(|\nabla u|)}{|\nabla u|}\nabla u, \nabla\psi\rangle \le \int \eta_\varepsilon(u)\varphi(|\nabla u|)|\nabla\psi|.$$

Letting $\varepsilon \to 0$, using Lebesgue convergence theorem and the boundedness of φ we get

$$\int_{U_k} b\big(f_k(u)\big)_+ l(|\nabla u|)\psi \le \int_{U_k} \varphi(|\nabla u|)|\nabla\psi| \le C_1 \int_M |\nabla\psi|. \tag{10.54}$$

Fix a diverging sequence $\{R_j\}$ such that $\{2R_j\}$ realizes the liminf in (10.52), and define $\psi(x) = \psi_j(x) = \gamma(r(x)/R_j)$, where $\gamma \in \mathrm{Lip}(\mathbb{R})$ satisfies

$$\gamma = 1 \quad on\ [0, 1), \qquad \gamma = 0 \quad on\ (2, \infty), \qquad \gamma(t) = 2 - t \quad on\ [1, 2].$$

Evaluating (10.54) with $\psi = \psi_j$ and letting $k \to \infty$ we obtain

$$\int_U b\big(f(u)\big)_+ l(|\nabla u|)\psi_j \le \frac{C_1}{R_j}|B_{2R_j}|.$$

The conclusion follows by letting $j \to \infty$, and the case of equality is handled as in Theorem 10.30.

$\square$

We next consider inequalities $(P_{\geq})$ under the validity of the Keller–Osserman condition

$$F^{-\frac{1}{\chi+1}} \in L^1(\infty),$$

when just a volume growth upper bound is imposed on M. The main result of this section, Theorem 10.33, improves on [181, 182] (see also [191, Thm. 1.3]). Although the proof is still based on the delicate iteration argument in [181, 182], the presence of a nontrivial gradient term l calls for new estimates, inspired by recent work in [87].

Towards this aim, we assume

$$\varphi(t) \leq Ct^{p-1} \qquad \text{for some } p > 1, \ C > 0 \ \text{ and } t \in \mathbb{R}^+. \tag{10.55}$$

Theorem 10.33 *Let M be a complete Riemannian manifold, and consider φ, b, f, l meeting assumptions (2.3), (2.5) and (10.55), for some $p > 1$. Assume that, for some $\mu, \chi, \omega \in \mathbb{R}$ with*

$$\chi \geq 0, \qquad \mu \leq \chi + 1, \qquad \omega > \chi \tag{10.56}$$

the following inequalities are satisfied:

$$b(x) \geq C\big(1 + r(x)\big)^{-\mu} \qquad \text{on } M,$$

$$f(t) \geq Ct^{\omega} \qquad\qquad \text{for } t \gg 1 \tag{10.57}$$

$$l(t) \geq C\frac{\varphi(t)}{t^{\chi}} \qquad\qquad \text{on } \mathbb{R}^+,$$

for some constant $C > 0$. Let $u \in \mathrm{Lip}_{\mathrm{loc}}(M)$ be a non-constant solution of $(P_{\geq})$ on M, and suppose that either

$$\mu < \chi + 1 \text{ and} \qquad \liminf_{r \to \infty} \frac{\log |B_r|}{r^{\chi+1-\mu}} < \infty \quad (= 0 \text{ if } \chi = 0);$$

or \hfill (10.58)

$$\mu = \chi + 1 \text{ and} \qquad \liminf_{r \to \infty} \frac{\log |B_r|}{\log r} < \infty \quad (\leq p \text{ if } \chi = 0).$$

Then, u is bounded above and $f(u^*) \leq 0$. In particular in case $f > 0$ on $\mathbb{R}^+$, (SL) holds.

Remark 10.34 In Euclidean space $\mathbb{R}^m$, and when the third in (10.57) is replaced with the stronger $l(t) \geq Ct^{p-1-\chi}$, Liouville type results covering some of the cases in Theorem 10.33 have been obtained by various authors (in some instances, even for more general quasilinear operators). Among them, we stress Thm 1 in [87], that considers the entire range (10.56). However, if $\mu = \chi + 1$, the authors need the further condition $p > m$

independently of the value of χ, a quite stronger requirement compared with the second in (10.58). Previous work in [91] considered the case $0 < \chi \leq p - 1, \omega > \chi$ and $\mu < \chi + 1$ under the restriction[2] $p \in (1, m)$, for operators close either to the p-Laplacian or to the mean curvature ones.

The existence of a Liouville theorem for $\mu = \chi + 1$ and $l(t) \equiv 1$ was conjectured by Mitidieri–Pohozaev in [162, Sect. 14 Ch. 1], and has previously been proved in [168] (for the p-Laplace operator) and [230] (for the mean curvature operator), in both cases on $\mathbb{R}^m$.

Theorem 10.33 is a consequence of Theorem 7.5 and of the next

Proposition 10.35 *Let M be a complete Riemannian manifold, and let φ satisfy (2.3), and (10.55) with $p > 1$. Fix $\mu, \omega, \chi \in \mathbb{R}$ satisfying*

$$\chi \geq 0, \qquad \mu \leq \chi + 1, \qquad \omega > \chi, \tag{10.59}$$

and assume either one of the following requirements:

$$\mu < \chi + 1 \ and \qquad \liminf_{r \to \infty} \frac{\log |B_r|}{r^{\chi + 1 - \mu}} < \infty$$

or $\tag{10.60}$

$$\mu = \chi + 1 \ and \qquad \liminf_{r \to \infty} \frac{\log |B_r|}{\log r} < \infty.$$

If $u \in \mathrm{Lip}_{\mathrm{loc}}(M)$ satisfies

$$\Delta_\varphi u \geq K(1 + r)^{-\mu} u^\omega \frac{\varphi(|\nabla u|)}{|\nabla u|^\chi} \qquad on \ \Omega_\eta = \{u > \eta\} \neq \emptyset, \tag{10.61}$$

for some $\eta > 0$, then u is bounded above.

Remark 10.36 Although we require no upper bound on $\varphi(t)/t^\chi$ in a neighbourhood of zero, the weak inequality (10.61) implicitly assumes the term

$$\frac{\varphi(|\nabla u|)}{|\nabla u|^\chi}$$

to be locally integrable on Ω_η.

Proof Suppose by contradiction that $u^* = \infty$. Fix $\gamma > \eta$, and take $\lambda \in C^1(\mathbb{R})$ such that

$$0 \leq \lambda \leq 1, \quad \lambda' \geq 0, \quad \lambda \equiv 0 \ \text{on} \ (-\infty, \gamma], \quad \lambda > 0 \ \text{on} \ (\gamma, \infty).$$

---

[2]See Corollaries 1.3 and 1.4 in [91]; the bound $p \in (1, m)$ is assumed at p.2904.

Let $\psi \in C_c^\infty(M)$ be a cut-off function, and let $\varsigma, \alpha > 1$ to be specified later. We insert the non-negative test function

$$\phi = \psi^\varsigma \lambda(u) u^\alpha \in \mathrm{Lip}_c(M)$$

in the weak definition of (10.61) to deduce, using $\lambda' \geq 0$ and (10.57),

$$K \int \psi^\varsigma \lambda \frac{u^{\alpha+\omega}}{(1+r)^\mu} \frac{\varphi(|\nabla u|)}{|\nabla u|^\chi} \leq -\int \frac{\varphi(|\nabla u|)}{|\nabla u|} \langle \nabla u, \nabla(\psi^\varsigma \lambda u^\alpha)\rangle$$

$$\leq \varsigma \int \psi^{\varsigma-1} \lambda u^\alpha \varphi(|\nabla u|)|\nabla \psi| \qquad (10.62)$$

$$- \alpha \int \psi^\varsigma \lambda u^{\alpha-1} \varphi(|\nabla u|)|\nabla u|.$$

We now divide the proof into several steps: □

Step 1: Basic Growth Estimates
The following inequalities hold:

– If $\mu < \chi + 1$, then for each $q > 0$ there exists $\alpha_q > 1$ and a constant $C_q > 0$ depending on p, q, χ, μ, ω such that, whenever $\alpha \geq \alpha_q$,

$$\int_{B_R \cap \Omega_\gamma} \lambda \frac{u^{\alpha+\omega}}{(1+r)^\mu} \frac{\varphi(|\nabla u|)}{|\nabla u|^\chi} \leq C_q \frac{|B_{2R}|}{R^q}.$$

– If $\mu = \chi + 1$, then there exists a constant $C > 0$ depending on p, χ, μ, ω such that

$$\int_{B_R \cap \Omega_\gamma} \lambda \frac{u^{\alpha+\omega}}{(1+r)^\mu} \frac{\varphi(|\nabla u|)}{|\nabla u|^\chi} \leq C \frac{|B_{2R}|}{R^p}. \qquad (10.63)$$

Proof of Step 1 The argument is an adaptation of Lemma 2.2 in [87], and rests on the use of the triple Young inequality to the first term on the right-hand side of (10.62): we need to find $z_1, z_2, z_3 > 1$ satisfying

$$\frac{1}{z_1} + \frac{1}{z_2} + \frac{1}{z_3} = 1 \qquad (10.64)$$

and $\tau, \bar{C} > 0$ such that

$$\varsigma \psi^{\varsigma-1} \lambda u^\alpha \varphi(|\nabla u|)|\nabla \psi| = \mathcal{J}_1^{\frac{1}{z_1}} \mathcal{J}_2^{\frac{1}{z_2}} \mathcal{J}_3^{\frac{1}{z_3}}, \qquad (10.65)$$

with

$$\mathcal{J}_1 = \frac{K}{2}\psi^{\varsigma}\lambda \frac{u^{\alpha+\omega}}{(1+r)^{\mu}}\frac{\varphi(|\nabla u|)}{|\nabla u|^{\chi}}$$

$$\mathcal{J}_2 = \alpha\psi^{\varsigma}\lambda u^{\alpha-1}\varphi(|\nabla u|)|\nabla u| \tag{10.66}$$

$$\mathcal{J}_3 = \bar{C}(1+r)^{\tau}\left[\frac{\varphi(|\nabla u|)}{|\nabla u|^{p-1}}\right]|\nabla\psi|^{z_3},$$

considering powers of u, $|\nabla u|$, r and ψ, to obtain (10.65) we need the following balancing:

$$(i) \text{ powers of } u: \qquad \alpha = \frac{\alpha+\omega}{z_1} + \frac{\alpha-1}{z_2}$$

$$(ii) \text{ powers of } |\nabla u|: \quad 0 = -\frac{\chi}{z_1} + \frac{1}{z_2} - \frac{p-1}{z_3}$$

$$(iii) \text{ powers of } r: \qquad 0 = -\frac{\mu}{z_1} + \frac{\tau}{z_3}$$

$$(iv) \text{ powers of } \psi: \qquad \varsigma - 1 = \frac{\varsigma}{z_1} + \frac{\varsigma}{z_2}.$$

To find z_1, z_2, z_3 note that, by (10.64), the equality for $|\nabla u|$ can be rewritten as

$$p - 1 = \frac{p-1-\chi}{z_1} + \frac{p}{z_2}.$$

Thus, solving the equations for u, $|\nabla u|$ with respect to z_1 and z_2, and then recovering z_3 from (10.64), we get

$$\frac{1}{z_1} = \frac{\alpha+p-1}{(\chi+1)(\alpha-1)+p(\omega+1)}, \qquad \frac{1}{z_2} = \frac{\chi\alpha+(p-1)\omega}{(\chi+1)(\alpha-1)+p(\omega+1)},$$

$$\frac{1}{z_3} = \frac{\omega-\chi}{(\chi+1)(\alpha-1)+p(\omega+1)}$$

(these are positive numbers less than 1 because of (10.59)), and from the last two equations,

$$\tau = \mu\frac{z_3}{z_1} = \mu\frac{\alpha+p-1}{\omega-\chi}, \qquad \varsigma = z_3.$$

The constant $\bar{C}$ is then uniquely determined by (10.65). Having found the right parameters, from the triple Young inequality

$$\mathcal{J}_1^{\frac{1}{z_1}}\mathcal{J}_2^{\frac{1}{z_2}}\mathcal{J}_3^{\frac{1}{z_3}} \leq \mathcal{J}_1 + \mathcal{J}_2 + \mathcal{J}_3,$$

and replacing into (10.65) and (10.66) we deduce

$$\varsigma\psi^{\varsigma-1}\lambda u^{\alpha}\varphi(|\nabla u|)|\nabla\psi| \leq \frac{K}{2}\psi^{\varsigma}\lambda\frac{u^{\alpha+\omega}}{(1+r)^{\mu}}\frac{\varphi(|\nabla u|)}{|\nabla u|^{\chi}} + \alpha\psi^{\varsigma}\lambda u^{\alpha-1}\varphi(|\nabla u|)|\nabla u|$$

$$+\bar{C}(1+r)^{\mu\frac{z_3}{z_1}}\left[\frac{\varphi(|\nabla u|)}{|\nabla u|^{p-1}}\right]^{z_3}|\nabla\psi|^{z_3}.$$

Inserting into (10.62) and using (10.55) we get

$$\frac{K}{2}\int\psi^{\varsigma}\lambda\frac{u^{\alpha+\omega}}{(1+r)^{\mu}}\frac{\varphi(|\nabla u|)}{|\nabla u|^{\chi}} \leq C_1\int(1+r)^{\mu\frac{z_3}{z_1}}|\nabla\psi|^{z_3}.$$

For large $R > 1$, we choose $\psi \in C_c^{\infty}(M)$ satisfying

$$0 \leq \psi \leq 1, \quad \psi \equiv 1 \text{ on } B_R, \quad \psi \equiv 0 \text{ on } M\backslash B_{2R}, \quad |\nabla\psi| \leq \frac{C}{R}, \qquad (10.67)$$

for an absolute constant C. Using (10.67) and the fact that $\lambda = 0$ when $u \leq \gamma$, we obtain

$$\frac{K}{2}\int_{B_R\cap\Omega_{\gamma}}\lambda\frac{u^{\alpha+\omega}}{(1+r)^{\mu}}\frac{\varphi(|\nabla u|)}{|\nabla u|^{\chi}} \leq \frac{K}{2}\int\psi^{\varsigma}\lambda\frac{u^{\alpha+\omega}}{(1+r)^{\mu}}\frac{\varphi(|\nabla u|)}{|\nabla u|^{\chi}}$$

$$\leq \frac{C_2}{R^{z_3}}\int_{B_{2R}}(1+r)^{\mu\frac{z_3}{z_1}} \qquad (10.68)$$

$$\leq C_3 R^{\mu\frac{z_3}{z_1}-z_3}|B_{2R}|.$$

The exponent of R in (10.68) can be written as

$$\mu\frac{z_3}{z_1} - z_3 = \frac{\alpha+p-1}{\omega-\chi}(\mu-\chi-1) - p. \qquad (10.69)$$

We examine the two cases, according to whether $\mu < \chi + 1$ or $\mu = \chi + 1$.

- If $\mu < \chi + 1$, then for any given $q > 0$ we can choose α_q sufficiently large that, for $\alpha \geq \alpha_q$,

$$\mu\frac{z_3}{z_1} - z_3 \leq -q.$$

Having fixed such α_q, from (10.68) we get

$$\int_{B_R\cap\Omega_{\gamma}}\lambda\frac{u^{\alpha+\omega}}{(1+r)^{\mu}}\frac{\varphi(|\nabla u|)}{|\nabla u|^{\chi}} \leq \frac{C_q}{K}\frac{|B_{2R}|}{R^q},$$

and the thesis follows.

- If $\mu = \chi + 1$, then by (10.69) the exponent of R in (10.68) is $-p$ independently of α, and we obtain

$$\int_{B_R \cap \Omega_\gamma} \lambda \frac{u^{\alpha+\omega}}{(1+r)^\mu} \frac{\varphi(|\nabla u|)}{|\nabla u|^\chi} \leq \frac{C_5}{K} \frac{|B_{2R}|}{R^p}.$$

as claimed.

□

Step 2: A Preliminary Inequality
We consider again (10.62), but we are going to choose $\alpha, \varsigma > \chi + 1$ in a way different to the one in Step 1.

Case 1: $\chi > 0$
We use Young's inequality with exponents $\chi + 1$ and $(\chi + 1)/\chi$ to remove the second term in the right-hand side of (10.62): for each $\varepsilon > 0$, we get

$$\varsigma \int \psi^{\varsigma-1} \lambda u^\alpha \varphi(|\nabla u|) |\nabla \psi|$$

$$\leq \frac{\varsigma}{(\chi+1)\,\varepsilon^{\chi+1}} \int \psi^{\varsigma-\chi-1} \lambda u^{\alpha+\chi} \frac{\varphi(|\nabla u|)}{|\nabla u|^\chi} |\nabla \psi|^{\chi+1} \tag{10.70}$$

$$+ \frac{\chi \varsigma \varepsilon^{\frac{\chi+1}{\chi}}}{\chi+1} \int \psi^\varsigma \lambda u^{\alpha-1} \varphi(|\nabla u|) |\nabla u|,$$

choosing ε such that

$$\frac{\chi \varsigma \varepsilon^{\frac{\chi+1}{\chi}}}{\chi+1} = \alpha, \quad \text{that is,} \quad \varepsilon = \left(\frac{\alpha(\chi+1)}{\chi \varsigma} \right)^{\frac{\chi}{\chi+1}},$$

and inserting (10.70) into (10.62), we obtain

$$K \int \psi^\varsigma \lambda \frac{u^{\alpha+\omega}}{(1+r)^\mu} \frac{\varphi(|\nabla u|)}{|\nabla u|^\chi} \leq \frac{C_1 \varsigma^{\chi+1}}{\alpha^\chi} \int \psi^{\varsigma-\chi-1} \lambda u^{\alpha+\chi} \frac{\varphi(|\nabla u|)}{|\nabla u|^\chi} |\nabla \psi|^{\chi+1} \tag{10.71}$$

for some constant $C_1 = C_1(\chi) > 0$.

Case 2: $\chi = 0$
In this case, (10.71) with $\chi = 0$ and $C_1 = 1$ directly follows from (10.62), getting rid of the second term on the right-hand side.

Step 3: Induction for $\mu < \chi + 1$

If $\mu < \chi + 1$, the following inductive relation holds:

$$\int_{B_R \cap \Omega_\gamma} \lambda (1+r)^{-\mu} \frac{\varphi(|\nabla u|)}{|\nabla u|^\chi} \leq 2^{-BR^\theta} \left[\int_{B_{2R} \cap \Omega_\gamma} \lambda (1+r)^{-\mu} \frac{\varphi(|\nabla u|)}{|\nabla u|^\chi} \right],$$

where

$$B = \frac{C_6 \gamma^{\omega - \chi}}{\omega - \chi}, \qquad \theta = \chi + 1 - \mu > 0, \tag{10.72}$$

and $C_6 = C_6(K, \omega, \chi, \mu)$ is a positive constant independent of γ and R.

Proof of Step 3 Fix $\xi > 1$ close enough to 1 in order to satisfy

$$\omega - \chi - (\chi + 1) \left(1 - \frac{1}{\xi} \right) > 0 \tag{10.73}$$

and, for $R \geq 2$ choose a cut-off function $\psi \in \mathrm{Lip}_c(B_{2R})$ such that

$$0 \leq \psi \leq 1, \qquad \psi \equiv 1 \text{ on } B_R, \qquad |\nabla \psi| \leq \frac{C}{R} \psi^{1/\xi}, \tag{10.74}$$

for some $C = C(\xi) > 0$. Note that this is possible since $\xi > 1$ (for instance, one can take the cut-off in (10.67), call it ψ_0, and consider $\psi = \psi_0^{\xi/(\xi-1)}$).

Choose α and ς in order to satisfy

$$\varsigma = \alpha + \omega, \qquad \alpha > \max \{\omega, \chi + 1 - \omega\}. \tag{10.75}$$

However, for the ease of notation we feel convenient to keep ς and α independent in the next computations. By Step 2, inequality (10.71) holds for each $\chi \geq 0$. Using then (10.74), and since $\{\nabla \psi \neq 0\} \subset B_{2R} \backslash B_R$, $R \geq 2$, from (10.71) we deduce

$$K \int \psi^\varsigma \lambda \frac{u^{\alpha + \omega}}{(1+r)^\mu} \frac{\varphi(|\nabla u|)}{|\nabla u|^\chi}$$

$$\leq \frac{C_2 \varsigma^{\chi+1}}{\alpha^\chi R^{\chi+1}} \int_{\{\nabla \psi \neq 0\}} \psi^{\varsigma - (\chi+1)\left(1 - \frac{1}{\xi}\right)} \lambda u^{\alpha + \chi} \frac{\varphi(|\nabla u|)}{|\nabla u|^\chi} \tag{10.76}$$

$$\leq \frac{C_3 \varsigma^{\chi+1}}{\alpha^\chi R^{\chi+1-\mu}} \int \psi^{\varsigma - (\chi+1)\left(1 - \frac{1}{\xi}\right)} \lambda \frac{u^{\alpha + \chi}}{(1+r)^\mu} \frac{\varphi(|\nabla u|)}{|\nabla u|^\chi}.$$

Since $\omega > \chi$, we can apply Hölder's inequality to the RHS with exponents

$$q = \frac{\alpha + \omega}{\omega - \chi}, \qquad q' = \frac{\alpha + \omega}{\alpha + \chi} \tag{10.77}$$

and get

$$\int \psi^{\varsigma - (\chi+1)\left(1-\frac{1}{\xi}\right)} \lambda \frac{u^{\alpha+\chi}}{(1+r)^{\mu}} \frac{\varphi(|\nabla u|)}{|\nabla u|^{\chi}} \leq \left(\int \psi^{\varsigma} \lambda \frac{u^{q'(\alpha+\chi)}}{(1+r)^{\mu}} \frac{\varphi(|\nabla u|)}{|\nabla u|^{\chi}} \right)^{1/q'}$$

$$\cdot \left(\int \psi^{\varsigma - (\chi+1)q(1-1/\xi)} \lambda (1+r)^{-\mu} \frac{\varphi(|\nabla u|)}{|\nabla u|^{\chi}} \right)^{1/q}.$$

Inserting into (10.76) we obtain

$$\int \psi^{\varsigma} \lambda \frac{u^{\alpha+\omega}}{(1+r)^{\mu}} \frac{\varphi(|\nabla u|)}{|\nabla u|^{\chi}}$$

$$\leq \left(\frac{C_4 \varsigma^{\chi+1}}{\alpha^{\chi} R^{\chi+1-\mu}} \right)^q \int \psi^{\varsigma - (\chi+1)q\left(1-\frac{1}{\xi}\right)} \lambda (1+r)^{-\mu} \frac{\varphi(|\nabla u|)}{|\nabla u|^{\chi}} \tag{10.78}$$

for some constant $C_4 = C_4(\chi, \mu, K) > 0$. Now, by (10.73), (10.75), and (10.77),

$$\varsigma - (\chi + 1) q \left(1 - \frac{1}{\xi} \right) = \frac{\alpha + \omega}{\omega - \chi} \left[\omega - \chi - (\chi + 1) \left(1 - \frac{1}{\xi} \right) \right] > 0,$$

hence the term with ψ on the right-hand side of (10.78) can be estimated from above with 1 on B_{2R}. Together with condition $\alpha > \omega$ in (10.75), this gives

$$\int \psi^{\varsigma} \lambda \frac{u^{\alpha+\omega}}{(1+r)^{\mu}} \frac{\varphi(|\nabla u|)}{|\nabla u|^{\chi}}$$

$$\leq \left(\frac{C_5 \alpha}{R^{\chi+1-\mu}} \right)^q \int_{B_{2R} \cap \Omega_\gamma} \lambda (1+r)^{-\mu} \frac{\varphi(|\nabla u|)}{|\nabla u|^{\chi}}.$$

Now, since $u \geq \gamma$ on the domain where $\lambda(u)$ is positive and not zero, using again the properties of ψ and the definition of q we finally infer

$$\int_{B_R} \lambda (1+r)^{-\mu} \frac{\varphi(|\nabla u|)}{|\nabla u|^{\chi}}$$

$$\leq \left(\frac{C_5 \alpha}{R^{\chi+1-\mu} \gamma^{\omega-\chi}} \right)^{\frac{\alpha+\omega}{\omega-\chi}} \int_{B_{2R}} \lambda (1+r)^{-\mu} \frac{\varphi(|\nabla u|)}{|\nabla u|^{\chi}}.$$

Choose α in such a way that

$$\frac{C_5\alpha}{R^{\chi+1-\mu}\gamma^{\omega-\chi}} = \frac{1}{2},$$

that is,

$$\alpha = \alpha(R) = \frac{\gamma^{\omega-\chi}}{2C_5}R^{\chi+1-\mu} = C_6\gamma^{\omega-\chi}R^{\chi+1-\mu}.$$

Since $\chi + 1 - \mu > 0$ by assumption, if R is big enough then α satisfies (10.75). Then, setting

$$\mathcal{H}(R) = \int_{B_R} \lambda(1+r)^{-\mu}\frac{\varphi(|\nabla u|)}{|\nabla u|^{\chi}},$$

we have

$$\mathcal{H}(R) \leq 2^{-\frac{\alpha+\omega}{\omega-\chi}}\mathcal{H}(2R) \leq 2^{-\frac{\alpha}{\omega-\chi}}\mathcal{H}(2R) = 2^{-BR^{\theta}}\mathcal{H}(2R), \tag{10.79}$$

where B, θ are as in (10.72). $\square$

Step 4: Iteration and Conclusion for $\mu < \chi + 1$

Fix R_0 large and $\bar{R} > 2R_0$ such that u is not constant on $\Omega_\gamma \cap B_{\bar{R}}$. Then, $\mathcal{H}(\bar{R}) > 0$. Consider $R_j = 2^j\bar{R}$, and let k be the integer satisfying $R_k < R \leq R_{k+1}$. Iterating (10.79) k-times and taking the logarithm, we get

$$\log\mathcal{H}(\bar{R}) \leq -\left(\sum_{j=0}^{k-1}\bar{R}^{\theta}2^{j\theta}\right)B\log 2 + \log\mathcal{H}(R_k)$$

$$\leq -\left(\sum_{j=0}^{k-1}R_k^{\theta}2^{(j-k)\theta}\right)B\log 2 + \log\mathcal{H}(R).$$

Now,

$$\sum_{j=0}^{k-1}R_k^{\theta}2^{(j-k)\theta} = \frac{R_{k+1}^{\theta}}{2^{\theta}}\sum_{j=0}^{k-1}2^{(j-k)\theta} = \frac{R_{k+1}^{\theta}}{2^{\theta}}\left(\frac{2^{-\theta}-2^{-(k+1)\theta}}{1-2^{-\theta}}\right) \geq R^{\theta}C_{\theta},$$

for some constant $C_\theta > 0$, thus

$$\log\mathcal{H}(\bar{R}) \leq -R^{\theta}BC_{\theta}\log 2 + \log\mathcal{H}(R),$$

or in other words,

$$\frac{\log \mathcal{H}(R)}{R^\theta} \geq \frac{\log \mathcal{H}(\bar{R})}{R^\theta} + BC_\theta \log 2. \tag{10.80}$$

By Step 1, for fixed $q = 2$ there exists α_2 such that, for $\alpha \geq \alpha_2$,

$$\mathcal{H}(R) \leq \frac{1}{\gamma^{\alpha+\omega}} \int_{B_r \cap \Omega_\gamma} \lambda \frac{u^{\alpha+\omega}}{(1+r)^\mu} \frac{\varphi(|\nabla u|)}{|\nabla u|^\chi} \leq \frac{C_q}{\gamma^{\alpha+\omega}} \frac{|B_{2R}|}{R^2}. \tag{10.81}$$

Now, choosing $\bar{R}$ large enough that $\alpha(\bar{R}) \geq \alpha_2$, plugging (10.81) into (10.80), letting $R \to \infty$ and using the definition of B, because of (10.60) we get

$$\frac{C_6 C_\theta \log 2}{\omega - \chi} \gamma^{\omega-\chi} \leq \liminf_{R \to \infty} \frac{\log |B_{2R}|}{R^\theta} < \infty.$$

However, the assumption $\omega > \chi$ leads to a contradiction provided that γ is chosen to be large enough. Therefore, $u^* < \infty$, concluding the proof.

Step 5: Iteration and Conclusion for $\mu = \chi + 1$
We begin again with (10.71), but we fix $\varsigma = \chi + 2$. Choose a cut-off ψ satisfying

$$\psi \equiv 1 \quad \text{on } B_R, \qquad \psi \equiv 0 \quad \text{on } M \backslash B_{2R}, \qquad |\nabla \psi| \leq \frac{2}{R}$$

and define

$$\mathcal{H}_u(R) \doteq \int_{B_R \cap \Omega_\gamma} \lambda \frac{u^{\alpha+\omega}}{(1+r)^\mu} \frac{\varphi(|\nabla u|)}{|\nabla u|^\chi}.$$

Since $\mu = \chi + 1$, we obtain

$$\mathcal{H}_u(R) \leq \frac{C_2}{K \alpha^\chi R^{\chi+1}} \int_{B_{2R} \cap \Omega_\gamma} \lambda u^{\alpha+\chi} \frac{\varphi(|\nabla u|)}{|\nabla u|^\chi}$$

$$\leq \frac{C_3}{K \alpha^\chi \gamma^{\omega-\chi}} \int_{B_{2R} \cap \Omega_\gamma} \lambda \frac{u^{\alpha+\omega}}{(1+r)^\mu} \frac{\varphi(|\nabla u|)}{|\nabla u|^\chi},$$

for some $C_3 = C_3(\chi) > 0$. Since $\omega > \chi$, for fixed $S > 0$ we can choose γ large enough to satisfy

$$\mathcal{H}_u(R) \leq 2^{-S} \mathcal{H}_u(2R).$$

Fix R_0 large, $\bar{R} > 2R_0$, $R_i = 2^i \bar{R}$. For $R > \bar{R}$, let $k \in \mathbb{N}$ be such that $R_k < R \leq R_{k+1}$. Iterating the above inequality k-times and taking the logarithm we deduce

$$\log \mathcal{H}_u(\bar{R}) \leq -kS \log 2 + \log \mathcal{H}_u(R_k) \leq -kS \log 2 + \log \mathcal{H}_u(R).$$

Dividing by $\log R$ and using the inequality

$$\log R \leq (k+1) \log 2 + \log \bar{R} \leq 2k \log 2 \qquad \text{for large enough } R,$$

we deduce

$$\frac{\log \mathcal{H}_u(\bar{R})}{\log R} \leq -\frac{Sk}{\log R} \log 2 + \frac{\log \mathcal{H}_u(R)}{\log R} \leq -\frac{S}{2} + \frac{\log \mathcal{H}_u(R)}{\log R}. \tag{10.82}$$

Now, because of Step 1, $\mathcal{H}_u(R) \leq C_\alpha |B_R|/R^p$, where the constant C_α depends on α. If $\{R_j\}$ is a sequence realizing the liminf in (10.60),

$$\limsup_{j \to \infty} \frac{\log \mathcal{H}_u(R_j)}{\log R_j} \leq \lim_{j \to \infty} \frac{\log |B_{R_j}|}{\log R_j} - p \doteq C_* < \infty.$$

Inserting into (10.82) and letting $j \to \infty$ we obtain

$$0 \leq -\frac{S}{2} + C_*,$$

that leads to a contradiction provided that γ (hence, S) is chosen large enough.

We are now ready to prove Theorem 10.33.

Proof (of Theorem 10.33) We first show that u is bounded above. If not, using (10.57) we deduce that, for $\eta > 0$ sufficiently large, u would be a non-constant solution of

$$\Delta_\varphi u \geq K(1+r)^{-\mu} u^\omega \frac{\varphi(|\nabla u|)}{|\nabla u|^\chi} \qquad \text{on } \Omega_\eta = \{u > \eta\} \neq \emptyset, \tag{10.83}$$

for some $K > 0$, contradicting Proposition 10.35. Next, we invoke Theorem 7.5 with $\sigma = 0$ to infer that $f(u^*) \leq 0$, concluding the proof. $\qquad \square$

Remark 10.37 When $\chi = 0$, it is interesting to observe that the upper bound p for the growth of $|B_r|$ in the second of (10.58) does not appear in (10.60). In other words, the bound is not needed to infer that $u^* < \infty$, but it serves to guarantee (WMP$_\infty$) and deduce $f(u^*) \leq 0$.

10.6   Applications: Yamabe and Capillarity Equations

Yamabe Type Equations

With the aid of Theorem 10.33, we are able to improve on various geometric corollaries of [182, Thm. 4.8]. By a way of example, we consider the following conformal rigidity result for manifolds with negative scalar curvature, first investigated by M. Obata in [173] (in the compact case) and S.T. Yau in [238]. The geometric conditions in their main theorems have later been substantially weakened in [182, Thm. 4.9], and our next corollary is a mild generalization of it.

Corollary 10.38 *Let* $(M, \langle , \rangle)$ *be a complete Riemannian manifold of dimension* $m \geq 2$ *whose scalar curvature* $R(x)$ *satisfies*

$$R(x) \leq -C\big(1 + r(x)\big)^{-\mu} \qquad on \ M,$$

for some constants $\mu \in \mathbb{R}, C \in \mathbb{R}^+$. *If either*

$$\mu < 2 \ and \qquad \liminf_{r \to \infty} \frac{\log |B_r|}{r^{2-\mu}} < \infty, \qquad or$$

$$\mu = 2 \ and \qquad \liminf_{r \to \infty} \frac{\log |B_r|}{\log r} < \infty,$$

then any conformal diffeomorphism of M *preserving* R *is an isometry.*

Proof Let $T : (M, \langle , \rangle) \to (M, \langle , \rangle)$ be a conformal diffeomorphism, and let $(,) = T^*\langle , \rangle = \lambda^2 \langle , \rangle$ be the conformally deformed metric, for $0 < \lambda \in C^\infty(M)$. If $m \geq 3$, writing $\lambda = u^{\frac{2}{m-2}}$ then it is well known that u solves Yamabe equation

$$\Delta u = \frac{R}{c_m} u - \frac{\bar{R}}{c_m} u^{\frac{m+2}{m-2}} \qquad on \ M,$$

where $\bar{R}$ is the scalar curvature of $(,)$, Δ is the Laplacian of the background metric $\langle , \rangle$, and $c_m = \frac{4(m-1)}{m-2}$. On the other hand, if $m = 2$, writing $\lambda = e^u$ it holds

$$2\Delta u = R - \bar{R}e^{2u} \qquad on \ M.$$

Therefore, if T preserves the scalar curvature,

$$\Delta u = -R(x)f(u), \qquad with \qquad f(u) = \begin{cases} \dfrac{1}{c_m}\left[u^{\frac{m+2}{m-2}} - u\right] & if \ m \geq 3, \\[2ex] \dfrac{1}{2}\left[e^{2u} - 1\right] & if \ m = 2. \end{cases}$$

We now apply Theorem 10.33 with $b(x) = -R(x)$, $\varphi(t) = t$ and $\chi = 1$ both to u and to $-u$ to deduce that u is bounded and $f(u^*) \leq 0 \leq f(u_*)$. Hence, $u \equiv 1$ if $m \geq 3$, respectively, $u \equiv 0$ if $m = 2$, and T is therefore an isometry. $\qquad\square$

For many other applications to Geometry, we refer the reader to [6, 182]. Next, we focus on the mean curvature operator.

The Capillarity Equation

As observed in the Introduction, global solutions $u : \mathbb{R}^m \to \mathbb{R}$ of the capillary equation

$$\text{div}\left(\frac{\nabla u}{\sqrt{1 + |\nabla u|^2}}\right) = \kappa(x)u$$

have been considered in [168, 225], with subsequent improvements in [86, 219]. Combing their results, u must vanish identically provided that

$$\kappa(x) \geq C\big(1 + r(x)\big)^{-\mu}$$

on $\mathbb{R}^m$, for some constants $C > 0$ and $\mu < 2$. In fact, in [86] the authors investigated a more general class of equations including

$$\text{div}\left(\frac{\nabla u}{\sqrt{1 + |\nabla u|^2}}\right) = \kappa(x)|u|^{\omega - 1}u \tag{10.84}$$

on $\mathbb{R}^m$, with $\omega > 0$ and $\kappa(x)$ enjoying (2.36), see also Section 5 in [195]. Applying the Corollary at p. 4387 in [86], $u \equiv 0$ on $\mathbb{R}^m$ whenever either

$$\begin{cases} \omega > 1, & \mu \leq 2, \qquad \text{or} \\ \omega \in (0, 1], & \mu < \omega + 1. \end{cases} \tag{10.85}$$

The upper bound $\mu < 2$ is readily recovered for the capillarity equation ($\omega = 1$). In a manifold setting, the case $\omega > 1$ and $\mu < 2$ was already considered in [182, Thm. 4.8]: with the aid of Theorem 10.33, we can improve on it by describing the full range $\omega > 0$. In particular, specifying the next theorem to the capillarity problem yields Theorem 2.32 in the Introduction.

Theorem 10.39 *Suppose that M is complete, fix $\omega > 0$ and let $\kappa \in C(M)$ satisfying*

$$\kappa(x) \geq C\big(1 + r(x)\big)^{-\mu} \qquad \text{on } M,$$

for some constants $C > 0$ and $\mu \in \mathbb{R}$. Then, the only solution of (10.84) on M is $u \equiv 0$ whenever one of the following cases occurs:

$$(i) \quad \omega > 1, \qquad \mu < 2 \qquad and \qquad \liminf_{r \to \infty} \frac{\log |B_r|}{r^{2-\mu}} < \infty;$$

$$(ii) \quad \omega > 1, \qquad \mu = 2 \qquad and \qquad \liminf_{r \to \infty} \frac{\log |B_r|}{\log r} < \infty; \qquad (10.86)$$

$$(iii) \quad \omega \in (0, 1], \quad \mu < \omega + 1 \quad and \quad \liminf_{r \to \infty} \frac{\log |B_r|}{r^{\omega+1-\mu-\varepsilon}} < \infty,$$

for some $\varepsilon > 0$.

Remark 10.40 Case (i) is due to [182, Thm. 4.8]. From (10.86), we readily deduce (10.85) in the Euclidean setting.

Proof Clearly, $u \equiv 0$ is the only constant solution. Suppose that (10.84) admits a non-constant solution, and, up to replacing u with $-u$, assume that $\{u > 0\} \neq \emptyset$. Set $p = 2$ and define $\chi = 1$ if $\omega > 1$, while $\chi = \omega - \varepsilon$ if $\omega \in (0, 1]$. Up to reducing ε, we can assume that $\chi \in (0, 1]$. Therefore, the boundedness of $t^\chi / \sqrt{1 + t^2}$ on $\mathbb{R}$ guarantees the existence of a constant $C_1 > 0$ depending on χ such that

$$\mathrm{div} \left(\frac{\nabla u}{\sqrt{1 + |\nabla u|^2}} \right) = \kappa(x) |u|^{\omega-1} u$$

$$\geq C_1 \kappa(x) |u|^{\omega-1} u \frac{|\nabla u|^{1-\chi}}{\sqrt{1 + |\nabla u|^2}} \qquad on \ \{u > 0\}.$$

By the pasting Lemma, $u_+ = \max\{u, 0\}$ satisfies the same inequality on the entire M. Since $\omega > \chi$, applying Theorem 10.33 with the choices $b(x) = C_1 \kappa(x)$, $f(t) = |t|^{\omega-1} t$ we deduce $u_+^* \leq 0$, thus $u \leq 0$ on M, contradiction. $\qquad \square$

10.7   Other Ranges of Parameters

In our investigation of problem $(P_\geq)$, we mostly assumed (10.57) in the parameter range

$$\chi \geq 0, \qquad \mu \leq \chi + 1.$$

The main reason for this choice was the possibility to obtain maximum principles at infinity for the operator $(bl)^{-1} \Delta_\varphi$. However, in the recent literature some interesting results in Euclidean space give subtle hints to grasp how Geometry comes into play for other ranges of χ, μ. To our knowledge, the problem is still completely open in a manifold setting.

Remark 10.41 (The Range $\mu > \chi + 1$) This case is considered in [86, 87, 195]. In particular, we quote [87, Thm. 3] where the authors establish a Liouville theorem under the restriction

$$\omega > \max\{\chi, 0\}, \qquad \frac{\mu - \chi - 1}{\omega - \chi} < \frac{p - m}{p - 1}, \tag{10.87}$$

see also Thm. 2 and Ex. 3 in [86]. Note that $\mu > \chi + 1$ may enjoy (10.87) only if $p > m$. Further results for large μ can be found in Theorems 4, 8 and 12 in [87], Thm. C in [86], Thms. 1.3 and 5.3 in [195].

Remark 10.42 (The Range $\chi < 0$) This corresponds to a gradient dependence l that is allowed to vanish with high order in $t = 0$, and we quote [64, Thm. 11.4]. There, the conclusions of Theorem 10.33 are shown to hold when (10.57) holds with $l(t) \geq Ct^{p-1-\chi}$ and $\omega = 0$, provided that

$$\mu < 1, \qquad -\left[\frac{1 - \mu}{m - 1}\right](p - 1) \leq \chi < 0. \tag{10.88}$$

Note that, as shown in Remark 11.8 of [64], when $\mu = 0$ the value $\frac{1-\mu}{m-1}(p - 1)$ in (10.88) is sharp. A similar bound also appears in Thms. 2 and 7 in [87]. Related interesting results, for possibly singular $b(x)$ and still in the range $\chi < 0$, are given in [144].

Bibliography

1. V. Agostiniani, M. Fogagnolo, L. Mazzieri, Sharp geometric inequalities for closed hypersurfaces in manifolds with nonnegative Ricci curvature. Invent. Math. (2020, online first)
2. V. Agostiniani, M. Fogagnolo, L. Mazzieri, Minkowski inequalities via nonlinear potential theory. Available at arXiv:1906.00322
3. G. Albanese, M. Rigoli, A Schwarz-type lemma for noncompact manifolds with boundary and geometric applications. Commun. Anal. Geom. 25(4), 719–749 (2017)
4. G. Albanese, L.J. Alías, M. Rigoli, A general form of the weak maximum principle and applications. Rev. Mat. Iberoam. 29, 1437–1476 (2013)
5. G. Albanese, L. Mari, M. Rigoli, On the role of gradient terms in coercive quasilinear differential inequalities on Carnot groups. Nonlinear Anal. 126, 234–261 (2015)
6. L.J. Alías, P. Mastrolia, M. Rigoli, *Maximum Principles and Geometric Applications*. Springer Monographs in Mathematics (Springer, Cham, 2016)
7. L.J. Alías, J. Miranda, M. Rigoli, A new open form of the weak maximum principle and geometric applications. Commun. Anal. Geom. 24(1), 1–43 (2016)
8. F.J. Almgren Jr., Some interior regularity theorems for minimal surfaces and an extension of Bernstein's theorem. Ann. Math. 85, 277–292 (1966)
9. S. Altschuler, L. Wu, Translating surfaces of the non-parametric mean curvature flow with prescribed contact angle. Calc. Var. Partial Differ. Equ. 2, 101–111 (1994)
10. L. Ambrosio, N. Fusco, D. Pallara, *Functions of Bounded Variation and Free Discontinuity Problems*. Oxford Mathematical Monographs (The Clarendon Press, Oxford University Press, New York, 2000)
11. P. Antonini, D. Mugnai, P. Pucci, Quasilinear elliptic inequalities on complete Riemannian manifolds. J. Math. Pures Appl. 87, 582–600 (2007)
12. D. Azagra, M. García-Bravo, Some remarks about the Morse-Sard theorem and approximate differentiability. Rev. Mat. Complut. 33(1), 161–185 (2020)
13. R. Azencott, Behavior of diffusion semi-groups at infinity. Bull. Soc. Math. France 102, 193–240 (1974)
14. C. Bandle, A. Greco, G. Porru, Large solutions of quasilinear elliptic equations: existence and qualitative properties. Boll. Un. Mat. Ital. B (7) 11(1), 227–252 (1997)
15. H. Bao, Y. Shi, Gauss maps of translating solitons of mean curvature flow. Proc. Am. Math. Soc. 142(12), 4333–4339 (2014)
16. V. Benci, D. Fortunato, L. Pisani, Solitons like solutions of a Lorentz invariant equation in dimension 3. Rev. Math. Phys. 10, 315–344 (1998)
17. R. Benedetti, C. Petronio, *Lectures on Hyperbolic Geometry*. Universitext (Springer, Berlin, 1992), p. xiv+330

18. S. Bernstein, Sur un théorème de géométrie et son application aux équations aux dérivées partielles du type elliptique. Commun. Soc. Math. de Kharkov **2**(15), 38–45 (1915–1917); German translation: Über ein geometrisches Theorem und seine Anwendung auf die partiellen Differentialgleichungen vom elliptischen Typus. Math. Z. **26**(1), 551–558 (1927)

19. L. Bers, *Mathematical Aspects of Subsonic and Transonic Gas Dynamics.* Surveys in Applied Mathematics, vol. 3 (Wiley, New York/Chapman & Hall, Ltd., London, 1958), xv+164 pp.

20. D. Bianchi, A.G. Setti, Laplacian cut-offs, porous and fast diffusion on manifolds and other applications. Calc. Var. Partial Differ. Equ. **57**, 1 (2018), 33 pp. Art. 4

21. B. Bianchini, G. Colombo, M. Magliaro, L. Mari, P. Pucci, M. Rigoli, Recent rigidity results for graphs with prescribed mean curvature. Math. Eng. **3**(5), 1–48 (2021)

22. B. Bianchini, L. Mari, M. Rigoli, On some aspects of oscillation theory and geometry. Mem. Am. Math. Soc. **225**(1056), vi + 195 (2013)

23. B. Bianchini, L. Mari, M. Rigoli, Yamabe type equations with sign-changing nonlinearities on the Heisenberg group, and the role of Green functions. Recent trends in Nonlinear Partial Differential Equations I. Evolution problems "Workshop in honour of Patrizia Pucci's 60th birthday". Contemporary Mathematics, vol. 594 (American Mathematical Society, Providence, RI, 2013), pp. 115–136

24. B. Bianchini, L. Mari, M. Rigoli, Yamabe type equations with sign-changing nonlinearities on non-compact Riemannian manifolds. J. Funct. Anal. **268**(1), 1–72 (2015)

25. B. Bianchini, L. Mari, M. Rigoli, Yamabe type equations with a sign-changing nonlinearity, and the prescribed curvature problem. J. Differ. Equ. **260**(10), 7416–7497 (2016)

26. R.L. Bishop, Decomposition of cut loci. Proc. Am. Math. Soc. **65**, 133–136 (1977)

27. L. Boccardo, T. Gallouet, J.L. Vazquez, Nonlinear elliptic equations in R^n without restriction on the data. J. Differ. Equ. **105**, 334–363 (1993)

28. E. Bombieri, E. Giusti, Harnack's inequality for elliptic differential equations on minimal surfaces. Invent. Math. **15**, 24–46 (1972)

29. E. Bombieri, E. De Giorgi, M. Miranda, Una maggiorazione a priori relativa alle ipersuperfici minimali non parametriche. Arch. Ration. Mech. Anal. **32**, 255–267 (1969)

30. E. Bombieri, E. De Giorgi, E. Giusti, Minimal cones and the Bernstein problem. Invent. Math. **7**, 243–268 (1969)

31. A. Borbely, A remark on the Omori-Yau maximum principle. Kuwait J. Sci. Eng. **39**(2A), 45–56 (2012)

32. A. Borbely, Stochastic completeness and the Omori-Yau maximum principle. J. Geom. Anal. **27**, 3228–3239 (2017)

33. S. Bordoni, R. Filippucci, P. Pucci, Nonlinear elliptic inequalities with gradient terms on the Heisenberg group. Nonlinear Anal. **121**, 262–279 (2015)

34. L. Brandolini, M. Magliaro, A note on Keller-Osserman conditions on Carnot groups. Nonlinear Anal. **75**, 2326–2337 (2012)

35. L. Brandolini, M. Magliaro, Liouville type results and a maximum principle for non-linear differential operators on the Heisenberg group. J. Math. Anal. Appl. **415**, 686–712 (2014)

36. H. Brezis, Semilinear equations in $\mathbb{R}^n$ without condition at infinity. Appl. Math. Optim. **12**, 271–282 (1984)

37. G. Caristi, E. Mitidieri, Nonexistence of positive solutions of quasilinear equations. Adv. Differ. Equ. **2**(3), 319–359 (1997)

38. G. Carron, Une suite exacte en $L2$-cohomology. Duke Math. J. **95**(2), 343–372 (1998)

39. J.-B. Casteras, I. Holopainen, J.B. Ripoll, On the asymptotic Dirichlet problem for the minimal hypersurface equation in a Hadamard manifold. Potential Anal. **47**(4), 485–501 (2017)

40. J.-B. Casteras, E. Heinonen, I. Holopainen, Solvability of minimal graph equation under pointwise pinching condition for sectional curvatures. J. Geom. Anal. **27**(2), 1106–1130 (2017)

41. J.-B. Casteras, I. Holopainen, J.B. Ripoll, Convexity at infinity in Cartan-Hadamard manifolds and applications to the asymptotic Dirichlet and Plateau problems. Math. Z. **290**(1–2), 221–250 (2018)

42. J.-B. Casteras, E. Heinonen, I. Holopainen, J. Lira, Asymptotic Dirichlet problems in warped products. Math.Z. **295**, 1–38 (2019)

43. J.-B. Casteras, E. Heinonen, I. Holopainen, Dirichlet problem for f-minimal graphs. J. Anal. Math. **138**(2), 917–950 (2019)

44. J.-B. Casteras, E. Heinonen, I. Holopainen, Existence and non-existence of minimal graphic and p-harmonic functions. Proc. R. Soc. Edinb. Sect. A Math. **150**, 1–26 (2020)

45. J. Cheeger, A lower bound for the smallest eigenvalue of the Laplacian, in *Problems in Analysis* (Papers Dedicated to Salomon Bochner, 1969) (Princeton University Press, Princeton, NJ, 1970), pp. 195–199

46. J. Cheeger, T.H. Colding, Lower bounds on Ricci curvature and the almost rigidity of warped products. Ann. Math. (2) **144**(1), 189–237 (1996)

47. J. Cheeger, T.H. Colding, On the structure of spaces with Ricci curvature bounded below. I. J. Differ. Geom. **46**(3), 406–480 (1997)

48. J. Cheeger, T.H. Colding, On the structure of spaces with Ricci curvature bounded below. II. J. Differential Geom. **54**(1), 13–35 (2000)

49. J. Cheeger, T.H. Colding, On the structure of spaces with Ricci curvature bounded below. III. J. Differ. Geom. **54**(1), 37–74 (2000)

50. S.Y. Cheng, S.T. Yau, Differential equations on Riemannian manifolds and their geometric applications. Commun. Pure Appl. Math. **28**(3), 333–354 (1975)

51. S.Y. Cheng, S.T. Yau, Maximal space-like hypersurfaces in the Lorentz-Minkowski spaces. Ann. Math. (2) **104**(3), 407–419 (1976)

52. Q. Chen, Y.L. Xin, A generalized maximum principle and its applications in geometry. Am. J. Math. **114**(2), 355–366 (1992)

53. L. Cherfils, Y. Ilyasov, On the stationary solutions of generalized reaction diffusion equations with $p\&q$-Laplacian. Commun. Pure Appl. Anal. **4**, 9–22 (2005)

54. S.S. Chern, On the curvatures of a piece of hypersurface in Euclidean space. Abh. Math. Sem. Univ. Hamburg **29**, 77–91 (1965)

55. J. Clutterbuck, O.C. Schnürer, F. Schulze, Stability of translating solutions to mean curvature flow. Calc. Var. Partial Differ. Equ. **29**(3), 281–293 (2007)

56. T.H. Colding, New monotonicity formulas for Ricci curvature and applications, I. Acta Math. **209**, 229–263 (2012)

57. T.H. Colding, W.P. Minicozzi II, Harmonic functions with polynomial growth. J. Differ. Geom. **45**, 1–77 (1997)

58. T.H. Colding, W.P. Minicozzi II, Ricci curvature and monotonicity for harmonic functions. Calc. Var. Partial Differ. Equ. **49**(3–4), 1045–1059 (2014)

59. G. Colombo, M. Magliaro, L. Mari, M. Rigoli, The Bernstein and half-space properties for minimal hypersurfaces under Ricci lower bounds. Available at arXiv:1911.12054

60. C.B.Croke, Some isoperimetric inequalities and eigenvalues estimates. Ann. Sci. É Norm. Sup. **13**, 419–435 (1980)

61. M. Dajczer, J.H.S. de Lira, Conformal killing graphs with prescribed mean curvature. J. Geom. Anal. **22**, 780–799 (2012)

62. M. Dajczer, P.A. Hinojosa, J.H.S. de Lira, Killing graphs with prescribed mean curvature. Calc. Var. Partial Differ. Equ. **33**, 231–248 (2008)

63. L. D'Ambrosio, E. Mitidieri, A priori estimates, positivity results, and nonexistence theorems for quasilinear degenerate elliptic inequalities. Adv. Math. **224**(3), 967–1020 (2010)

64. L. D'Ambrosio, E. Mitidieri, A priori estimates and reduction principles for quasilinear elliptic problems and applications. Adv. Differ. Equ. **17**(9–10), 935–1000 (2012)

65. L. D'Ambrosio, E. Mitidieri, Liouville theorems for elliptic systems and applications. J. Math. Anal. Appl. **413**, 121–138 (2014)

66. L. D'Ambrosio, A. Farina, E. Mitidieri, J. Serrin, Comparison principles, uniqueness and symmetry results of solutions of quasilinear elliptic equations and inequalities. Nonlinear Anal. **90**, 135–158 (2013)

67. E.B. Davies, Heat kernel bounds, conservation of probability and the Feller property. Festschrift on the occasion of the 70th birthday of Shmuel Agmon. J. Anal. Math. **58**, 99–119 (1992)

68. E. De Giorgi, Una estensione del teorema di Bernstein. Ann. Scuola Norm. Sup. Pisa (3) **19**, 79–85 (1965)

69. E. De Giorgi, Errata-Corrige: "Una estensione del teorema di Bernstein". Ann. Scuola Norm. Sup. Pisa Cl. Sci. (3) **19**(3), 463–463 (1965)

70. L. De Pascale, The Morse-Sard theorem in Sobolev spaces. Indiana Univ. Math. J. **50**, 1371–1386 (2001)

71. C.H. Derrick, Comments on nonlinear wave equations as model elementary particles. J. Math. Phys. **5**, 1252–1254 (1964)

72. Q. Ding, Liouville type theorems for minimal graphs over manifolds. Available at arXiv:1911.10306

73. Q. Ding, J. Jost, Y. Xin, Minimal graphic functions on manifolds of nonnegative Ricci curvature. Commun. Pure Appl. Math. **69**(2), 323–371 (2016)

74. M.P. Do Carmo, *Riemannian Geometry*. Mathematics: Theory and Applications (Birkäuser Boston Inc., Boston, MA, 1992)

75. M.P. Do Carmo, H.B. Lawson Jr., On Alexandrov-Bernstein theorems in hyperbolic space. Duke Math. J. **50**(4), 995–1003 (1983)

76. M.P. Do Carmo, C.K. Peng, Stable complete minimal surfaces in $\mathbb{R}^3$ are planes. Bull. Am. Math. Soc. **1**, 903–906 (1979)

77. N. do Espírito Santo, S. Fornari, J.B. Ripoll, The Dirichlet problem for the minimal hypersurface equation in $M \times \mathbb{R}$ with prescribed asymptotic boundary. J. Math. Pures Appl. (9) **93**(2), 204–221 (2010)

78. J. Dodziuk, Maximum principle for parabolic inequalities and the heat flow on open manifolds. Indiana Univ. Math. J. **32**(5), 703–716 (1983)

79. D.M. Duc, J. Eells, Regularity of exponentially harmonic functions. Int. J. Math. **2**, 395–408 (1991)

80. K. Ecker, G. Huisken, Mean curvature evolution of entire graphs. Ann. Math. (2) **130**(3), 453–471 (1989)

81. J. Eells, L. Lemaire, Some properties of exponentially harmonic maps, in *Partial Differential Equations, Part 1, 2 (Warsaw, 1990)*. Banach Center Publications 27, Part 1, 2 (Polish Academy of Sciences of Institute of Mathematics, Warsaw, 1992), pp. 129–136

82. J.H. Eschenburg, E. Heintze, Comparison theory for Riccati equations. Manuscr. Math. **68**, 209–214 (1990)

83. A. Farina, Liouville-type theorems for elliptic problems, in *Handbook of Differential Equations*, ed. by M. Chipot. Stationary Partial Differential Equations, vol. 4 (Elsevier, Amsterdam, 2007), pp. 60–116

84. A. Farina, A Bernstein-type result for the minimal surface equation. Ann. Scuola Norm. Sup. Pisa XIV **5**, 1231–1237 (2015)

85. A. Farina, A sharp Bernstein-type theorem for entire minimal graphs. Calc. Var. Partial Differ. Equ. **57**(5), 5 pp. (2018). Art. 123

86. A. Farina, J. Serrin, Entire solutions of completely coercive quasilinear elliptic equations. J. Differ. Equ. **250**(12), 4367–4408 (2011)

87. A. Farina, J. Serrin, Entire solutions of completely coercive quasilinear elliptic equations, II. J. Differ. Equ. **250**(12), 4409–4436 (2011)

88. P. Felmer, M. Montenegro, A. Quaas, A note on the strong maximum principle and the compact support principle. J. Differ. Equ. **246**, 39–49 (2009)

89. P. Felmer, A. Quaas, B. Sirakov, Solvability of nonlinear elliptic equations with gradient terms. J. Differ. Equ. **254**, 4327–4346 (2013)

90. A. Figalli, A simple proof of the Morse-Sard theorem in Sobolev spaces. Proc. Am. Math. Soc. **136**(10), 3675–3681 (2008)

91. R. Filippucci, Nonexistence of positive weak solutions of elliptic inequalities. Nonlinear Anal. **8**, 2903–2916 (2009)

92. R. Filippucci, P. Pucci, M. Rigoli, Non-existence of entire solutions of degenerate elliptic inequalities with weights. Archive Rat. Mech. Anal. **188**, 155–179 (2008); Erratum, **188**, 181 (2008)

93. R. Filippucci, P. Pucci, M. Rigoli, On weak solutions of nonlinear weighted p-Laplacian elliptic inequalities. Nonlinear Anal. **70**, 3008–3019 (2009)

94. R. Filippucci, P. Pucci, M. Rigoli, On entire solutions of degenerate elliptic differential inequalities with nonlinear gradient terms. J. Math. Anal. Appl. **356**(2), 689–697 (2009)

95. R. Filippucci, P. Pucci, M. Rigoli, Nonlinear weighted p-Laplacian elliptic inequalities with gradient terms. Commun. Cont. Math. **12**(3), 501–535 (2010)

96. R. Finn, *Equilibrium Capillary Surfaces* (Springer, New York, 1986)

97. D. Fischer-Colbrie, R. Schoen, The structure of complete stable minimal surfaces in 3-manifolds of non negative scalar curvature. Comm. Pure Appl. Math. **XXXIII**, 199–211 (1980)

98. H. Flanders, Remark on mean curvature. J. Lond. Math. Soc. **41**(2), 364–366 (1966)

99. W.H. Fleming, On the oriented Plateau problem. Rend. Circolo Mat. Palermo **9**, 69–89 (1962)

100. M. Fogagnolo, L. Mazzieri, A. Pinamonti, Geometric aspects of p-capacitary potentials. Ann. Inst. H. Poincaré Anal. Non Linéaire **36**(4), 1151–1179 (2019)

101. D. Gilbarg, N. Trudinger, *Elliptic Partial Differential Equations of Second Order*, 3rd edn. (Springer, New York, 1998)

102. E. Giusti, *Minimal Surfaces and Functions of Bounded Variation*. Monographs in Mathematics, vol. 80 (Birkhäuser Verlag, Basel, 1984), xii+240 pp.

103. M. Ghergu, V. Radulescu, Nonradial blow-up solutions of sublinear elliptic equations with gradient term. Commun. Pure Appl. Anal. **3**(3), 465–474 (2004)

104. M. Ghergu, C. Niculescu, V. Radulescu, Explosive solutions of elliptic equations with absorption and nonlinear gradient term. Proc. Indian Acad. Sci. Math. Sci. **112**, 441–451 (2002)

105. A. Greco, On the existence of large solutions for equations of prescribed mean curvature. Nonlinear Anal. **34**(4), 571–583 (1998)

106. R. Greene, H.H. Wu, *Function Theory on Manifolds Which Possess a Pole*. Lecture Notes in Mathematics, vol. 699 (Springer, New York, 1979)

107. A. Grigor'yan, Analytic and geometric background of recurrence and non-explosion of the Brownian motion on Riemannian manifolds. Bull. Am. Math. Soc. **36**, 135–249 (1999)

108. B. Guan, J. Spruck, Hypersurfaces of constant mean curvature in hyperbolic space with prescribed asymptotic boundary at infinity. Am. J. Math. **122**, 1039–1060 (2000)

109. Y. Haitao, A compact support principle for a class of elliptic differential inequalities. Nonlinear Anal. **58**, 103–119 (2004)

110. E.K. Haviland, A note on unrestricted solutions of the differential equation $\Delta u = f(u)$. J. Lond. Math. Soc. **26**, 210–214 (1951)

111. E. Hebey, Optimal Sobolev inequalities on complete Riemannian manifolds with Ricci curvature bounded below and positive injectivity radius. Am. J. Math. **118**(2), 291–300 (1996)

112. E. Hebey, *Nonlinear Analysis on Manifolds: Sobolev Spaces and Inequalities*. Courant Lecture Notes in Mathematics, vol. 5 (New York University, Courant Institute of Mathematical Sciences, New York/American Mathematical Society, Providence, RI, 1999), x+309 pp.

113. H.-J. Hein, Weighted Sobolev inequalities under lower Ricci curvature bounds. Proc. Am. Math. Soc. **139**(8), 2943–2955 (2011)

114. E. Heinonen, Survey on the asymptotic Dirichlet problem for the minimal surface equation. Available at arXiv:1909.08437

115. J. Heinonen, T. Kilpelainen, $\mathscr{A}$-superharmonic functions and supersolutions of degenerate elliptic equations. Ark. Mat. **26**, 87–105 (1988)

116. J. Heinonen, T. Kilpeläinen, O. Martio, *Nonlinear Potential Theory of Degenerate Elliptic Equations*. Unabridged Republication of the 1993 Original (Dover Publications Inc., Mineola, NY, 2006), p. xii+404

117. E. Heinz, Über die Lösungen der Minimalflächengleichung. Nachr. Akad. Wiss. Göttingen. Math.-Phys. Kl. Math.-Phys.-Chem. Abt. **1952**, 51–56 (1952)

118. E. Heinz, Uber Flächen mit eindeutiger projektion auf eine ebene, deren krümmungen durch ungleichungen eingschränkt sind. Math. Ann. **129**, 451–454 (1955)

119. D. Hoffman, J. Spruck, Sobolev and isoperimetric inequalities for Riemannian submanifolds. Commun. Pure Appl. Math. **27**, 715–727 (1974)

120. I. Holopainen, Nonlinear potential theory and quasiregular mappings on Riemannian manifolds. Ann. Acad. Sci. Fenn. Ser. A I Math. Dissertationes **74**, 45 pp. (1990)

121. I. Holopainen, Positive solutions of quasilinear elliptic equations on Riemannian manifolds. Proc. Lond. Math. Soc. **65**, 651–672 (1992)

122. I. Holopainen, Volume growth, Green's functions, and parabolicity of ends. Duke Math. J. **97**(2), 319–346 (1999)

123. I. Holopainen, J.B. Ripoll, Nonsolvability of the asymptotic Dirichlet problem for some quasilinear elliptic PDEs on Hadamard manifolds. Rev. Mat. Iberoam. **31**(3), 1107–1129 (2015)

124. E. Hopf, *Elementäre Bemerkungen über die Lösungen partieller Differentialgleichungen zweiter Ordnung vom elliptischen Typus* (Sitzungsberichte Preussische Akademie der Wissenschaften, Berlin, 1927), pp. 147–152

125. E. Hopf, On S. Bernstein's theorem on surfaces $z(x, y)$ of nonpositive curvature. Proc. Am. Math. Soc. **1**, 80–85 (1950)

126. E.-P. Hsu, Heat semigroup on a complete Riemannian manifold. Ann. Probab. **17**(3), 1248–1254 (1989)

127. E.-P. Hsu, *Stochastic Analysis on Manifolds*. Graduate Studies in Mathematics, vol. 38 (American Mathematical Society, New York, 2002)

128. D. Impera, S. Pigola, A.G. Setti, Potential theory for manifolds with boundary and applications to controlled mean curvature graphs. J. Reine Angew. Math. **733**, 121–159 (2017)

129. T. Iwaniec, J.J. Manfredi, Regularity of p-harmonic functions on the plane. Rev. Mat. Iberoam. **5**(1–2), 1–19 (1989)

130. M. Kanai, Rough isometries, and combinatorial approximations of geometries of non-compact Riemannian manifolds. J. Math. Soc. Jpn. **37**, 391–413 (1985)

131. L. Karp, Differential inequalities on complete Riemannian manifolds and applications. Math. Ann. **272**(4), 449–459 (1985)

132. T. Kato, Schrödinger operators with singular potentials. Israel J. Math. **13**, 135–148 (1972)

133. R.Z. Khas'minskii, *Ergodic Properties of Recurrent Diffusion Processes and Stabilization of the Solution of the Cauchy Problem for Parabolic Equations*. Teor. Verojatnost. i Primenen., Akademija Nauk SSSR (Teorija Verojatnosteĭ i ee Primenenija 5, 1960), pp. 196–214

134. J.B. Keller, Electrohydrodynamics I. The equilibrium of a charged gas in a container. Indiana Univ. Math. J. **5**(4), 715–724 (1956)

135. J.B. Keller, On solutions of $\Delta u = f(u)$. Commun. Pure Appl. Math. **10**, 503–510 (1957)

136. S. Kichenassamy, L. Véron, Singular solutions of the p-Laplace equation. Math. Ann **275**(4), 599–615 (1986)

137. B. Kotschwar, L. Ni, Local gradient estimates of p-harmonic functions, $1/H$-flow, and an entropy formula. Ann. Sci. École Norm. Supér. (4) **42**(1), 1–36 (2009)

138. T. Kura, The weak supersolution-subsolution method for second order quasilinear elliptic equations. Hiroshima Math. J. **19**(1), 1–36 (1989)

139. A. Lair, A. Wood, Large solutions of semilinear elliptic equations with nonlinear gradient terms. Int. J. Math. Sci. **22**(4), 869–883 (1999)

140. J.-M. Lasry, P.-L. Lions, Nonlinear elliptic equations with singular boundary conditions and stochastic control with state constraints. I. The model problem. Math. Ann. **283**(4), 583–630 (1989)

141. V.K. Le, On some equivalent properties of sub- and supersolutions in second order quasilinear elliptic equations. Hiroshima Math. J. **28**, 373–380 (1998)

142. J.L. Lewis, Capacitary functions in convex rings. Arch. Rational Mech. Anal. **66**(3), 201–224 (1977)

143. J.L. Lewis, On critical points of p harmonic functions in the plane. Electron. J. Differ. Equ. **1994**(3), 4 pp. (1994)

144. X. Li, F. Li, Nonexistence of solutions for singular quasilinear differential inequalities with a gradient nonlinearity. Nonlinear Anal. **75**(5), 2812–2822 (2012)

145. P. Li, L. Karp, The heat equation on complete Riemannian manifolds. Unpublished

146. P. Li, S.T. Yau, On the parabolic kernel of the Schrödinger operator. Acta Math. **156**(3–4), 153–201 (1986)

147. M. Magliaro, L. Mari, P. Mastrolia, M. Rigoli, Keller–Osserman type conditions for differential inequalities with gradient terms on the Heisenberg group. J. Differ. Equ. **250**, 2643–2670 (2011)

148. J. Malý, W.P. Ziemer, *Fine Regularity of Solutions of Elliptic Partial Differential Equations.* Mathematical Surveys and Monographs, vol. 51 (American Mathematical Society, Providence, RI, 1997)

149. C. Mantegazza, *Lecture Notes on Mean Curvature Flow.* Progress in Mathematics, vol. 290 (Birkhäuser/Springer Basel AG, Basel, 2011). xii+166 pp.

150. C. Mantegazza, A.C. Mennucci, Hamilton-Jacobi equations and distance functions on Riemannian manifolds. Appl. Math. Opt. **47**(1), 1–25 (2002)

151. C. Mantegazza, G. Mascellani, G. Uraltsev, On the distributional Hessian of the distance function. Pacific J. Math. **270**(1), 151–166 (2014)

152. P. Marcellini, On the definition and the lower semicontinuity of certain quasiconvex integrals. Ann. Inst. H. Poincaré Anal. Non Linéaire **3**, 391–409 (1986)

153. L. Mari, L.F. Pessoa, Maximum principles at infinity and the Ahlfors-Khas'minskii duality: an overview, in *Contemporary Research in Elliptic PDEs and Related Topics.* Springer INdAM Series, vol. 33 (Springer, Cham, 2019), pp. 419–455

154. L. Mari, L.F. Pessoa, Duality between Ahlfors-Liouville and Khas'minskii properties for nonlinear equations. Commun. Anal. Geom. **28**(2), 395–497 (2020)

155. L. Mari, D. Valtorta, On the equivalence of stochastic completeness, Liouville and Khas'minskii condition in linear and nonlinear setting. Trans. Am. Math. Soc. **365**(9), 4699–4727 (2013)

156. L. Mari, M. Rigoli, A.G. Setti, On the $1/H$-flow by p-Laplace approximation: new estimates via fake distances under Ricci lower bounds. Available at arXiv:1905.00216

157. L. Mari, M. Rigoli, A.G. Setti, Keller-Osserman conditions for diffusion-type operators on Riemannian Manifolds. J. Funct. Anal. **258**(2), 665–712 (2010)

158. O. Martio, G. Porru, Large solutions of quasilinear elliptic equations in the degenerate case. Complex analysis and differential equations (Uppsala, 1997). Acta Univ. Upsaliensis Skr. Uppsala Univ. C Organ. Hist. **64**, 225–241 (1999)

159. E.J. Mickle, A remark on a theorem of Serge Bernstein. Proc. Am. Math. Soc. **1**, 86–89 (1950)

160. V. Miklyukov, V. Tkachev, Denjoy-Ahlfors theorem for harmonic functions on Riemannian manifolds and external structure of minimal surfaces. Commun. Anal. Geom. **4**(4), 547–587 (1996)

161. V. Minerbe, Weighted Sobolev inequalities and Ricci flat manifolds. Geom. Funct. Anal. **18**(5), 1696–1749 (2009)

162. E. Mitidieri, S.I. Pokhozhaev, A priori estimates and the absence of solutions of nonlinear partial differential equations and inequalities. Tr. Mat. Inst. Steklova **234**, 1–384 (2001)

163. J. Moser, On Harnack's theorem for elliptic differential equations. Commun. Pure Appl. Math. **14**, 577–591 (1961)

164. R. Moser, The inverse mean curvature flow and p-harmonic functions. J. Eur. Math. Soc. **9**(1), 77–83 (2007)

165. W.F. Moss, J. Piepenbrink, Positive solutions of elliptic equations. Pacific J. Math. **75**, 219–226 (1978)

166. K. Motomiya, On functions which satisfy some differential inequalities on Riemannian manifolds. Nagoya Math. J. **81**, 57–72 (1981)

167. A. Naber, D. Valtorta, Sharp estimates on the first eigenvalue of the p-Laplacian with negative Ricci lower bound. Math. Z. **277**(3–4), 867–891 (2014)

168. Y. Naito, H. Usami, Entire solutions of the inequality $div(A(|\nabla u|)\nabla u) \geq f(u)$. Math. Z. **225**, 167–175 (1997)

169. B. Nelli, H. Rosenberg. Minimal surfaces in $\mathbb{H}^2 \times \mathbb{R}$. Bull. Braz. Math. Soc. (N.S.) **33**(2), 263–292 (2002)

170. L. Ni, Mean value theorems on manifolds. Asian J. Math. **11**(2), 277–304 (2007)

171. J.C.C. Nitsche, Elementary proof of Bernstein's theorem on minimal surfaces. Ann. Math. (2) **66**, 543–544 (1957)

172. J.C.C. Nitsche, *Lectures on Minimal Surfaces*, vol. 1 (Cambridge University Press, Cambridge, 1989), xxvi+563 pp.

173. M. Obata, Conformal transformations of compact Riemannian manifolds. Illinois J. Math. **6**, 292–195 (1962)

174. H. Omori, Isometric immersions of Riemannian manifolds. J. Math. Soc. Jpn. **19**, 205–214 (1967)

175. R. Osserman, *On the inequality $\Delta u \geq f(u)$*. Pacific J. Math. **7**, 1641–1647 (1957)

176. R. Osserman, *A Survey of Minimal Surfaces*, 2nd edn. (Dover Publications, Inc., New York, 1986), p. vi+207

177. P. Petersen, *Riemannian Geometry* (Springer, New York, 1997)

178. S. Pigola, A.G. Setti, The Feller property on Riemannian manifolds. J. Funct. Anal. **262**(5), 2481–2515 (2012)

179. S. Pigola, M. Rigoli, A.G. Setti, Maximum principles and singular elliptic inequalities. J. Funct. Anal. **193**(2), 224–260 (2002)

180. S. Pigola, M. Rigoli, A.G. Setti, A remark on the maximum principle and stochastic completeness. Proc. Am. Math. Soc. **131**(4), 1283–1288 (2003)

181. S. Pigola, M. Rigoli, A.G. Setti, Volume growth, "a priori" estimates, and geometric applications. Geom. Funct. Anal. **13**(6), 1303–1328 (2003)

182. S. Pigola, M. Rigoli, A.G. Setti, Maximum principles on Riemannian manifolds and applications. Mem. Am. Math. Soc. **174**(822), x+99 (2005)

183. S. Pigola, M. Rigoli, A.G. Setti, Maximum principles at infinity on Riemannian manifolds: an overview. Matemática Contemporânea **31**, 81–128 (2006)

184. S. Pigola, M. Rigoli, A.G. Setti, Some non-linear function theoretic properties of Riemannian manifolds. Rev. Mat. Iberoam. **22**(3), 801–831 (2006)

185. S. Pigola, M. Rigoli, A.G. Setti, *Vanishing and Finiteness Results in Geometric Analysis. A Generalization of the Böchner Technique*. Progress in Mathematics, vol. 266 (Birkäuser, Basel, 2008), xiv+282 pp.

186. S. Pigola, A.G. Setti, M. Troyanov, The topology at infinity of a manifold supporting an $L^{q,p}$-Sobolev inequality. Expo. Math. **32**(4), 365–383 (2014)

187. Y. Pinchover, K. Tintarev, Ground state alternative for p-Laplacian with potential term. Calc. Var. Partial Differ. Equ. **28**, 179–201 (2007)

188. A. Pogorelov, On the stability of minimal surfaces. Soviet Math. Dokl. **24**, 274–276 (1981)

189. M.H. Protter, H.F. Weinberger, *Maximum Principles in Differential Equations*. Corrected reprint of the 1967 original (Springer, New York, 1984)

190. P. Pucci, V. Radulescu, The maximum principle with lack of monotonicity. Electron. J. Qual. Theory Differ. Equ. (58), 11 pp. (2018)

191. P. Pucci, M. Rigoli, Entire solutions of singular elliptic inequalities on complete manifolds. Disc. Cont. Dyn. Sys. **20**(1), 115–137 (2008)

192. P. Pucci, J. Serrin, The strong maximum principle revisited. J. Differ. Equ. **196**, 1–66 (2004); Erratum, J. Differ. Equ. **207**, 226–227 (2004)

193. P. Pucci, J. Serrin, Dead cores and bursts for quasilinear singular elliptic equations. SIAM J. Math. Anal. **38**, 259–278 (2006)

194. P. Pucci, J. Serrin, *The Maximum Principle*. Progress in Nonlinear Differential Equations and their Applications, vol. 73 (Birkhäuser Verlag, Basel, 2007), x+235 pp.

195. P. Pucci, J. Serrin, A remark on entire solutions of quasilinear elliptic equations. J. Differ. Equ. **250**(2), 675–689 (2011)

196. P. Pucci, J. Serrin, H. Zou, A strong maximum principle and a compact support principle for singular elliptic inequalities. J. Math. Pures Appl. **78**, 769–789 (1999)

197. P. Pucci, M. García-Huidobro, R. Manásevich, J. Serrin, Qualitative properties of ground states for singular elliptic equations with weights. Ann. Mat. Pura Appl. **185**(suppl. 4), 205–243 (2006)

198. P. Pucci, M. Rigoli, J. Serrin, Qualitative properties for solutions of singular elliptic inequalities on complete manifolds. J. Differ. Equ. **234**(2), 507–543 (2007)

199. V. Radulescu, Singular phenomena in nonlinear elliptic problems: from boundary blow-up solutions to equations with singular nonlinearities, in *Handbook of Differential Equations: Stationary Partial Differential Equations*, vol. 4, ed. by M. Chipot (North-Holland Elsevier Science, Amsterdam, 2007), pp. 483–591

200. A. Ratto, M. Rigoli, Elliptic differential inequalities with applications to harmonic maps. J. Math. Soc. Jpn. **45**, 321–337 (1993)

201. A. Ratto, M. Rigoli, A.G. Setti, On the Omori-Yau maximum principle and its application to differential equations and geometry. J. Funct. Anal. **134**, 486–510 (1995)

202. R. Redheffer, Maximum principles and duality. Monatsh. Math. **62**, 56–75 (1958)

203. R. Redheffer, On the inequality $\Delta u \geq f(u, |\operatorname{grad} u|)$. J. Math. Anal. Appl. **1**, 277–299 (1960)

204. R. Redheffer, Nonlinear differential inequalities and functions of compact support. Trans. Am. Math. Soc. **220**, 133–157 (1976)

205. M. Rigoli, M. Salvatori, M. Vignati, Volume growth and p-subharmonic functions on complete manifolds. Math. Z. **227**(3), 367–375 (1998)

206. M. Rigoli, M. Salvatori, M. Vignati, A Liouville type theorem for a general class of operators on complete manifolds. Pacific J. Math. **194**(2), 439–453 (2000)
207. M. Rigoli, M. Salvatori, M. Vignati, Some remarks on the weak maximum principle. Rev. Mat. Iberoamericana **21**(2), 459–481 (2005)
208. M. Rigoli, M. Salvatori, M. Vignati, On the compact support principle on complete manifolds. J. Differ. Equ. **246**, 870–894 (2009)
209. M. Rigoli, M. Salvatori, M. Vignati, The compact support principle for differential inequalities with gradient terms. Nonlinear Anal. **72**, 4360–4376 (2010)
210. J. Ripoll, M. Telichevesky, Regularity at infinity of Hadamard manifolds with respect to some elliptic operators and applications to asymptotic Dirichlet problems. Trans. Am. Math. Soc. **367**(3), 1523–1541 (2015)
211. J. Ripoll, M. Telichevesky, On the asymptotic plateau problem for CMC hypersurfaces in hyperbolic space. Bull. Braz. Math. Soc. (N.S.) **50**(2), 575–585 (2019)
212. H. Rosenberg, F. Schulze, J. Spruck, The half-space property and entire positive minimal graphs in $M \times \mathbb{R}$. J. Differ. Geom. **95**, 321–336 (2013)
213. I. Salavessa, Graphs with parallel mean curvature. Proc. Am. Math. Soc. **107**, 449–458 (1989)
214. L. Saloff-Coste, *Aspects of Sobolev-Type Inequalities*. (English summary). London Mathematical Society Lecture Note Series, vol. 289 (Cambridge University Press, Cambridge, 2002), x+190 pp.
215. R. Schoen, S.-T. Yau, *Lectures on Differential Geometry*, vol. 1 (International Press Cambridge, Cambridge, 1994)
216. J. Serrin, Local behaviour of solutions of quasilinear equations. Acta Math. **111**, 247–302 (1964)
217. J. Serrin, Isolated singularities of solutions of quasilinear equations. Acta Math. **113**, 219–240 (1965)
218. J. Serrin, Entire solutions of nonlinear Poisson equations. Proc. Lond. Math. Soc. **24**(3), 348–366 (1972)
219. J. Serrin, Entire solutions of quasilinear elliptic equations. J. Math. Anal. Appl. **352**(1), 3–14 (2009)
220. J. Serrin, The Liouville theorem for homogeneous elliptic differential inequalities. J. Math. Sci. (N.Y.) **179**(1), 174–183 (2011). Problems in Mathematical Analysis, vol. 61
221. L.M. Sibner, R.J. Sibner, A non-linear Hodge-de-Rham theorem. Acta Math. **125**, 57–73 (1970)
222. L. Simon, Entire solutions of the minimal surface equation. J. Differ. Geom. **30**, 643–688 (1989)
223. L. Simon, The minimal surface equation, in *Geometry, V*. Encyclopaedia of Mathematical Sciences, vol. 90 (Springer, Berlin, 1997), 239–272
224. J. Simons, Minimal varieties in Riemannian manifolds. Ann. Math. **88**, 62–105 (1968)
225. V.G. Tkachev, Some estimates for the mean curvature of nonparametric surfaces defined over domains in $\mathbb{R}^n$. Ukr. Geom. Sb. **35**, 135–150 (1992); Translated in: J. Math. Sci. (N. Y.) **72**, 3250–3260 (1994)
226. P. Tolksdorf, Regularity of a more general class of quasilinear elliptic equations. J. Differ. Equ. **51**, 126–150 (1984)
227. F. Toumi, Existence of blowup solutions for nonlinear problems with a gradient term. Int. J. Math. Math. Sci. (2006), 11 pp. Art. ID 80605
228. M. Troyanov, Parabolicity of manifolds. Siberian Adv. Math. **9**(4), 125–150 (1999)
229. M. Troyanov, Solving the p-Laplacian on manifolds. Proc. Am. Math. Soc. **128**(2), 541–545 (2000)
230. H. Usami, Nonexistence of positive entire solutions for elliptic inequalities of the mean curvature type. J. Differ. Equ. **111**, 472–480 (1994)

231. N.T. Varopoulos, The Poisson kernel on positively curved manifolds. J. Funct. Anal. **44**(3), 359–380 (1981)

232. L. Véron, Singularities of solutions of second order quasilinear equations, in *Nonlinear Diffusion Equations and Their Equilibrium States*. Pitman Research Notes in Mathematics Series, vol. 353 (Birkhäuser, Boston, MA, 1996), p. viii+377

233. X.-J. Wang, Convex solutions to the mean curvature flow. Ann. Math. (2) **173**(3), 1185–1239 (2011)

234. L. Wang, A Bernstein type theorem for self-similar shrinkers. Geom. Dedicata **151**, 297–303 (2011)

235. X. Wang, L. Zhang, Local gradient estimate for p-harmonic functions on Riemannian manifolds. Commun. Anal. Geom. **19**(4), 759–771 (2011)

236. F.E. Wolter, Distance function and cut loci on a complete Riemannian manifold. Arch. Math. (Basel) **32**, 92–96 (1979)

237. S.T. Yau, A general Schwarz lemma for Kähler manifolds. Am. J. Math. **100**(1), 197–203 (1978)

238. S.T. Yau, Remarks on conformal transformations. J. Differ. Geom. **8**, 369–381 (1973)

239. S.T. Yau, Harmonic functions on complete Riemannian manifolds. Commun. Pure Appl. Math. **28**, 201–228 (1975)

240. V.V. Zhikov, Averaging of functionals of the calculus of variations and elasticity theory. Izv. Akad. Nauk SSSR Ser. Mat. **50**, 675–710 (1986)

241. V.V. Zhikov, S.M. Kozlov, O.A. Oleinik, *Homogenization of Differential Operators and Integral Functionals*. Translated from the Russian by G.A. Yosifian [G.A. Iosifýan] (Springer, Berlin, 1994), xii+570 pp.

Index

Printed in the United States
By Bookmasters